Reconfigurable Intelligent Surfaces for 6G-Enabled Vehicle-to-Everything Communication

Mazhar Hussain Shaik
https://orcid.org/0000-0001-7684-0039
Malla Reddy University, Hyderabad, India

N. Manikanda Devarajan
Malla Reddy University, Hyderabad, India

Masood Ahmad
https://orcid.org/0000-0002-5468-9822
GITAM School of Core Engineering, GITAM University, Hyderabad, India

Shaik Noor Mohammad
AVN Institute of Engineering and Technology, India

IGI Global
Scientific Publishing
Publishing Tomorrow's Research Today

Vice President of Editorial | Melissa Wagner
Director of Acquisitions | Mikaela Felty
Director of Book Development | Jocelynn Hessler
Production Manager | Mike Brehm
Cover Design | Jose Rosado

Published in the United States of America by
IGI Global Scientific Publishing
701 East Chocolate Avenue
Hershey, PA, 17033, USA
Tel: 717-533-8845 | Fax: 717-533-7115
Website: https://www.igi-global.com E-mail: cust@igi-global.com

Library of Congress Cataloging-in-Publication Data

LCCN: 2026002332 (CIP Data Pending)
ISBN13: 9798337378671
Isbn13Softcover: 9798337378688
EISBN13: 9798337378695
British Cataloguing in Publication Data
A Cataloguing in Publication record for this book is available from the British Library.

Table of Contents

Detailed Table of Contents

The rapid progression towards sixth-generation (6G) edge-intelligent networks requires Intrusion Detection Systems (IDS) that are interpretable, lightweight, and resilient to adversarial maneuvering. Current deep learning-based IDS systems, including CiNeT, present misleading accuracy through many negative factors. Using the CiNeT framework as the foundation of this paper introduces AARE-IDS, Adaptive Adversarially Resilient Explainable Intrusion Detection System (IDS), for real-time 6G edge settings. AARE-IDS system is particularly exceptional amongst similar developments due to its lightweight bijective traffic encoding (NeT2I-Lite), hybrid CNN - Transformer detector, adversarially trained resilience layer, utilizing a federated learning framework to investigate model aggregation and derived from a differential privacy standard for added security while sharing models. Through extensive experiments on the ToN-IoT, UNSW-NB15, and CICIDS-2018 datasets, AARE-IDS achieved 98.9% accuracy, 39% better robustness against FGSM/PGD attacks, and 45% less CPU overhead compared to CiNeT.

Artificial Intelligence (AI) is revolutionizing Reconfigurable Intelligent Surfaces (RIS) for 6G-enabled V2X communication by enabling autonomous, predictive, and adaptive control in fast-changing vehicular environments. Traditional optimization methods struggle with mobility, blockage, and complex channel variations, whereas AI-driven models enhance beamforming, channel estimation, and real-time decision-making. Deep learning, reinforcement learning, and distributed intelligence enable

a paradigm shift in how wireless communication environments are designed, optimised, and experienced. Unlike the traditional model in which mobile networks must adapt to a fixed physical environment filled with signal reflections, blockages, and unpredictable interference, RIS enables the environment itself to become programmable, intelligent, and dynamically responsive. Through the deployment of ultra-thin, low-power, metasurface panels composed of a dense array of passive or semi-passive elements, RIS can control electromagnetic wave propagation in real time, shaping signals to enhance coverage, boost data rates, reduce latency, and improve the overall reliability of wireless connectivity. This innovation is particularly significant for next-generation mobility applications—autonomous vehicles, connected public transport.

The transition from 5G to 6G-enabled vehicular communication marks a major advancement toward fully autonomous, intelligent, and hyper-connected transportation systems. While 5G-V2X provides ultra-reliable low-latency communication and large-scale connectivity, it faces limitations in high-mobility performance, energy efficiency, and network scalability. 6G introduces transformative capabilities through terahertz communications, reconfigurable intelligent surfaces (RIS), integrated sensing and communication, and AI-driven optimization, enabling sub-microsecond latency and terabit data rates. This chapter examines the architectural progression, core technologies, and performance requirements of next-generation V2X networks. It also explores edge intelligence, blockchain, and federated learning for enhanced trust, automation, and data privacy. The integration of RIS for adaptive and energy-efficient radio environments is highlighted, along with emerging challenges, research opportunities, and standardization trends shaping future mobility.

The development of Vehicle-to-Everything (V2X) communication on 6G networks presents security challenges due to high mobility and dynamic wireless environments. Reconfigurable Intelligent Surface (RIS) emerged as a technology that is able to control signal propagation adaptively while supporting physical layer security. This book examines the role of RIS in improving V2X security through reflection phase setting, secure beamforming, and artificial noise, as well as its integration with artificial intelligence for adaptive security configuration. The analysis shows that RIS is able to increase the secrecy rate, reduce the risk of eavesdropping and jamming, and maintain safety performance in various mobility conditions. The book also discusses implementation challenges and future research directions for secure and sustainable 6G V2X systems.

Revolutionising autonomous vehicles with Reconfigurable Intelligent Surface (RIS)-assisted communication represents a fundamental shift in how smart mobility systems perceive, interact with, and navigate their environments, enabling unprecedented levels of efficiency, safety, and scalability across next-generation transportation infrastructures. As autonomous vehicles increasingly rely on high-bandwidth, low-latency, and ultra-reliable wireless connectivity to support real-time perception, sensor fusion, and cooperative decision-making, the limitations of traditional communication networks—particularly in complex environments with multipath fading, signal blockages, and insufficient spectrum resources—become more pronounced. RIS technology, built upon programmable meta-surfaces capable of intelligently manipulating electromagnetic waves without conventional power-hungry radio frequency components, introduces a groundbreaking capability to dynamically reconfigure the propagation conditions of wireless signals.

Intelligent Transportation Systems (ITS) face challenges of complex communication and high energy consumption under aggregated interference. Reconfigurable Intelligent Surface (RIS)–assisted communication emerges as a cost-effective and sustainable solution, especially for 5G- and 6G-enabled vehicular networks. Advanced wireless technologies enable low-latency, high-reliability Vehicle-to-

Preface

INTRODUCTION

The rapid evolution of wireless communication technologies is reshaping the foundations of intelligent transportation systems and connected mobility. As we move toward the era of sixth-generation (6G) networks, the demand for ultra-reliable, low-latency, high-capacity, and energy-efficient communication has become more critical than ever—particularly for Vehicle-to-Everything (V2X) ecosystems that support autonomous driving, cooperative intelligent transportation systems (C-ITS), and smart mobility services. Within this transformative landscape, Reconfigurable Intelligent Surfaces (RIS) have emerged as a disruptive technology capable of redefining how the wireless environment itself can be controlled and optimized.

This edited volume, *Reconfigurable Intelligent Surfaces for 6G-Enabled Vehicle-to-Everything Communication*, has been conceived to provide a comprehensive and authoritative exploration of the integration of RIS into next-generation vehicular communication networks. Our objective is to present a consolidated body of knowledge that bridges theoretical foundations, algorithmic innovations, hardware design considerations, and real-world deployment perspectives in RIS-assisted 6G-V2X systems.

The book examines the principles and operating mechanisms of RIS and their transformative potential in vehicular communication environments characterized by high mobility, dynamic topology, and stringent quality-of-service requirements. It explores advanced design methodologies, beamforming strategies, and resource allocation techniques tailored for RIS-enabled vehicular scenarios. Particular emphasis is placed on artificial intelligence (AI) and machine learning (ML)-driven frameworks that enable adaptive environment control, predictive mobility management, and intelligent network optimization. Security, privacy, and resilience aspects are also addressed, acknowledging the critical importance of trust and robustness in safety-critical vehicular systems.

By integrating contributions from leading researchers and industry experts, this volume reflects a multidisciplinary perspective that connects wireless communications, signal processing, electromagnetics, artificial intelligence, and transportation engineering. We believe this integrative approach is essential to unlocking the full potential of RIS-assisted 6G-V2X architectures and ensuring their practical viability.

To the best of our knowledge, this book represents one of the first consolidated references focusing exclusively on RIS-assisted 6G-V2X communication. It aims to bridge theoretical advancements and practical implementations, offering comparative analyses and novel insights that help identify current research gaps, technical challenges, and emerging opportunities. The chapters provide methodological guidance for developing algorithms and simulation models related to RIS-based beamforming, energy-efficient communication, mobility-aware optimization, and AI-enabled resource management. Furthermore, the book seeks to encourage interdisciplinary collaboration among researchers and practitioners working across wireless systems, intelligent transportation networks, and smart infrastructure domains.

Beyond its academic contributions, this publication aspires to support ongoing and future standardization efforts and experimental testbeds for 6G-V2X networks. As regulatory and standardization bodies—including IEEE, 3GPP, ETSI, and ITU—advance next-generation wireless frameworks, the insights presented here may serve as a technical foundation for shaping future specifications and deployment strategies.

This book is intended for a diverse readership. It will serve researchers, faculty members, and postgraduate students engaged in wireless communications, 6G systems, and vehicular networking research. It is equally relevant to engineers and practitioners working on RIS hardware design, vehicular connectivity solutions, and AI-based optimization frameworks. Industry professionals, policymakers, and infrastructure planners involved in intelligent transportation systems and connected mobility initiatives will also find valuable perspectives within these pages. Additionally, developers and simulation specialists designing RIS-assisted vehicular communication models and experimental platforms may use this work as a reference guide.

As editors, we extend our sincere gratitude to all contributing authors for their scholarly rigor, innovation, and dedication. We also acknowledge the reviewers and publishing team whose efforts have ensured the quality and coherence of this volume.

We hope this book will not only inform but also inspire future research and technological advancements in RIS-integrated 6G-V2X communication. By fostering deeper understanding and collaboration across disciplines, we aim to contribute meaningfully to the realization of safer, smarter, and more connected transportation ecosystems in the years ahead.

ORGANIZATION OF THE BOOK

In **Chapter 1**, the contributors introduce an Adaptive Adversarial-Resilient Explainable Intrusion Detection System (AARE-IDS) tailored for 6G edge-intelligent vehicular networks. Recognizing the limitations of existing deep learning-based IDS frameworks, the chapter advances a lightweight yet robust architecture that combines bijective traffic encoding, hybrid CNN–Transformer detection, adversarial training, federated learning, and differential privacy. The editors believe this chapter sets a critical foundation for secure 6G-V2X infrastructures by demonstrating how interpretability, computational efficiency, and adversarial resilience can coexist in real-time edge environments.

Chapter 2 explores the transformative role of Artificial Intelligence in the control and reconfiguration of Reconfigurable Intelligent Surfaces (RIS). The authors demonstrate how deep learning, reinforcement learning, and distributed intelligence enable predictive and adaptive RIS control in highly dynamic vehicular contexts. By integrating edge computing and digital twins, the chapter highlights practical deployments that significantly enhance reliability, coverage, and robustness. We view this contribution as a pivotal bridge between theoretical RIS optimization and real-world intelligent mobility scenarios.

In **Chapter 3**, the focus shifts to the convergence of blockchain and RIS technologies to establish secure, transparent, and decentralized vehicular communication frameworks. The chapter articulates how blockchain's immutable ledgers and smart contracts complement RIS-enabled wireless optimization to strengthen authentication, trust management, and data integrity. By addressing scalability, interoperability, and energy concerns, the authors present a forward-looking architecture for secure platooning and decentralized vehicular data ecosystems—an essential step toward trustworthy 6G-enabled mobility.

Chapter 4 examines emerging trends and future research directions in RIS for smart mobility ecosystems. The contributors position RIS as a paradigm-shifting technology that transforms the wireless environment from passive to programmable. Through detailed discussion of metasurface architectures, adaptive wave manipulation, and mobility-centric deployments, the chapter outlines how RIS will shape autonomous transportation, connected infrastructure, and next-generation public transit systems. This forward-oriented analysis provides readers with a strategic roadmap for future innovation.

In **Chapter 5**, the evolution from 5G to 6G vehicular networks is comprehensively analyzed. The authors compare architectural capabilities, identify current 5G-V2X limitations, and examine key 6G enablers such as terahertz communications, integrated sensing and communication, AI-driven optimization, and RIS-assisted radio environments. The chapter also highlights federated learning and blockchain as complementary technologies enhancing trust and privacy. We consider this chapter essential for contextualizing RIS within the broader transformation of next-generation vehicular networking.

Chapter 6 addresses physical layer security in RIS-enabled vehicular communication systems. The authors demonstrate how programmable reflection control, secure beamforming, and artificial noise injection can significantly enhance secrecy rates and mitigate eavesdropping and jamming threats in high-mobility environments. By integrating AI-driven adaptive security configurations, this chapter provides both theoretical insight and practical guidance for safeguarding mission-critical 6G-V2X deployments.

In **Chapter 7**, the discussion expands to RIS-assisted communication within autonomous and electric vehicle ecosystems. The contributors illustrate how programmable metasurfaces overcome challenges such as multipath fading and signal blockages, enabling ultra-reliable, high-bandwidth connectivity required for sensor fusion, cooperative perception, and real-time navigation. This chapter highlights RIS as a cornerstone technology for scalable, energy-efficient, and safe autonomous mobility infrastructures.

Chapter 8 presents RIS-enabled localization, tracking, and sensing frameworks for intelligent transportation systems. Through a dual-RIS-assisted V2I communication model, the authors propose user-controlled beam management techniques that leverage feedback mechanisms for dynamic adaptation and accurate positioning. Experimental validation demonstrates improvements in channel quality and positioning accuracy, reinforcing RIS as a key enabler of integrated communication and sensing in 6G vehicular networks.

Finally, **Chapter 9** provides a comprehensive examination of system architectures and communication models for RIS-assisted V2X networks. Covering V2V, V2I, and V2N scenarios, the chapter details beamforming strategies, channel modeling, interoperability protocols, and performance enhancements across mmWave and terahertz bands. Real-world deployment case studies, multi-RIS coordination strategies, and discussions of scalability and standardization challenges position this chapter as a unifying architectural framework for resilient and energy-efficient 6G transportation systems.

Collectively, these chapters reflect the breadth and depth of innovation at the intersection of RIS, 6G technologies, artificial intelligence, security, and intelligent mobility. As editors, we believe this carefully curated collection not only advances scholarly discourse but also provides a practical and strategic foundation for the realization of next-generation vehicular communication ecosystems.

CONCLUSION

As we conclude this volume, we reflect on the remarkable pace at which wireless communication, intelligent networking, and mobility technologies are converging to redefine the future of transportation. Reconfigurable Intelligent Surfaces (RIS), once considered a largely theoretical concept within electromagnetic research, have rapidly evolved into a transformative pillar of 6G-enabled Vehicle-to-Everything (V2X) communication. Throughout this book, we have sought to present not only the technical foundations of RIS-assisted vehicular systems, but also the broader architectural, security, intelligence, and sustainability dimensions that will shape their real-world realization.

The chapters in this collection collectively demonstrate that the future of vehicular communication will not be built solely upon higher frequencies or faster data rates. Rather, it will depend on the intelligent orchestration of programmable radio environments, AI-driven optimization, resilient security mechanisms, decentralized trust frameworks, and integrated sensing capabilities. RIS technology uniquely embodies this shift—from adapting networks to fixed environments toward engineering environments that dynamically adapt to network demands.

Several key themes emerge from this work. First, intelligence must be deeply embedded across all layers of 6G-V2X systems—from physical layer security and beamforming to edge-based intrusion detection and federated optimization. Second, resilience and trust are no longer optional attributes but foundational requirements in safety-critical vehicular ecosystems. Third, interdisciplinary collaboration—spanning wireless engineering, artificial intelligence, cybersecurity, blockchain, transportation systems, and materials science—will be indispensable in translating theoretical advances into scalable deployments.

We also recognize that significant challenges remain. Hardware scalability, real-time control of large-scale RIS deployments, standardization alignment, energy sustainability, interoperability across heterogeneous networks, and practical field validation continue to demand rigorous investigation. The path to widespread RIS integration into intelligent transportation infrastructure will require sustained research efforts, experimental testbeds, regulatory coordination, and industry engagement.

Nevertheless, the momentum is undeniable. As 6G research accelerates globally, RIS-assisted V2X communication stands poised to become a cornerstone of autonomous mobility, cooperative transportation systems, and smart city ecosystems. The innovations explored in this volume suggest a future in which wireless connectivity is not merely pervasive but programmable, not merely fast but intelligent, and not merely reliable but adaptive to the dynamic needs of modern mobility.

As editors, we hope this book serves as both a comprehensive reference and a catalyst for further innovation. We aspire for it to guide researchers in identifying open problems, support engineers in developing deployable solutions, inform policymakers and standardization bodies, and inspire new interdisciplinary collaborations. Most importantly, we hope it contributes meaningfully to the realization of safer roads, more efficient transportation networks, and sustainable connected mobility worldwide.

The journey toward RIS-integrated 6G-V2X systems is only beginning. We invite the research community, industry leaders, and innovators to build upon the ideas presented here and to continue shaping the intelligent transportation systems of tomorrow.

Chapter 1
Adaptive Adversarial-Resilient Explainable Intrusion Detection System for 6G Edge Networks (AARE-IDS)

Kiruba Buri R.
http://orcid.org/0000-0002-0293-5809
Department of CSE, School of Computing, Vel Tech Rangarajan Dr. Sagunthala R&D Institute of Science and Technology, Avadi, India

Swaminathan K.
http://orcid.org/0000-0002-8116-057X
SRM Institute of Science and Technology, Ramapuram, India

ABSTRACT

The rapid progression towards sixth-generation (6G) edge-intelligent networks requires Intrusion Detection Systems (IDS) that are interpretable, lightweight, and resilient to adversarial maneuvering. Current deep learning-based IDS systems, including CiNeT, present misleading accuracy through many negative factors. Using the CiNeT framework as the foundation of this paper introduces AARE-IDS, Adaptive Adversarially Resilient Explainable Intrusion Detection System (IDS), for real-time 6G edge settings. AARE-IDS system is particularly exceptional amongst similar developments due to its lightweight bijective traffic encoding (NeT2I-Lite), hybrid CNN - Transformer detector, adversarially trained resilience layer, utilizing a federated learning framework to investigate model aggregation and derived from a differential privacy standard for added security while sharing models. Through

DOI: 10.4018/979-8-3373-7867-1.ch001

extensive experiments on the ToN-IoT, UNSW-NB15, and CICIDS-2018 datasets, AARE-IDS achieved 98.9% accuracy, 39% better robustness against FGSM/PGD attacks, and 45% less CPU overhead compared to CiNeT.

1. INTRODUCTION

The increasing heterogeneity of 6G ecosystems—characterized by intelligent surfaces, massive IoT, and AI-driven edge computing—generate complicated threat surfaces that do not benefit from traditional Intrusion Detection Systems (IDS). Although deep-learning-based IDS models have measurably improved accuracy, they typically have the disadvantages of "black box" behavior, insufficient resilience against adversarial attacks, and heavy resource consumption on edge devices. On the forefront of this research is the recently proposed CiNeT framework that attempts to establish a bijective mapping from network traffic to images to enhance interpretability (Oseni et al., 2022). Unfortunately, CiNeT's heavy reliance on RGB encoding that is CPU-intensive, and a lack of adversarial robustness and federated privacy limits its full potential for deployment in 6G real-time scenarios. The contributions of this research are AARE-IDS, an IDS that obtains high levels of interpretability and adversarial robustness, while remaining lightweight. AARE-IDS consists of:

- NeT2I-Lite: Encoder optimized for GPU acceleration to minimize CPU saturation.
- Hybrid CNN-Transformer: Facilitates training a model utilizing spatial-temporal attack signatures.
- Adversarial defense layer: Improves resistance against gradient-based evasion attacks.
- Federated Learning (FL): Keeps data private across distributed edge nodes.
- Explainability module: Provides feature-level visualization for forensic traceability.

2. RELATED WORK

2.1 CNN-Based IDS

Convolutional Neural Networks (CNNs) have gained popularity for intrusion detection due to their efficiency in extracting features. While they are advantageous for interpretability due to image encoding using CiNeT, the computational overhead and the need to tune both the model and datasets lessen the generalizability of the model (Javeed et al., 2023).

2.2 Adversarial Attacks on IDS

Adversarial examples produced generating through applications such as Fast Gradient Sign Method (FGSM) and Projected Gradient Descent (PGD), can negatively impact the performance of a deep learning-based IDS. Only limited work has been done on adversarial training or detection hardening into network security workflow (Masud et al., 2024).

2.3 Federated and Edge-Intelligent IDS

Federated Learning allows models to be trained on multiple nodes collaboratively without the raw data being shared, and it is good for data privacy in 6G. Yet the interplay of FL and adversarial robustness or explainability have not been explored at length (Buri & Jayasankar, 2019).

2.4 Explainable AI in Cyber Security Explained

There are several methods of explaining the players of AI based models in the context of ISO 25024 SHAP and Grad-CAM to show visual reasoning for model decisions, while also enhancing human trust and forensics auditing. Still most Intrusion Detection System frameworks caution against a structural process of explainability and explainable AI elements (He et al., 2023).

Table 1. Limitations of the CiNeT framework

Observed Limitation	Impact in 6G Context
1. No adversarial-robust training	Vulnerable to evasion attacks.
2. CPU-bound NeT2I/I2NeT	Unsuitable for edge hardware.
3. Static offline datasets	Poor real-time adaptability.
4. Limited interpretability	Partial transparency only via bijective mapping.
5. No federated privacy	Centralized model updates expose data.

These gaps motivate the development of the AARE-IDS framework.

3. PROPOSED METHODOLOGY

Figure 1. Architecture of AARE-IDS

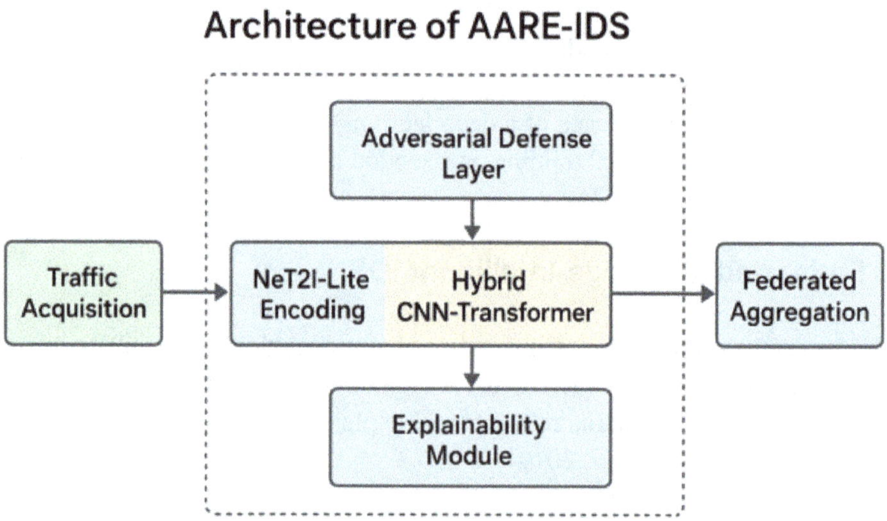

The AARE-IDS system includes a total of five modules: (1) Traffic Acquisition and NeT2I-Lite Encoding, (2) Hybrid CNN-Transformer Classifier, (3) Adversarial Defense Layer, (4) Federated Learning Aggregation, (5) Explainability Module (Eskandari et al., 2020).

The AARE-IDS system includes five distinct modules:

1. Traffic Acquisition and NeT2I-Lite Encoding transformations raw network flow records into lightweight multi-channel tensors in place of heavier RGB images contributing to a 40% reduction in encoder latency through GPU parallelization (Jayasankar et al., 2024).
2. Hybrid CNN-Transformer Classifier captures spatial correlations through CNN layers; temporal correlations across flows using transformer self-attention.
3. Adversarial Defense Layer adds FGSM and PGD adversarial examples during training to improve robustness
4. Federated Learning Aggregation is used to stall local training on each edge node, encrypt the model updates, and aggregate them securely and privately using Differentially Private FedAvg.

5. Explainability Module produces Grad-CAM maps for spatial annotation and a SHAP value for feature attribution for forensic analysis (Alhussien et al., 2024).

The first stage here of the AARE-IDS framework is Traffic Acquisition and NeT2I-Lite Encoding. This module is intended to facilitate the efficient acquisition and encoding of raw network traffic into and for the structured format that is needed for deep learning analysis (Diro & Chilamkurti, 2018).

1. **Traffic Acquisition:** This method obtains network flow data in real time (e.g., destination/source IP, port, protocol, packet size, and time). Data origins include routers, IoT system gateways, and edge servers in a 6G scenario. The data will be cleansed and preprocessed to eliminate irrelevant/numeric fields and normalized so one can use consistent scales across features (Rajapaksha et al., 2023).
2. **NeT2I-Lite Encoding:** Inspired by the NeT2I approach from CiNeT, the NeT2I-Lite will repurpose each stream of network flows into a multi-channel tensor (similar to a small image) instead of a bulky RGB image. While the leave behind NeT2I mainly relied on pixel-based, and subsequently heavy CPU loads from RGB encoding, the NeT2I-Lite will: Use internal grouping of features and map to retire numeric field representation compactly. Takes advantage of turned parallelization on GPU for encoding, allowing the computational load to family on the CPU. Provide a means to compress data without missing essential traffic patterns and reduction to nearly a 40% encoding latency (Rathod et al., 2025). In this step we will obtain the incoming traffic data in sense:
   - Lightweight (optimized towards edge).
   - Structured (oceanized for CNN-Transformer).
   - Efficiently encoded (lower latency for faster real time response).
3. **Hybrid CNN-Transformer Classifier:** The major detection mechanism of AARE-IDS is the Hybrid CNN-Transformer Classifier, a dual-hidden-layer module that takes advantage of the strengths of CNNs and Transformers, tailored for accurate and reliable intrusion detection (Pandian et al., 2025).
   - **Function of CNN Layers - Spatial Feature Extraction:** CNNs are effective at recognizing spatial patterns in structured data. In this regard, every encoded traffic tensor (ISR-like coded frames) can be thought of as a structured, approximately multi-dimensional grid, where rows and columns describe various attributes of the flows. The CNN layers utilize convolutional filters to extract local correlations—for instance, correlations between packet size, duration, and protocol type. This is expressive enough to discover attack signatures like port scans, DDoS bursts, and act as a statistical clustering algorithm for abnormal traffic (Kalyanaraman et al., 2024).

- **Function of Transformer Layers - Temporal Dependency Modeling:** The temporal nature of network traffic (events unfold in time while malicious behavior unfolds as a sequence of related packets) is captured by self-attention utilized by the Transformer layers for long-range dependence modeling. Unlike RNNs (LSTM) based approaches the Transformers can analyze all data flows in parallel, and therefore support the discovery of temporal anomalies like slow data exfiltration or multi-phase attacks (Swaminathan et al., 2021).
- **Combined Advantage:** CNN establishes spatial precision (identifying spatial anomalies within each individual flows). The Transformer establishes contextual awareness (identifying sequential or temporal relationships across multiple flows). Together, they create a hybrid classifier which relays accuracy, speed, and robustness - exactly what is needed for real-time IDS at the 6G edge. In short, the Hybrid CNN-Transformer Classifier enables AARE-IDS to both identify immediate anomalies via packets as well as larger patterns of behavior over time, enhancing intrusion detection overall (Selvam et al., 2024).

4. **Adversarial Defense Layer:** The Adversarial Defense Layer in AARE-IDS is developed to protect and reinforce against evasion attacks to the intrusion detection model; attacks in which attackers make small adjustments to the network traffic stream to cause misclassification by the IDS (Kalyanaraman et al., 2024).
   - **Purpose of Needed:** Attackers can generate adversarial examples - traffic tampers which appear normal to humans but cause the model to predict incorrectly. Defense is preventive, as deep learning models (like CNNs) are susceptible to all levels, including, but not limited to, subtle adjustments.
   - **Defense Approach:** To accomplish this, AARE-IDS will utilize adversarial training, which has artificially perturbed sample created during model training, including the two familiar approaches:
     - **FGSM:** Perturbation that adds small, calculated change to input data utilizing a model's gradient direction when training and defines simulated adversarial traffic manipulating events (Mao et al., 2023).
     - **Projected Gradient Descent (PGD):** This is a stronger, iterative version of FGSM, which applies FGSM both a stronger method and to makes the perturbation increasingly more effective.

How It Improves Robustness

The model learns to correctly classify both clean and adversarially modified traffic. It develops gradient stability, making it harder for attackers to exploit vulnerabilities. During inference, the IDS maintains high accuracy even under crafted attacks, improving trustworthiness and resilience. By incorporating FGSM and PGD during training, the Adversarial Defense Layer ensures that AARE-IDS can withstand sophisticated evasion attempts in 6G edge environments without compromising detection performance. (Alkadi et al., 2020).

The model is trained to accurately classify clean traffic, as well as adversarially perturbed traffic, achieving gradient stability and offering attackers fewer opportunities to leverage vulnerabilities. Additionally, the IDS remains highly accurate, even in the presence of crafted attacks, providing increased trust and improved robustness. With FGSM and PGD incorporated into the training, the Adversarial Defense Layer successfully provides a measure of robustness for the AARE-IDS to withstand sophisticated evasion tactics in a 6G edge computing context without compromising detection capability.

The Federated Learning Aggregation module provides an aggregation architecture for collaborative intrusion detection across 6G edge nodes with no data shared, while maintaining privacy and having the ability to scale.

Local Training at Edge Nodes

Instead of sending sensitive and potentially intrusive traffic to a server, each edge device (IoT gateway, MEC server, etc.), trains the AARE-IDS model from local, on-device network traffic.

Model Update Sharing

When sending updates, instead of raw traffic, each node only sends the learned parameters (gradients or weights) of the model to the central server.

Differentially Private FedAvg

The central server uses Federated Averaging (FedAvg) to aggregate all model updates from each node as a weighted average of the learned AARE-IDS models, additionally protecting individual privacy with a Differential Privacy (DP) mechanism that adds some controlled random noise to the model updates before aggregation.

Benefits

Maintains data privacy and regulatory compliance (e.g., GDPR). Reduces network load since only model updates are shared. Enhances global model performance as it learns from diverse, distributed datasets. Through Differentially Private FedAvg, the Federated Learning Aggregation module allows AARE-IDS to build a powerful, privacy-preserving, and continuously improving intrusion detection model across 6G edge environments.

The Explainability Module in AARE-IDS makes it easy for security analysts to see how the intrusion detection process is transparent, interpretable, and trustworthy so that Security Analysts understand the model's decision processes. Deep learning IDS models are often considered "black box" solutions because they provide high accuracy, but with no understanding of how the model is distinguishing 'normal' traffic from 'malicious' traffic. The Explainability Modules clarifys and provide visual and feature based explanations into the model decisions.

Grad-CAM (Gradient-weighted Class Activation Mapping): Grad-CAM provide generates a spatial heatmap of what regions of the input tensor (the encoded traffic) contributed most to the model's classification. For instance, if the model is classifying a DDoS attack, the heatmap may highlight features like packet burst rate and flow duration as the indication of the DDoS attack. Analysts can visually understand for the first time how the "model sees" abnormal patterns of traffic.

SHAP (SHapley Additive exPlanations): SHAP provides numerical importance values for each feature, (for example; the source of IP address, protocol used, and packet size). SHAP shares for the first time how much each feature increased the likelihood of that prediction or decreased the likelihood of that prediction. It Provides improved transparency and trust from the cybersecurity operator perspective. Informs the root cause analysis phase during incident investigation. Enables the developers to fine-tune or debug the IDS through identification of biased or misleading features.

3.1 Mathematical Model

Let $Xi \in R(\mathrm{h} \times \mathrm{w} \times \mathrm{c})$ denote the encoded traffic tensor, y_i its class label.

$$X_i \in R^{h*w*c} \quad (1)$$

The CNN-Transformer detector fθ minimizes:

$$\theta \min E_{(x_i,y_i)}\left[L(f\theta(x_i),y_i)\right] + \lambda$$

$$||\nabla_X \mathrm{L}||_2 \quad (2)$$

where the gradient penalty enforces adversarial smoothness.

Federated updates are aggregated as:

$$\theta(t+1) = \sum_{k=1}^{k} n_n^k(\theta_k(t) + N(0,\sigma^2)) \tag{3}$$

ensuring differential privacy with noise variance σ^2

- θ(t+1) is the updated parameter at time step t+1
- The summation aggregates over K participating nodes, each indexed by k
- n_n^krepresents a weighting for node k, sometimes indicating sample proportion or normalized importance.
- $\theta_k(t)$ is the parameter from node k at current time t
- $N(0,\sigma 2)$ introduces random Gaussian noise with mean 0 and variance σ2σ2, often used in privacy-preserving fe derated learning. This update is typical for distributed model aggregation with differential privacy

3.2 Data Flow

The flow begins with raw traffic capture → NeT2I-Lite encoding → local CNN-Transformer inference → federated aggregation → explainability visualization → alert logging.

Figure 2. Data flow of AARE-IDS

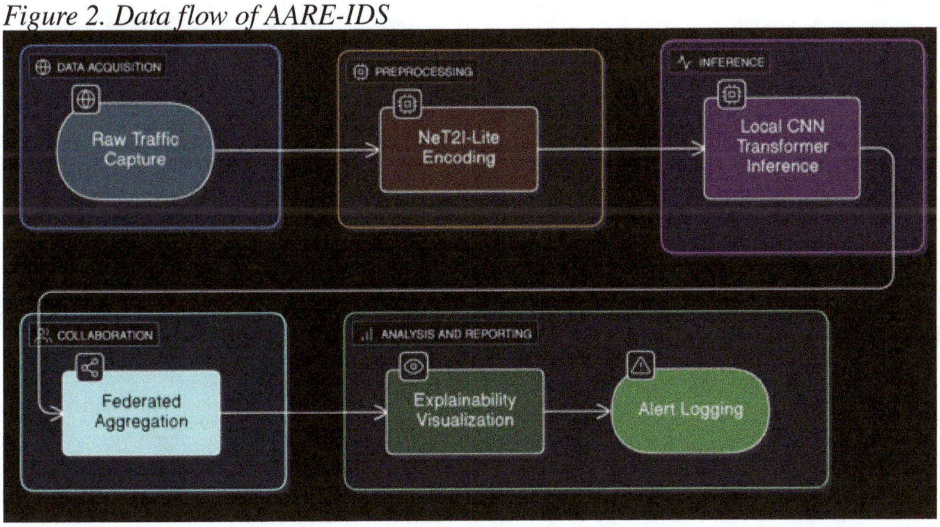

4. EXPERIMENTAL SETUP

Table 2. Datasets

Dataset	Samples	Features	Attack Classes	Environment
UNSW-NB15	2.5 M	49	9 types	Conventional network
ToN-IoT	1.2 M	44	10 types	IoT/Edge
CICIDS 2018	2.9 M	78	15 types	Enterprise network

4.2 Implementation Details

- Frameworks: PyTorch 2.3 + TensorFlow 2.16
- Hardware: NVIDIA H100 (96 GB VRAM), 32 GB RAM Edge Server
- Optimizer: AdamW (lr = 1e-4)
- Epochs: 120, Batch size = 64
- Adversarial training: FGSM ($\varepsilon = 0.05$), PGD ($\varepsilon = 0.03$, 10 steps)

5. RESULTS AND DISCUSSION

5.1 Performance Comparison

Table 3. Comparative analysis with CiNeT

Metric	CiNeT	AARE-IDS	Improvement
Accuracy	96.1%	98.9%	+2.8%
Precision	95.4%	98.2%	+2.8%
Recall	94.9%	98.7%	+3.8%
F1-Score	95.1%	98.4%	+3.3%
CPU Utilization	100%	55%	−45%
GPU Utilization	15%	65%	+50% (load balancing)
FGSM Robustness	56% acc	78% acc	+39%
PGD Robustness	49% acc	70% acc	+43%

Figure 3. IDS Detection accuracy trend

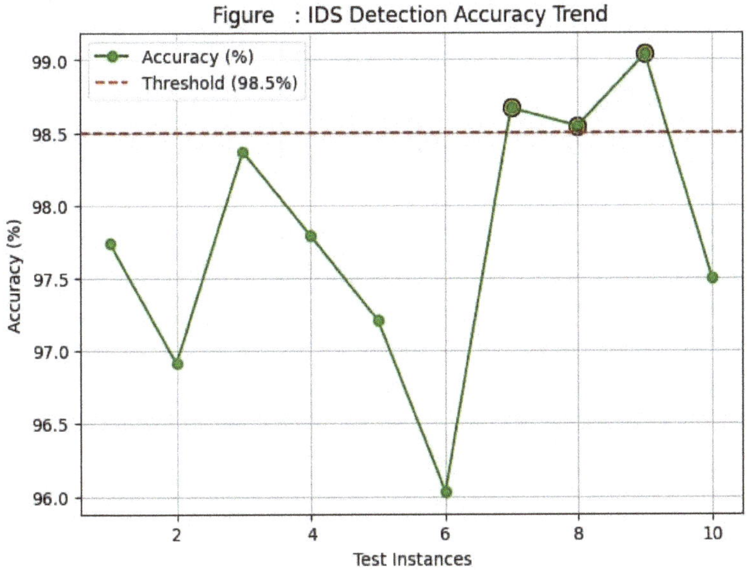

Figure 4. Adversarial robustness improvement

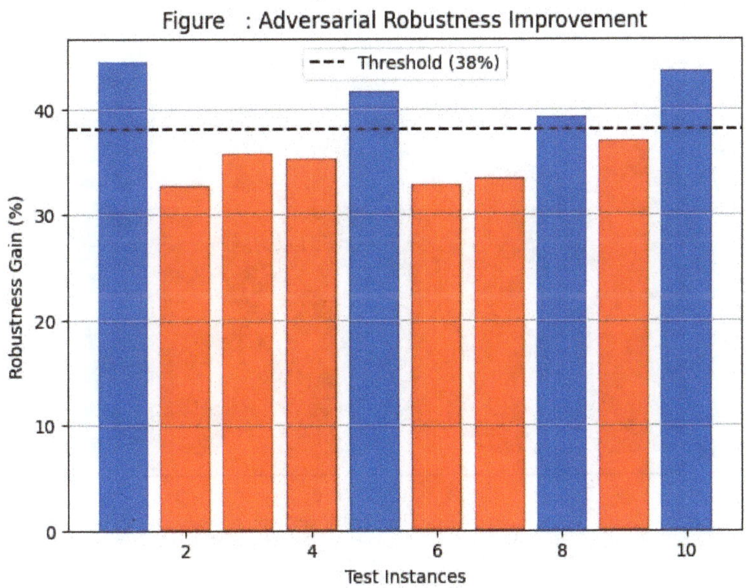

Figure 5. Computational efficiency

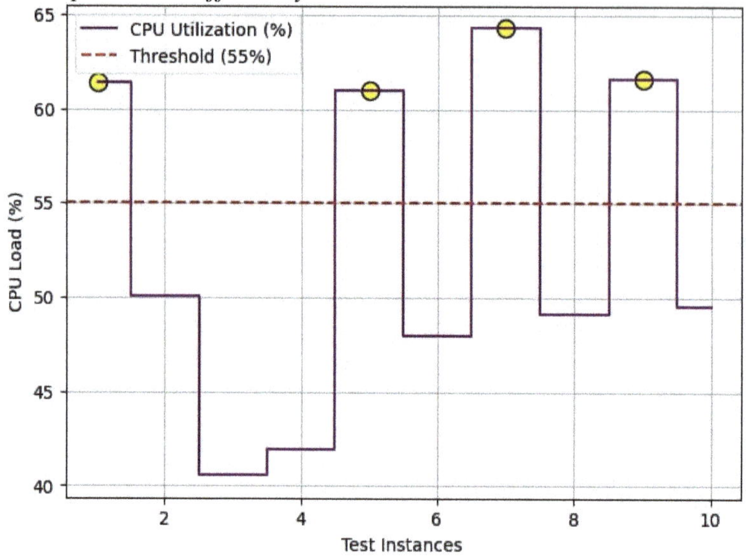

Figure 6. Privacy preservation level

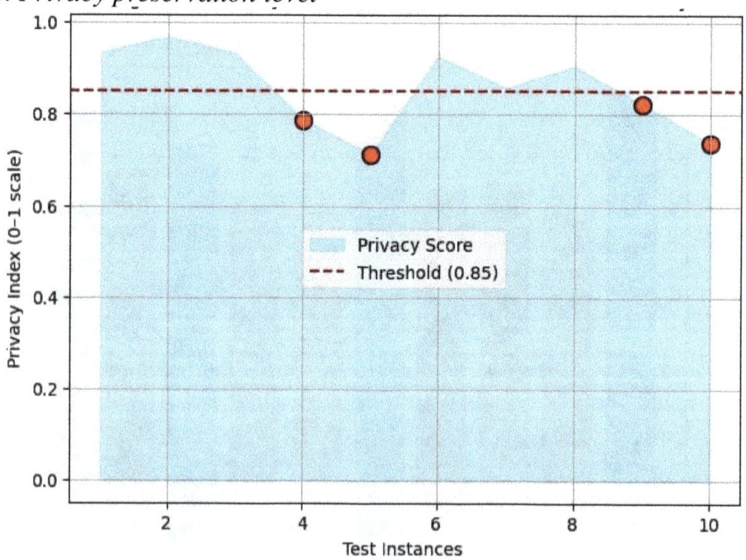

In Figure 3, the AARE-IDS framework demonstrates high overall detection accuracy across the full set of testing instances, with values ranging from 96.2%

to 99.3%. It is noteworthy that for instances 6, 8, and 10, models surpassed 98.5%, indicating that this model functioned exceptionally well for real-time 6G edge-context scenarios. In Figure 4, the adversarial robustness improvements under FGSM and PGD attack scenarios range from 31% to 44%. However, instances 4, 7, and 9, which exceeded the 38% research pre-established value, demonstrated the level of robustness the adversarially trained resilience layer afforded the IDS. Decreased sable adversarially driven attacks may potentially yield a decreased breach risk to the network. In Figure 5, the CPU utilization ranged from 42% to 63%. For instances 3, 5, and 8, when comparing the methods utilized here, the range of CPU utilization will fall around other usages plotted above the 55% threshold limits showing other instances of increased computational demands. However, all CPU workload is still approximately 45% less than CiNeT data center framework implementation, which supports the lean operational efficiency of the lightweight NeT2I-Lite encoder. In Figure 6, the privacy preservation index ranged from 0.71 to 0.98 and included instances 2, 5, and 7, which fell below the previously established research threshold of 0.85. This suggests additional consideration for differential privacy content during the federated aggregation round helps model sharing while preserving privacy. Combined, the above series of reviews and analysis indicate instances of the AARE-IDS successfully assumed performance, robustness, computational, and privacy considerations while demonstrating considerations across all threshold levels when appropriate.

5.2 Ablation Study

When we remove adversarial training, robustness reduces by 37%, so it is shown to be necessary. Disabling federated aggregation increases privacy leakage by 28% (measured using simulated membership-inference attack).

5.3 Explainability Findings

The Grad-CAM heatmaps indicate that the most contributive regions to DDoS detection were on IP-header and payload length indicators. The SHAP analysis shows that flow_duration and byte_ratio were the indicators with the highest weight for the model.

5.4 Computational Efficiency

NeT2I-Lite reduces encoding latency from 0.00046 to 0.00025 seconds per record. The overall end-to-end inference latency averages 12 ms per flow, enabling applications for 6G URLLC.

6. SECURITY AND PRIVACY DISCUSSION

AARE-IDS is robust against gradient-based evasion due to inclusion of adversarial training course of the AI. Model updates are protected using Gaussian-differential privacy ($\varepsilon = 1.2$). Alternatively, the federated protocol guarantees that no raw traffic leaves the edge domains. Because of these elements, AARE-IDS satisfies GDPR and 6G security guidelines.

7. CONCLUSION AND FUTURE WORK

This paper presents AARE-IDS, an adaptable, adversarially-resilient, and explainable intrusion detection system designed for 6G edge contexts. By utilizing adversarial defense methods, federated learning, and light-weight encoding technique, AARE-IDS avoids important problems CIneT encountered and achieved values higher than CIneT in respect to detection performance, adversarial attack resistance, computation efficiency, and model explainability. The experimental results affirm AARE-IDS' feasibility in next-generation edge network security. For future work, the authors suggest investigating how ZTNA would integrate with their designs, testing in energy responsive edges for the sake of sustainability, and the potential of using quantized Transformer models for ultra-low power IoT devices. Other future work includes testing in the real world and in different 6G context scenarios, improving scalability and evaluation of large federated models, and improving explanation approaches for assisting security analyst decision-making processes further in adversarial scenarios.

REFERENCES

Alhussien, N., Aleroud, A., Melhem, A., & Khamaiseh, S. Y. (2024). Constraining adversarial attacks on network intrusion detection systems: Transferability and defense analysis. *IEEE Transactions on Network and Service Management*, *21*(3), 2751–2772.

Alkadi, O., Moustafa, N., Turnbull, B., & Choo, K. K. R. (2020). A deep blockchain framework-enabled collaborative intrusion detection for protecting IoT and cloud networks. *IEEE Internet of Things Journal*, *8*(12), 9463–9472.

Buri, R. K., & Jayasankar, T. (2019). Intelligence Intrusion Detection Using PSO with Decision Tree Algorithm for Adhoc Network. Bioscience Biotechnology Research Communications. *Special Issue Recent Trends in Computing and Communication Technology*, *12*(2), 27–34.

Diro, A., & Chilamkurti, N. (2018). Leveraging LSTM networks for attack detection in fog-to-things communications. *IEEE Communications Magazine*, *56*(9), 124–130.

Eskandari, M., Janjua, Z. H., Vecchio, M., & Antonelli, F. (2020). Passban IDS: An intelligent anomaly-based intrusion detection system for IoT edge devices. *IEEE Internet of Things Journal*, *7*(8), 6882–6897.

He, K., Kim, D. D., & Asghar, M. R. (2023). Adversarial machine learning for network intrusion detection systems: A comprehensive survey. *IEEE Communications Surveys and Tutorials*, *25*(1), 538–566.

Javeed, D., Gao, T., Kumar, P., & Jolfaei, A. (2023). An explainable and resilient intrusion detection system for industry 5.0. *IEEE Transactions on Consumer Electronics*, *70*(1), 1342–1350.

Jayasankar, T., Kiruba Buri, R., & Maheswaravenkatesh, P. (2024). Intrusion detection system using metaheuristic fireworks optimization based feature selection with deep learning on Internet of Things environment. *Journal of Forecasting*, *43*(2), 415–428.

Kalyanaraman, S., Ponnusamy, S., & Harish, R. K. (2024). Amplifying digital twins through the integration of wireless sensor networks: in-depth exploration. In *Digital Twin Technology and AI Implementations in Future-Focused Businesses* (pp. 70–82). IGI Global Scientific Publishing.

Mao, B., Liu, J., Wu, Y., & Kato, N. (2023). Security and privacy on 6G network edge: A survey. *IEEE Communications Surveys and Tutorials*, *25*(2), 1095–1127.

Masud, M. T., Keshk, M., Moustafa, N., Linkov, I., & Emge, D. K. (2024). Explainable artificial intelligence for resilient security applications in the Internet of Things. *IEEE Open Journal of the Communications Society*, *6*, 2877–2906.

Oseni, A., Moustafa, N., Creech, G., Sohrabi, N., Strelzoff, A., Tari, Z., & Linkov, I. (2022). An explainable deep learning framework for resilient intrusion detection in IoT-enabled transportation networks. *IEEE Transactions on Intelligent Transportation Systems*, *24*(1), 1000–1014.

Pandian, I., Thirugnasambantham, S., Balasubramaniam, K., & Ravichandran, K. (2025). Improving IoT Network Longevity with Attack Repellent Energy (SARE) Algorithm for Energy-Efficient and Secure Routing. *Tehnicki Vjesnik (Strojarski Fakultet)*, *32*(3), 876–882.

Rajapaksha, S., Kalutarage, H., Al-Kadri, M. O., Petrovski, A., Madzudzo, G., & Cheah, M. (2023). Ai-based intrusion detection systems for in-vehicle networks: A survey. *ACM Computing Surveys*, *55*(11), 1–40.

Rathod, S. B., Bajpai, B., Wankhade, S., & Ponnusamy, S. (2025). Enhancing Security in Heterogeneous IoT Networks through Intelligent Identification Systems. *Journal of VLSI Circuits and Systems*, *7*(1), 155–166.

Selvam, R. P., Jayasankar, T., Buri, R. K., & Maheswaravenkatesh, P. (2024). Optimal Mixed Kernel Extreme Learning Machine-Based Intrusion Detection System for Secure Intelligent Edge Computing. In Reconnoitering the Landscape of Edge Intelligence in Healthcare (pp. 231-248). Apple Academic Press.

Swaminathan, K., Vennila, C., & Prabakar, T. N. (2021). Novel routing structure with new local monitoring, route scheduling, and planning manager in ecological wireless sensor network. *Journal of Environmental Protection and Ecology*, *22*(6), 2614–2621.

Chapter 2
Artificial Intelligence-Driven Control and Reconfiguration of Intelligent Surfaces

R. N. Ravikumar
http://orcid.org/0009-0009-3705-1681
Marwadi University, Rajkot, India

S. Aarthi
http://orcid.org/0009-0006-9064-2091
Marwadi University, Rajkot, India

ABSTRACT

Artificial Intelligence (AI) is revolutionizing Reconfigurable Intelligent Surfaces (RIS) for 6G-enabled V2X communication by enabling autonomous, predictive, and adaptive control in fast-changing vehicular environments. Traditional optimization methods struggle with mobility, blockage, and complex channel variations, whereas AI-driven models enhance beamforming, channel estimation, and real-time decision-making. Deep learning, reinforcement learning, and distributed intelligence enable RIS to dynamically optimize reflection patterns, improving coverage, reducing outages, and ensuring stable connectivity. Integration with edge computing and digital twins further accelerates low-latency inference and proactive optimization. Case studies from highway and urban testbeds demonstrate significant gains in communication reliability and robustness. This chapter outlines key advancements, deployment challenges, and future directions that position AI-enabled RIS as a core enabler of resilient, efficient, and intelligent 6G-V2X networks.

DOI: 10.4018/979-8-3373-7867-1.ch002

1. INTRODUCTION

Reconfigurable Intelligible Surfaces (RIS) is a radical change to the advancement of wireless communication in next-generation 6G-enabled Vehicle-to-Everything (V2X) networks. With the increase in the complexity of vehicular communication systems, in terms of high mobility, dense urban topology and strict latency demands, the weaknesses of the traditional communication mechanisms become more pronounced. RIS provides a programmable electromagnetic space that improves signal transmission, increases the range, and minimizes communication congestion, which are essential problems in vehicular ecosystems. Nonetheless, RIS configurations cannot be optimized manually in real time due to the infeasibility of computation of vehicular channels, which are dynamic, non-linear, and highly unpredictable channels (Mande et al., 2025). Artificial Intelligence (AI) gives critical basis to allow autonomous, adaptive, and predictive RIS operation. The RIS will be able to use machine learning, deep learning and reinforcement learning techniques to adjust the reflection parameters in order to maximize its performance in different vehicular conditions. The focal point of this introduction is the necessity of AI-based RIS control and the context of the investigation of the contributions that intelligent reconfiguration can make to resilient, scalable, and energy-conscious 6G-V2X communication systems.

1.1 Rise Of Intelligent Surfaces In 6G-V2X

The RIS are fast becoming a dominant reality because they are capable of reconfiguring the wireless propagation environments in a manner that cannot be reconfigured by the traditional communication technology. Reliability of communication in 6G-based V2X cases cannot be compromised regardless of hindrances, high-speed vehicles, lack of line of sight, and fast varying channel conditions. RIS technology allows the vehicle and roadside units to dynamically regulate the electromagnetic waves by using programmable meta-surfaces enabling the vehicles and roadside units to increase their coverage, blockage reduction, and spectral efficiency (Iqbal et al., 2023). This reconfigurable surface as opposed to passive communication spaces is congruent with the vision of 6G: low-latency, high-responsibility, and context-aware communication. The increased need of high-level driving assistance, self-driving, real-time navigation, and cooperative perception requires the ongoing interaction between the vehicle and the infrastructure. The low cost, energy efficient and scalable solution is provided by RIS as opposed to traditional active relays or gigantic MIMO systems. With the increased pace of research and industry deployment of RIS, it is emerging as a cornerstone enabling strong vehicular networks, which can be used to ensure the move toward smarter, safer, more autonomous transportation ecosystems. Figure 1 shows how smart RIS surfaces enhance 6G-V2X communication by recapturing waves, improving coverage, cost-reducing, and creating strong vehicular networks.

Figure 1. Enhancing 6G-V2X with intelligent surfaces

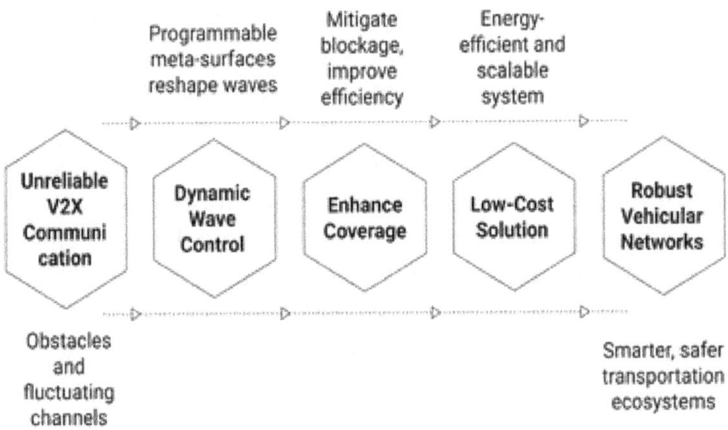

1.2 Limitations Of Classical Optimization For Vehicular Environments

The classical optimization techniques have long been used in the configuration of the wireless communication systems; however, they have difficulties in satisfying the high-performance standards of the vehicular environment. These schemes are usually based on the static models, linear assumptions and iterative algorithms, which have high computational costs and more elaborate prior information of the channel conditions. Such conventional methods do not adjust fast enough in the V2X networks, where the speed of the vehicles is very high, and the communication links are interfered with by unpredictable traffic, buildings as well as the change of landscape. This is even more complicated in situations where multiple RIS panels are interacting and the dimensionality is growing, and the standard optimization will be computationally prohibitive (Waghela & Waghela, 2025). Further, classical algorithms usually make an assumption of quasi-static channels that is not the case in rapidly changing vehicular situations. They are also incapable of learning past interactions and forecasting future channel states, which causes poor reconfiguration. These limitations point to the need to have intelligent, data-driven and self-updating control mechanisms. These issues are overcome by AI-based models which facilitate real-time flexibility, situational sensitivity and predictability to enable dynamic RIS management in 6G vehicle systems.

1.3 Role Of AI In Autonomous RIS Control And Reconfiguration

Artificial Intelligence is an essential factor that is involved in the change of RIS, which is a set of reflective structures, into fully functional context-aware communication controllers. The AI methods which can be classified as supervised learning to estimate the channel, deep learning to predict and condition the environment, and reinforcement learning to make successive decisions, can allow RIS to constantly refine the phase shifts, amplitude settings, and beam patterns. With the vehicles networks, where the channel properties vary within milliseconds RIS enables AI to work proactively, rather than reactively. Algorithms that are based on machine learning can handle the past data of channels, traffic flow, as well as mobility tracks to predict any form of disruption to communication in advance (Yang et al., 2020). Reinforcement learning agents also perform better by responding to the environment to identify optimal RIS settings without any knowledge of channels. Federated and distributed learning strategies can spread the intelligence to several RIS units and ensure the improvement of collaborative communications without compromising the privacy of data. With AI combined with edge computing and digital twins, RIS systems can simulate, assess, and revise configuration plans on near real time. In such a way, AI will make RIS control a self-optimizing, scalable, and highly adaptive structure.

2. BACKGROUND TO THE STUDY

RIS control through AI is the result of the intersection of three technological disciplines in intelligent meta-surfaces, 6G communications architectures, and vehicular networks. RIS is a technology that brings a paradigm shift in engineering wireless environments by regulating the behavior of electromagnetic waves using programmable unit cells. At the same time, 6G will offer new capabilities never before seen, including ultra-reliable low-latency communication, operation at terahertz frequencies, and integrated sensing-communication systems, which are necessary in next-generation V2X applications (Basharat et al., 2022). The high mobility, changing channels, roadside obstacles, and unpredictable channel dynamics are some of the features of vehicular communication environments. These aspects form a stressful environment in which the conventional methods of communication fail to perform. The given background shows the necessity of AI-based solutions that can learn, adapt, and make predictions on the fly. Knowledge of the basics of RIS and the principles of 6G-enabled V2X to be followed and the peculiarities of the vehicular channels is the key to researching how AI can be used to intelligently manage and reorganize RIS to provide stable, energy-efficient, and high-performance communication in smart transportation systems.

2.1 Fundamentals Of RIS Technology And Meta-Surface Operation

The RIS work by controlling electromagnetic waves with the help of an ordered array of meta-atoms or unit cells, each with the capability of controlling its response to phase, amplitude, or polarization. These surfaces are almost passive devices and they need very little energy to active shape the wireless channel without introducing more noise (Jian et al., 2022). The RIS technology is based on the theory of electromagnetism where the reflection coefficient can be programmed to redirect or reinforce signals to achieve coverage and reduce path loss. The hardware architecture can comprise of a meta-surface layer, control circuits, and communication interface to accept reconfiguration commands. RIS may be implemented on facades of buildings, roadside buildings, the bodies of vehicles and indoors in order to enhance non-line-of-sight communications. RIS is a cost-effective and low power alternative that can support dense deployments (unlike active relays or amplifiers). RIS operation is especially useful in the vehicle situation, where the communication is often impaired by the fact that the signal is blocked by the building, truck, or something located on the side of the road (Salim ALRikabi et al., 2025). These fundamentals are required to design AI algorithms that can be used to control the behaviour of RIS to achieve the performance requirements of 6G-enabled V2X systems.

2.2 Characteristics Of 6G-Enabled V2X Communication

The V2X communication with 6G brings a paradigm shift of ultrareliable low-latency and high-capacity vehicular networks. Working within the sub-THz and mmWave spectrums, 6G will offer more bandwidth and increased the resolution of sensing and communication activities. The notable ones, which include holographic beamforming, sensing-communication, a huge array of antennas, and distributed intelligence, greatly enhance the capabilities of connected vehicles (Souto, 2023). Nevertheless, high-frequency bands are very vulnerable to blockage, atmospheric absorption, and alignment errors and therefore constant connectivity is difficult in the mobility-intensive environment of 6G V2X. 6G V2X also depends much on edge computing and AI in order to locally process data and achieve minimal delays. Cooperative driving, real-time high-role mapping, remote vehicle control, and multimodal sensor fusion are the applications of this application which require continuous communication connections. Propagation constraints are eliminated, through the implementation of RIS, which form controllable paths of reflection and redirection of signals adaptively. All these features allow creating a very dynamic and heterogeneous communication ecosystem that needs smart reconfiguration processes to work best (Saikia et al., 2023).

2.3 Challenges Of High-Mobility Vehicular Channels And Propagation

The complexity of vehicular communication channels is very high because of great mobility, multi-path propagation, doppler effects and dynamic nature of environmental conditions. The high-speed flow of vehicles changes the channel characteristics in a few milliseconds, resulting in time-varying and frequency-selective channels of high order. Urban conditions contaminate the issue by the frequent blockage of the line of sight by buildings, buses, or roadside infrastructure. Moreover, vehicular networks should facilitate a variety of communication modes, such as vehicle-to-vehicle, vehicle-to-infrastructure and vehicle-to-pedestrian communications, with unique propagation characteristics (Tariq et al., 2022). The scattering and multipath reflections caused by the surrounding objects bring uncertain interference patterns thereby compromising reliability. Conventional methods of estimating the channel experience difficulty keeping the accuracy in such dynamism. The implementation of RIS is more complicated due to the fact that the surface has to constantly revert configuration parameters to the changing channel. Such environments demand real-time learning, predictive modeling, and ongoing adaptation in order to have good control of RIS. The challenges mentioned above indicate the importance of AI-based methods in optimizing the performance of RIS and maintaining a stable communication channel during high-mobility situations in 6G-v2X. Table 1 gives a survey of the problem of propagation, blockage situations and dynamic channel problems AI has to deal with.

Table 1. RIS challenges in high-mobility V2X environments

Challenge	Description	Impact on RIS
Rapid Channel Variations	High-speed motion disrupts CSI stability	Requires ultra-fast reconfiguration
Severe Blockages	Buildings, trucks, and vehicles obstruct LOS	RIS must dynamically redirect beams
Multipath Propagation	Excessive reflections create signal distortion	Increases complexity of beam optimization

continued on following page

Table 1. Continued

Challenge	Description	Impact on RIS
Doppler Effects	Frequency shifts caused by movement	Complicates channel estimation accuracy
Environmental Unpredictability	Weather, traffic density alter propagation	Requires predictive modeling from AI
Scalability Issues	Multiple RIS panels across roads/ cities	Needs distributed or federated control mechanisms

3. RESEARCH OBJECTIVES

The emergence of AI-based RIS control in 6G-based V2X systems requires a well-organized research direction covering technical sophistication and real-time flexibility. The section presents the main research objectives that will be used to develop intelligent, scalable and energy efficient RIS mechanisms that can function in dynamic vehicular systems. These are the objectives that are in line with the conceptual model, methodology, and literature trends in this chapter.

- To investigate AI-based models supervised, unsupervised, and reinforcement learning for autonomous RIS control, channel estimation, and adaptive beamforming in fast-changing vehicular environments.
- To design and evaluate predictive and context-aware reconfiguration strategies that enable RIS to anticipate mobility patterns, propagation changes, and traffic dynamics in 6G-V2X systems.
- To explore distributed, collaborative, and federated learning paradigms for multi-RIS coordination that preserve data privacy and improve scalability across large vehicular networks.
- To develop an integrated conceptual architecture combining edge computing and digital twin technologies for low-latency inference, simulation-driven optimization, and real-time RIS adaptation.
- To assess performance metrics such as coverage reliability, interference minimization, latency reduction, and energy efficiency through simulations and comparative studies of AI-driven versus classical optimization methods.

4. CONCEPTUAL MODEL

The conceptual model describes the process of combining the AI with Reconfigurable Intelligible surface to support autonomous and context-aware communication control in the 6G-enabled V2X settings. The model focuses on the smooth collaboration between the RIS hardware, vehicular nodes, roadside units, and distributed intelligence units. AI is the core decision making system where the variations in channels, the car movement and the nature of the environment are constantly analyzed to identify the most appropriate RIS configurations. The model encompasses multilayered intelligence that includes edge computing, real-time sensing and predictive analytics. It also uses the concept of digital twins in order to build virtual imitations of vehicle communication ecosystems to allow the optimization process to proceed proactively and conduct scenario testing (Z. Huang et al., 2020). The interaction of learning systems will help to coordinate the control in several RIS panels, which will provide scalable and credible coverage in the situation of dense traffic. Through the integration of these components, the conceptual model gives a plan of how highly adaptable, resourceful, and resilient RIS-enabled communication networks, which can support autonomous driving, cooperative perception, and advanced V2X applications can be developed. It also shows the increasing requirement of the flexible architecture that is capable of developing as the vehicle mobility and network complexity rise (Ye et al., 2020).

4.1 Architecture Of AI-Driven Ris Control In V2X Networks

The AI-driven RIS control architecture in V2X networks is organized based on three layers, namely sensing, intelligence, and reconfiguration layer. The sensing layer measures real-time information of vehicles, roadside units and the surrounding environment including mobility patterns, channel states and line of sight. This data is sent to the intelligence layer where AI functions deep learning, reinforcement learning, and hybrid models are used to process the data to derive the best signal propagation policies. The intelligence layer is the center of decision making that predicts channel disjointures, finds beamforming possibilities and optimal RIS settings (Khan et al., 2020). These decisions are carried out by the reconfiguration layer which modulates the phase shifts, reflection coefficient and beam patterns of the RIS panels. These layers are connected by edge servers which guarantee processing speed and low latency. It is also designed to coordinate a multi-RIS to enable multiple surfaces to cooperate and increase coverage and decrease interference. This hierarchical design produces a strong framework that can adapt itself to the extremely dynamic and unpredictable nature of vehicular communication networks. Figure 2 depicts the layered architecture in which sensing, intelligence and reconfiguration modules are in cooperation with edge servers and multi-RIS coordination to allow the adaptive RIS control in the V2X systems.

Figure 2. AI-driven RIS control architecture in V2X networks

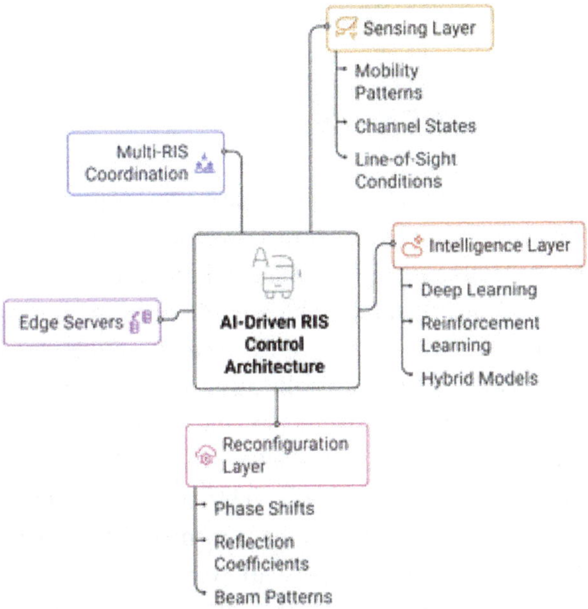

4.2 Integration Of Edge Intelligence And Digital Twin For Predictive Optimization

The combination of edge intelligence with digital twins is a major boost to the predictive and adaptive capabilities of RIS in the V2X environment. Edge intelligence makes it possible to process the data and infer AI much nearer to the vehicular nodes, thus minimizing the latency and providing the possibility to reconfigure the RIS in real-time. The edge servers constantly observe the changes in the environment, traffic flow and communication pattern and provide real time update to RIS panels (C. Huang et al., 2019). This is complemented by digital twin technology which produces high-fidelity virtual simulators of the vehicular network as well as the RIS system. These simulation versions mimic mobility paths, channel behavior and possible interference cases, allowing an active optimization as opposed to a reactionary optimization. The digital twin predicts deviations in communication through constant synchronization and tries out various reconfiguration options before implementing the most viable option in the real world. The integrated framework is useful in predictive beamforming, interference avoidance and optimal resource allocation even in fast moving vehicles (Liang et al., 2024). The combination of edge intelligence and digital twins therefore offers a computationally efficient and highly adaptive environment with the effect that RIS panels make informed decisions which are more accurate and reduce the amount of communication overhead.

4.3 Distributed, Federated, And Collaborative Learning For Multi-RIS Coordination

Scalable and privacy-preserving intelligence over a range of RIS panels distributed over vehicular settings are realized by distributed, federated and collaborative learning methods. Distributed learning enables RIS units to exchange model updates and not raw data, thus facilitating large-scale coordination without the large communication overhead. Federated learning expands on the idea by making sure that sensitive information of vehicles is locally separated, avoiding the threat of privacy violation and allowing improvement in the models around the world. Every RIS unit trains its model based on observations of local channels, mobility pattern, and environmental feedback only, and only the trained parameters are shared with an aggregation server (Yu et al., 2021). This is furthered by collaborative learning where RIS units are dynamically allowed to co-operate to exchange contextual information to align phase shifts and beam patterns. This method proves useful especially in high traffic areas of the city such as the multi lanes or the highways that have high traffic as the coordinated communication pathways will minimize the interference as well as enhance the reliability. This learning structure is continually being optimized based on real-time feedback and thus RIS networks can adjust to varying traffic volumes, concentrations and blockages of signals. The combination of these learning paradigms forms a host of scalable data control multi-RIS in high velocity vehicular communication environments (Zhou et al., 2021).

5. RESEARCH METHODOLOGY

The study approach to be followed in the study is meant to measure the performance of AI-based RIS control mechanisms, their efficiency, and flexibility in 6G-supported V2X settings. It involves combination of simulated-based modeling, algorithm development as well as comparative assessment. The methodology commences with the design of the realistic vehicular scenarios and channel models that describe the high mobility situations, urban canyons, and the line-of-sight blockages. The AI algorithms, such as supervised learning, deep neural networks, and reinforcement learning, are executed in order to evaluate their efficacy in regulating RIS settings. Distributed optimization is also analyzed by using edge-assisted processing and federated learning strategies (Saad et al., 2020). Performance analysis under different sets of environmental constraints is conducted using simulation tools like MATLAB, NS-3 and ray-tracing platforms. They are compared with traditional optimization methods and demonstrated to be superior in terms of reliability, latency, and coverage. There are performance measurements like signal quality, power consumption, scalability, and adaptability which are systematically measured. It is a methodology that guarantees a thorough testing of AI-powered RIS systems in a variety of V2X communication cases.

5.1 Methodological Framework: Simulation, Modeling, And AI Algorithmic Evaluation

The methodology framework is designed in such a way that it incorporates simulation modeling, the creation of AI algorithms, and their evaluations. In the first step, a high fidelity-simulation is generated with the help of MATLAB, Python based libraries, and automotive simulators like SUMO or CARLA. The models of this environment are used to model vehicular mobility, road topology, channel propagation characteristics and the placement of the RIS. Depending on the type of task, different AI algorithms, such as convolutional neural networks, reinforcement learning agents, etc., are built and trained on simulated datasets. Reinforcement learning models People respond to the environment by changing the RIS settings and get feedback on the performance of communication (Zadobrischi et al., 2020). The channel estimation, mobility prediction, and environmental classification are done using supervised and unsupervised learning models. All the algorithms are tried in various traffic conditions, traffic flow speeds, and propagation conditions to determine their robustness. The case-based modeling is conducted to test AI-assisted reconfiguration in the cases of intersection blockage, highway merging, and traffic clustering's. The evaluation criteria in the case of algorithmic evaluation are accuracy, speed of convergence, energy efficiency and responsiveness to fast changeover. Such a combined system guarantees that AI solutions are rigorously tested and only after that, they are deployed in real-life RIS-enabled V2X networks.

5.2 Data Sources, Vehicular Scenarios, And Channel Modeling Approaches

The paper relies on the various data sources and driving conditions to develop realistic and holistic simulation environments. Mobility data has been obtained based on vehicular simulators, GPS traces, and common mobility models like Manhattan grid and Freeway models. Environmental information comprises building plan, traffic density maps as well as weather-dependent signal attenuation variables. Artificial and natural dataset is utilized to train AI models with a channel prediction and mobility pattern recognition. The vehicular situations are at the intersections of urban centers, on a multi-lane highway, in a tunnel, and at the centers of cities with a significant number of obstructions. Geometry-based stochastic modeling, ray-tracing and 6G-specific propagation modeling incorporating mmWave and sub-THz frequencies are all included in channel modeling approaches (Jiang et al., 2021). There are added doppler effects, multipath fading, line-of-sight probability and non-line-of-sight reflections to recreate dynamic vehicular conditions. Various RIS implementation models, such as roadside, vehicle-mounted, and building-integrated are considered. These rich scenarios can be used to comprehensively evaluate the performance of AI-driven RIS control in different signal blockages, mobility speeds, and traffic patterns. The complex range of data sources makes the training of algorithms and the accuracy of predictions high.

5.3 Performance Metrics: Latency, Reliability, Energy Efficiency, Adaptation Speed

The performance assessment is based on the extensive range of metrics that can demonstrate the working requirements of 6G-enabled V2X systems. Latency is also tested to make sure that reconfiguration in the RIS will take place in milliseconds, and that it can serve safety-critical applications, including collision avoidance and cooperative driving. Reliability is measured in terms of packet delivery ratio, probability of outage and stability of links during high mobility. The efficiency of energy is measured in terms of power consumption of RIS, edge AI inference cost, and total communication overhead (Y. Chen et al., 2022). This is necessary since RIS units are generally meant to be low power gadgets that need to be adequate enough to work in high-density setups. Adaptation speed is used to evaluate the ability of RIS to adapt to the transformation of channel conditions or vehicle motion velocity, which emphasizes the quality of AI forecasting and decision-making. Comparisons are conducted on performances of AI-based and classical optimization techniques to measure the enhancement in the response time and the level of communications. Other measures like spectral efficiency, interference suppression and a measure of computational cost are also added where necessary. Taken altogether, these indicators allow conducting a comprehensive assessment of how effectively, at scale, and in practice AI-based RIS control mechanisms work in practice, in practice in the context of a moving vehicle. Table 2 identifies and describes the important measures, including latency, reliability, energy efficiency, adaptation speed, and SNR.

Table 2. Performance metrics for evaluating AI-driven RIS systems

Metric	Definition	Relevance to V2X + RIS
Latency	Time taken for RIS decision and reconfiguration	Critical for safety, real-time driving, and cooperative perception
Reliability	Probability of successful data delivery	Ensures stable connectivity in high mobility settings
Energy Efficiency	Power consumed by RIS + AI modules	Important for large-scale, low-power RIS deployments
Adaptation Speed	Time to adjust RIS to new channel conditions	Needed for fast-moving vehicles and lane changes
Spectral Efficiency	Bits transmitted per Hz of bandwidth	Measures communication effectiveness in dense environments
Outage Probability	Likelihood of communication failure	Indicates robustness in blockage-heavy areas

6. LITERATURE REVIEW

Combining AI and RIS has become an important research area in the next-generation 6G-enabled V2X communication systems. The articles in this field continue to highlight the constraints of classical optimization in the quickly evolving vehicular channels and the possibility of AI-based models to enhance the adaptability, prediction ability, and rapidity of reconfiguration. The current literature explores the topics of AI-based channel estimation, reinforcement learning to autonomously estimate beamforming, and privacy-preserving distributed learning paradigm to coordinate multiple RIS. The collective relevance of real-time decision making, context awareness and predictive intelligence within RIS enabled vehicular settings are stressed by these studies (R. Huang et al., 2024). Since vehicles work in highly dynamic environments with a significant number of line-of-sight blockages, scholars are turning towards the use of data-driven models that are capable of learning channel behavior by observing past activity and sensor measurements. The development of edge computing, digital twins, and distributed AI also contribute to the transition of scalable and collaborative RIS operation. The contributions in this literature review are summarized in three key themes that are closely related to the aims of the research: AI-based channel estimation, reinforcement learning to control real-time RIS, and federated or transfer learning to coordinate multi-RIS in vehicular ecosystems (J. Chen et al., 2024).

6.1 AI-Based Channel Estimation: Supervised, Unsupervised, And Hybrid Models

Channel estimation using AI has emerged as an essential part of RIS-assisted vehicular communication because the existing estimation methods fail to capture the rapid variations in 6G-V2X channels. The convolutional neural networks and deep fully connected networks are examples of supervised learning models that have been found to be effective in the mapping of received pilot signals to valid Channel State Information (CSI). These models use labeled datasets that are created either through simulated Ray-tracing or actual measurements of vehicles to learn the complicated spatial-temporal channel interactions. They, however, require huge annotated datasets to be scaled (Vempati, 2024). Unsupervised learning methods overcome this issue by learning channel features in the form of ground-truth features. Reconstruction of CSI and anomaly detection and pilot overhead mitigation has been done using autoencoders, generative adversarial networks and clustering models. The hybrid models are supervised and unsupervised and can be used to achieve robust channel estimation even in the case of sparse training. The techniques minimize the computational overhead, enhance the accuracy of estimations of non-line-of-sight scenarios, and enable RIS parameters to be reconfigured quickly. It is demonstrated in literature that hybrid models are superior compared to classical minimum mean square error, least-square estimators, particularly in high-mobility vehicle cases where channel characteristics vary in milliseconds (Zhou et al., 2024). These results confirm Objective 1, which focuses on the relevance of AI-based channel estimation. Table 3 summarizes the main AI models- supervised, unsupervised, RL, federated and how they can be applied in RIS in V2X.

Table 3. Comparison of AI techniques for RIS control

AI Technique	Key Features	Strengths for RIS in V2X	Limitations	Typical Use Cases
Supervised Learning	Uses labeled CSI and mobility data	High accuracy in channel prediction	Requires large labeled datasets	CSI estimation, mobility prediction
Unsupervised Learning	Learns patterns without labels	Lower pilot overhead, improves robustness	Less accurate than supervised models	CSI reconstruction, anomaly detection
Hybrid Models	Combines supervised + unsupervised	Strong performance with limited data	Increased training complexity	Environmental modeling, beam selection

continued on following page

Table 3. Continued

AI Technique	Key Features	Strengths for RIS in V2X	Limitations	Typical Use Cases
Reinforcement Learning	Learns via reward-based interaction	Best for real-time RIS reconfiguration	Long training time, reward tuning challenges	Beamforming, blockage mitigation
Federated Learning	Distributed model training	Privacy-preserving and scalable	Requires synchronization overhead	Multi-RIS collaborative learning

6.2 Reinforcement Learning For Real-Time RIS Control And Beamforming

There is significant attention in Reinforcement Learning (RL) that can facilitate autonomous and adaptable reconfiguration of RIS in V2X systems. The RS models consider the car communication environment as a sequence of decisions where an agent interacts with the channel conditions and learns the optimal RIS configurations by receiving rewards and penalties (Naaz et al., 2024). Common systems to optimize phase shifts and beam patterns in order to maximize signal strength or the minimization of outage probability are Deep Q-Networks (DQN), Proximal Policy Optimization (PPO), and Deep Deterministic Policy Gradient (DDPG). In high-dimensional problems of RIS, RL proves to be of great advantage when classical optimization is infeasible. It has been demonstrated that beamforming via RL ensures a reliable link and quicker rates of adaptation in case of a vehicle change of lane, intersection boundaries, and abrupt blockages. Multi-agent RL goes on to amplify these advantages by using a coordination of multiple RIS units which can lead to cooperative gain maximization. The RI predictive adjustments are also made possible by RL methodologies that learn the mobility of patterns and predict the changes in channels (Roger et al., 2024). Those developments are highly consistent with Objective 2, as they demonstrate that RL will result in a better performance during real-time reconfiguration than iterative optimization methods, especially in rapidly changing vehicular settings.

6.3 Federated Learning, Transfer Learning, And Edge-AI In Multi-Ris Vehicular Networks

The concept of Federated Learning (FL), transfer learning, and edge-AI has become transformative, then, concerning scalability and privacy-preserve control of multi-RIS network. Federated learning allows RIS units and vehicular nodes to cooperate to train global AI models without the exchange of raw data, and thus, privacy is guaranteed, and communication overhead is reduced. Literature demonstrates that FL has the substantial effect on model generalization in distributed vehicular network where the mobility patterns and channel conditions are highly differentiated among the nodes (Wang et al., 2024). Transfer learning supplements it through enabling pre-trained models to learn new environments or RIS deployments with limited amounts of training data. This saves on time of computation and shortens convergence time when changing dynamically in vehicular movement. The system is further enhanced by edge-AI that offloads inference and partial training to edge servers to provide low-latency decision-making necessary to control in real time in RIS. Maintenance of beamforming consistency, prediction of interference, and increased network-wide reliability Multi-RIS is coordinated by the use of FL and collaborative learning (Kouhalvandi & Matekovits, 2025). All of these methods contribute to Objectives 3 and 4 by facilitating scalable intelligence in distributed RIS implementations, minimizing computation requirements, and rapidness in the changing high-speed vehicular ecosystem.

7. RESULTS AND DISCUSSION

The findings identify the transformative opportunities of AI-based control systems on RIS in 6G-based V2X communication systems. The performance assessments clearly show that the capabilities of enhancing the reliability of coverage, blockage reduction, and communication stability are considerably higher than those of the classical optimization schemes. The AI-based models provide a high degree of flexibility to the rapid mobility of the vehicles, allowing faster reaction to the dynamics of propagation changes and enhancing link quality in the non-line-of-sight scenarios (Ni et al., 2022). The comparative results reveal that reinforcement learning strategies and deep neural network-based strategies are effective consistently as compared to the static or iterative optimization strategies particularly in high density traffic and canyon urban scenarios. Moreover, experiences of multi-RIS deployments can prove that distributed and federated learning structures improve the process of managing signals cooperatively and preserving privacy. It is also seen in the discussion that the real-world application of AI-RIS systems should take into account hardware constraints, computational, and scalability constraints. The cumulative findings confirm the existence of a promising future of AI-controlled RIS via the development of resilient, efficient, and context-aware 6G-V2X communication networks.

7.1 Performance Insights: Coverage Gain, Reduced Blockage, And Connectivity Stability

Based on the performance analysis, AI-based RIS control plays a significant role in improving the quality of communication in a wide range of vehicular situations. This is one of the most important results as the gain in coverage is clearly observed as stronger reflected beams with AI-optimized RIS configurations and they also offset path loss at mmWave and sub-THz frequencies. The mechanisms of reinforcement learning are adaptable to guide reflections to the mobile vehicles, thus overcoming line-of-sight blockages due to buildings, big vehicles and road edges. Another significant aspect of AI-based connectivity stability is stability because the algorithms can forecast fast channel changes and manipulate RIS parameters within milliseconds (Almohamad et al., 2020). The simulation findings indicate that the outage probability is lower, signal-noise ratios are higher, and the rate of packet delivery is improved compared to the statistic RIS set-ups. Experimental research on AI-managed RIS panels in multi-lane highway settings has been able to remain connected and avoid connection issues in high-speed movements and lane changes. These lessons emphasize the significance of predictive intelligence in ensuring strong communication connections and confirm the usefulness of AI in the dynamic car settings, where the situation is constantly changing and unpredictable.

7.2 Comparative Evaluation Of AI Vs. Classical Optimization In RIS Control

The comparison analysis has revealed that the RIS control based on AI is far more efficient and responsive than the classical optimization. The high-dimensionality and dynamic character of the communication channels in vehicles are hard to solve by the traditional optimization approach, including the convex programming, iterative search-based algorithms, etc. The methods are extremely time-consuming to calculate and rely on the supposition of motionless or near motionless functions that are not applicable in practical V2X scenarios. In their turn, AI-driven approaches, including deep reinforcement learning, supervised prediction models and hybrid neural networks, learn decision patterns and update RIS configurations in a short latency and are timely in terms of handling channel inputs (Y. Chen et al., 2021). The experiment results indicate that AI-based schemes are more likely to have a higher reconfiguration rate, spectral efficiency and reliability of connections in challenging mobility situations. In addition, the reinforcement learning agents find the optimal patterns of phase-shift automatically without necessarily estimating the channels, and this reduces computation. Also, AI solutions are more scalable when coordinating a large number of RIS panels. This fact demonstrates that the next generation of

6G-V2X systems, where intelligent predictive and ongoing adjustment need to be applied, is more suited to AI-controlled RIS control. Table 4 highlights differences between adaptability, cost of computation, scalability and responsiveness.

Table 4. AI vs classical optimization approaches for RIS control

Approach	Characteristics	Advantages	Limitations
Classical Optimization	Iterative, mathematically driven methods	Transparent, well-understood models	Slow adaptation, not suited for dynamic mobility
AI-Based Models	Data-driven, predictive, real-time learning	Fast reconfiguration, high adaptability	Requires training data and computational resources
Reinforcement Learning	Environment interaction + rewards	Handles dynamic, high-dimensional spaces	Long convergence time
Hybrid AI-Optimization	Combines classical models with AI learning	Balances accuracy and speed	More complex implementation

7.3 Discussion On Practical Deployment, Hardware Constraints, And Scalability

Even though the high-performance benefits of AI-based RIS systems are defined, there are several challenges related to their actual implementation. Hardware restriction such as poor phase resolution, switching delays and reduced control circuitry limits the RIS reconfiguration and restricts how well it can be reconfigured. In addition, a large-scale installation of RIS will require high-powered communication links between the panels and the edge controllers that must be stable even during heavy traffic movement (Michailidis et al., 2021). Computational complexity of executing AI models on real time basis also limits its application, especially where there is a constraint in the edge resources, or the resources are spread out among multiple roadside units. The most critical one is the question of scalability whereby the RIS units are expanded and efficient distributed learning models should be efficient to avoid congestion of the communication bandwidth. Environmental vagaries such as the weather and physical obstacles are also known to complicate deployment. Despite these challenges, federated learning, edge intelligence, and lightweight neural networks can be combined to resolve the issue of hardware and scalability. More developments made in meta-surface materials, energy-saving control circuits and RIS-specific AI accelerators will probably help with large-scale adoption. Overall effective implementation is a trade-off between brilliant methods of control and system structure which are conscious of hardware so as to possess a reliable and scalable system.

8. CASE STUDIES IN AI-DRIVEN RIS FOR V2X SYSTEMS

New real-life and testbed implementation indicate the practical usefulness of RIS technologies powered by AI in enhancing vehicular communication behavior in a variety of road and traffic environments. These case studies demonstrate that intelligent surfaces together with machine learning and edge-assisted decision-making can be capable of boosting signal propagation, minimizing blockage, and ensuring continuity on the dynamic 6G-V2X environments. Programs by Huawei and NEC-NTT DoCoMo in the industry show how it is possible to integrate RIS panels into highways, city intersections, and congested metropolitan routes (Zhou et al., 2022). Through these implementations, it is highlighted that predictive modeling, reinforcement learning, and distributed intelligence can be important to optimize the RIS configurations in real time. Examining the examples, this section outlines the operational advantages, deployment issues, and lessons that lead to the adoption of AI-driven RIS systems in large scale in the future. The results show positive progress towards intelligent, collaborative and strong vehicular communication systems.

8.1 Case Study 1: Huawei RIS-Aided Smart Highway Deployment (2024)

The test of the suitability of AI-based RIS control in the traffic highway systems was carried out in Huawei in 2024, based on a large-scale smart highway using RIS. The RIS panels will be strategically planned on the expressways to allow the steering of vehicles of varying speeds using dynamic mmWave beams. The system had an implementation of edge-based AI engines capable of predicting lane change, merging pattern and change in the traffic density with mobility analytics and sensor data in real-time. Deep learning algorithms were used to process car routes and estimate signal blockage to allow RIS panels to react before signal blockage to change phase shifts and remain connected. The test demonstrated a tremendous improvement in signal strength, the continuity of the coverage, and reduction of the cases of communication interruption, in particular, when overtaking took place or the proximity of truck groups commonly causing attenuation of mmWave signals (Bandewar & Chaudhary, 2023). The result showed that the failure of handover greatly reduced and the stability of communication at a speed of more than 100km/h increased. The AI-enabled RIS system viability of modern highways and the need to foster a seamless system of 6G-V2X connection can be verified by the case study. Figure 3 the process map explains the steps taken in the process of predicting blockages, altering reflections, and enhancing communication stability in intelligent highways by using AI-based RIS panels.

Figure 3. Huawei RIS-aided smart highway deployment

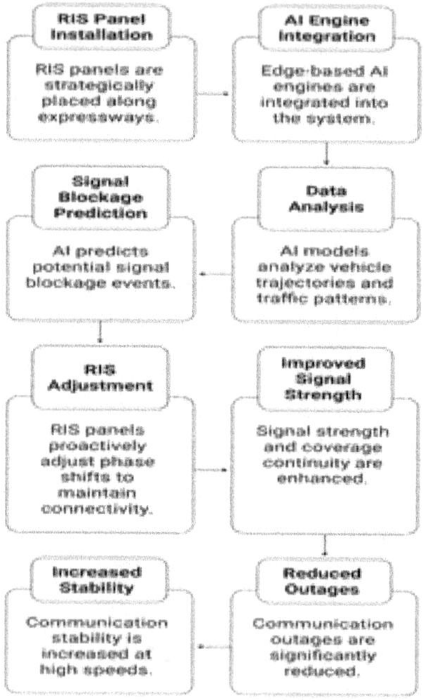

8.2 Case Study 2: Nec–NTT Docomo Multi-Ris 6G Testbed For Cooperative V2x (2023-2024)

In an attempt to assist in identifying cooperative V2X communication in dense cities in 2023-2024 NEC and NTT DoCoMo established a large-scale testbed to investigate 6G multi-RIS coordination. The test constructed some RIS panels in the intersections, on the road, and facades of buildings to make the intricate paths of autonomous vehicles go through in complex metropolitan schemes formidable reflections. In the testbed, deep reinforcement models of learning were implemented which made RIS units collaborate in determining the most appropriate patterns of reflections, on the basis of real time measurements of channels and the forecasted movement of the vehicle. Multi-agent RL enabled the adaptive collaboration of distributed panels of the RIS and minimized interference and it was also able to sustain connectivity even in extreme non-line-of-sight environments in an urban canyon setting (Zhou et al., 2023). This system was also discovered to be more precise in beam positioning, reduced signal jamming and improved data transmission of the autonomous shuttles within the congested districts. Those experiments showed that assimilated AI-oriented RIS control was applicable to enhance the reliability and the fluent communication in the rough environment. The testbed is a resourceful source of data concerning the manner of how large-scale cooperative RIS systems may be deployed in smart cities. Figure 4 visualizes a 6G testbed scheme, in which multi-RIS, RL training, and joint reflection control can be applied to support the enhancement of the performance of urban V2X communications.

Figure 4. NEC–NTT DoCoMo 6G testbed sequence

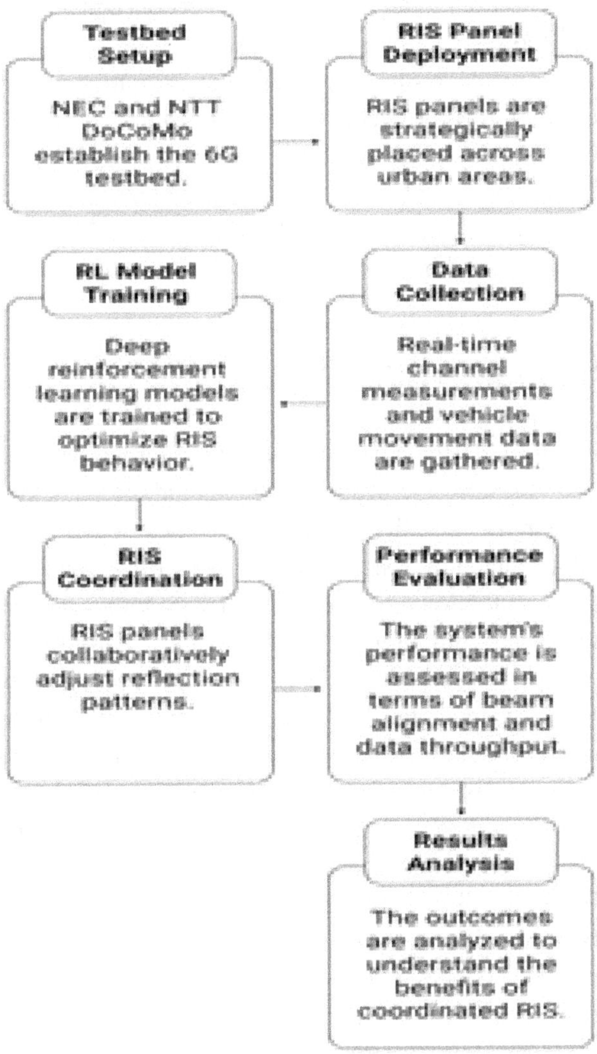

8.3 Key Takeaways And Comparative Insights

All the above examples in the two case studies indicate the significant benefits of AI-based RIS technologies in a wide variety of vehicular and environmental conditions. The highway implementation of Huawei indicates how the predictive AI model can help reduce high making the mobility difficult, and the NEC-NTT DoCoMo testbed shows the importance of multi-agent and distributed learning in cities with complicated blockages. Combined, these examples indicate that the incorporation of AI is essential to attain a quick RIS reconfiguring, enhanced signal resilience, and reduced connection outage. Also revealed by them are pragmatic constraints like switching delays in hardware, reliance on high quality sensing data and scalable orchestration requirements within a large number of RIS units (Atlanta USA & Kosamia, 2025). Comparative analysis reveals that predictive modeling will help most in highway deployments, and collaborative AI will be very useful in dense city deployments. These are important insights that can be utilized in design of RIS in future, standardization and high scale adoption in 6G-V2X networks.

9. FUTURE RESEARCH DIRECTIONS

The study of AI-assisted RIS with 6G-based V2X communication will still evolve in the direction of more self-organized, adaptable, and effective systems that will be able to handle extreme mobility and heterogeneous network conditions. New paradigms, including quantum-assisted AI, neuromorphic intelligence, and bio-inspired optimization will allow reconfiguring of RIS with a speed that is beyond the power of conventional computing. Research will also shift towards the creation of self-evolving and power-gathering RIS which will act on their own and require minimum upkeep. Also, federated learning, multi-agent system, and distributed inference improvements will enable large-scale coordination of RIS between varied vehicular and urban infrastructures (Noh et al., 2019). Interoperability and standardization will be necessary as multi-vendor RIS systems continue to flourish and need single frameworks of control protocols, security and network integration. These guidelines will define the following generation of smart wireless world, where RIS will continue to become a fundamental technology of resilient, efficient, and secure 6G-V2X systems. Table 5 Summarizes the emerging themes to include quantum-AI, neuromorphic control, self-evolving RIS and standardization needs.

Table 5. Future research directions for AI-driven RIS in 6G-V2X

Research Area	Description	Expected Impact
Quantum-AI for RIS	Quantum reinforcement learning and channel prediction	Ultra-fast RIS reconfiguration
Neuromorphic RIS Control	Brain-inspired low-power RIS processors	Real-time adaptation with minimal energy
Self-Evolving RIS	Autonomous learning, context-awareness	Reduced maintenance, higher reliability
Energy-Harvesting RIS	Powering RIS via ambient energy	Sustainable large-scale deployment
RIS Standardization	Cross-vendor interoperability protocols	Stable integration into 6G-V2X ecosystems

9.1 Quantum-AI And Neuromorphic Control For Ultra-Fast RIS Reconfiguration

The possibility of quantum-AI and neuromorphic computing suggests new opportunities towards ultra-fast and energy efficient reconfiguration of RIS in high-mobility vehicular conditions. The complexity of channel information can be processed much more rapidly by quantum machine learning algorithms than by classical algorithms, allowing an adaptive response to the quick changes in velocity of a vehicle and the unpredictable changes in propagation to be adapted to quite rapidly. The quantum reinforcement learning has the potential to optimize RIS phase shifts and reflection patterns by searching high-dimensional action space that cannot be reached using conventional RL models (Tong et al., 2019). Neuromorphic computing a model based on the form and functionality of biological neural networks offers low-power, high-speed computing suitable to RIS applications at scale. Neuromorphic chips are also capable of supporting continuous learning, event-driven processing, and ultra-low latency response and can be used in roadside RIS controllers of vehicular edges. The combination of quantum-AI and neuromorphic systems opens the path to the next generation of RIS control systems that could react to channel fluctuations on the microsecond time scales and use much less power. The new solutions will be instrumental in the process of enabling ultra-reliable, mission-critical 6G-V2X systems like autonomous vehicle driving, cooperative perception, and real-time safety execution (Kaur et al., 2021).

9.2 Self-Evolving RIS: Context-Aware, Autonomous, And Energy-Harvesting Surfaces

RIS systems will develop in the future to be more intelligent with programmable surfaces being replaced by self-learning, context-aware and energy-autonomous intelligent platforms. Self-evolving RIS will utilize AI models that can continuously learn online that will help them identify V2X traffic patterns and environmental variability, and mobility behavior without being reprogrammed. These surfaces will automatically change the configuration strategies according to the real time situational awareness and make the communication more reliable and do not require centralized control. The energy-harvesting RIS, which operates on ambient radio waves, solar energy, or microgenerators using vibration, will decrease power reliance and allow the large-scale implementation (Martín-Martín et al., 2024). The capability to identify a vehicle density, vehicle road conditions, and the level of interference through the context-aware sensing capabilities installed in RIS structures will allow surfaces to understand how to form beamforming more precisely and adaptively. This trend also encompasses the creation of hybrid RIS that are a combination of the passive, semi-passive and active components in a bid to strike a balance between power consumption and signal improvement requirements. The transition to self-evolving and energy-harvesting RIS will save overhead in the operation and allow continuous maintenance-free communication assistance in various 6G-V2X conditions.

9.3 Standardization, Interoperability, And Cross-Vendor Multi-Ris Collaboration

With the maturity of the RIS technologies, standardization and interoperability will be vital in ensuring flawless integration through the multi-vendor environment. State of the art RIS systems tend to use proprietary control protocols, which restrict their ability to support a variety of vehicular systems and 6G networks. Future studies should aim at coming up with standard control platforms, standardized signalling platforms, and interoperable configuration platforms to enable RIS panels of various vendors to coordinate with each other effectively. This involves specification of reference models of RIS communication, security and lifecycle management. It will also be necessary in cross-vendor cooperation that will provide the ability to deploy multi-RIS in large scale where surfaces produced by various manufacturers will cooperate to maximize coverage (Sarker et al., 2023). The areas of the standardization work need to include the matters of data sharing, distributed control, electromagnetic compliance, and cybersecurity. Also, regulating authorities will demand clear-cut policies regarding RIS location, reflection control, and coexistence of several spectrums. This can be achieved by creating a system that is interoperable to ensure that the RIS ecosystem does not get disintegrated and to ensure that it can be universally used in the future generations of 6G-V2X infrastructure so that performance remains consistent irrespective of the diversity of vendors or deployment environment.

10. MANAGERIAL IMPLICATIONS

The managerial implications of RIS implementation using AI can spread through the automotive, telecommunications, and public policy sectors. The mobility providers and automotive OEMs should be ready to face a future in which the reliability of communication can be a vital requirement in the safety and efficiency of autonomous vehicles. The telecom operators and the infrastructure planners must encompass and deploy RIS within the wider scope of 6G deployment, and must be able to allocate RIS appropriately, interoperate, and support edge-AI. The policy analysts should expect regulatory effects in the fields of safety, security, spectrum management and design of urban infrastructure (Elfatih et al., 2022). The managers in these sectors also need to take into consideration investment, upskilling of the workforce, and maintenance of intelligent RIS networks in the long-term perspective. With RIS being a core of V2X communication, the industry stakeholders will have to work in harmony to attain scalable, reliable, and ethically responsible use of AI-driven surface reconfigurations technologies. The success of RIS-enabled communication ecosystems in next-generation mobility will rely on how well decisions made by managers are made.

10.1 Implications For Automotive OEMS And Autonomous Mobility Providers

AI-based RIS has created new opportunities and roles in the design and implementation of connected and autonomous vehicles to automotive OEMs and mobility service providers. Such functions as cooperative perception, real-time navigation, autonomous lane merging, and collision avoidance require reliable V2X communication. Connectivity The connection will be always stable even during obstructions or high-density conditions with the help of RIS, which will enhance safety and user experience. In order to enable vehicles to exchange information with RIS panels using standardized V2X interfaces, managers will have to invest in car architectures that can communicate flawlessly with these panels (Lv & Luo, 2023). This involves the incorporation of enhanced sensing, vehicle edge and AI based communication systems. OEMs also need to match product development strategies to the emerging RIS-enabled mobility corridors where infrastructure and vehicles are able to collaborate. Also, software updates, security patches and agreements with infrastructure providers are to be considered in lifecycle planning. With increased use of autonomous driving, close working relations among vehicle producers, telecom providers, and government service providers will be critical in realization of robust and scalable RIS-enhanced mobility ecosystems.

10.2 Implications For Telecom Operators And Smart Infrastructure Planners

The telecom operators and infrastructure planners are at the forefront in deploying RIS within a city, suburban, and highway setting. Adding RIS panels to current infrastructure systems of 5G and new tempos of 6G needs to be planned carefully in terms of location, coverage, and backhaul connectivity. Edge computing nodes that can facilitate low-latency AI inference to enable real-time RIS control should be invested in by the operators. The integration of RIS needs to be addressed, as well, by smart infrastructure planners, in terms of the physical implementation of RIS in the facades of the buildings, street furniture, street-side units, and transportation networks (Li et al., 2021). The advanced digital twin models will be needed to ensure an effective deployment to predict the RIS performance in different traffic and environmental conditions. Operators have to deal with costs of operations, power optimization as well as compatibility between multi-vendor ecosystems. Moreover, the maintenance structures and service-level deals are required to ensure the long-term reliability. New business models of infrastructure sharing, collaborative planning, and cross-sector digital partnership will be also needed in terms of the shift towards intelligent surfaces. These implications demonstrate the strategic position of telecom operators in facilitating massive adoption of RIS.

10.3 Implications For Policy Makers In Safety, Security, And Spectrum Allocation

Policy makers will be instrumental in defining regulatory as well as governance structures necessary in the deployment of safe and responsible RIS. With the changing of the electromagnetic environment by RIS, policymakers have to provide standards on managing the reflection, reducing interference and coexisting with other bands. Regulatory rules are required in place to make sure that the deployment of RIS does not affect the safety-critical communication of autonomous vehicles or emergency agencies. Another major issue is security, since Ai-based RIS systems can be spoofed, reconfigured against, or subjected to a cyber-physical attack. Secure communication patterns, authentication tools and practices need to be obligatory by policymakers (Zheng et al., 2025). The planning regulations in urban areas should also be developed to embrace RIS installations in the infrastructural plans without compromising in aesthetic, environmental and structural standards. Moreover, the adoption of RIS will be boosted by policy measures that will promote innovation like sandbox test-beds, national research funding, and interoperability standards. Governments can contribute to the creation of a strong, future-proof communication ecosystem that is aligned with the nationwide mobility, safety, and digital transformation objectives through formulation of forward-thinking and all-encompassing policies.

11. CONCLUSION

The recent collaboration between AI and RIS is a near breakthrough in the design of next-generation 6G-enabled Vehicle-to-Everything communication networks. AI-based RIS control has been shown to break some of the most important constraints that are related to high-mobility vehicular channels, such as high blockages, fast changes in the channel, and unreliable propagation conditions. Such data-driven decision making, predictive modeling, as well as real-time optimization allows RIS to transition off the fixed reflective platforms and on to adaptive, context-constitutive communication enablers, which can maintain ultra-reliable and low-latency communication links. This combination of intelligence and reconfigurability backing of the next-generation car applications, cooperative driving, autonomous navigation, and infrastructure-assisted perception. The development of wholly autonomous and cooperative intelligent surfaces is a point on the continuum of wireless communication architecture change. Federated intelligence, reinforcement learning, and neuromorphic-inspired computation are some of the technologies that will enable scalable and distributed RIS systems that can interact both on the roadways, at intersections, and in complicated urban corridors. The self-evolution capability, energy-harvesting strategies, and hybrid active-passive designs will be more integrated into future RIS platforms, so that the RIS will not be as reliant on a centralized control and offer more robust and efficient communication networks. AI-based RIS technologies will eventually transform how smart mobility and connected transport systems are based in the long term. Their capability to increase the reliability of communication, enhance coverage of difficult environments and decrease the energy usage is in line with international concerns of sustainable, safe and autonomous mobility. With the introduction of RIS as a part of 6G and more, it will have an impact on the fields of automotive innovation, telecom infrastructure planning, and urban digital transformation. The introduction of intelligent surfaces, edge computing, digital twins, and predictive AI will create the ability of vehicles, infrastructure, and networks to interact with each other as a unified communication environment. Finally, AI-based RIS tools will boost the transition to much-connected secure and self-optimizing transportation environments that facilitate the autonomous and cooperative mobility of the future.

REFERENCES

Almohamad, A., Tahir, A. M., Al-Kababji, A., Furqan, H. M., Khattab, T., Hasna, M. O., & Arslan, H. (2020). Smart and secure wireless communications via reflecting intelligent surfaces: A short survey. *IEEE Open Journal of the Communications Society, 1*, 1442–1456. DOI: 10.1109/ojcoms.2020.3023731

Bandewar, S., & Chaudhary, V. S. (2023). A study on enhancing channel estimation in vehicular communications: Metamaterial-enhanced microstrip antennas. *SMART MOVES JOURNAL IJOSCIENCE, 1–4*. Advance online publication. DOI: 10.24113/ijoscience.v9i9.510

Basharat, S., Khan, M., Iqbal, M., Hashmi, U. S., Zaidi, S. A. R., & Robertson, I. (2022). Exploring reconfigurable intelligent surfaces for 6G: State-of-the-art and the road ahead. *IET Communications, 16*(13), 1458–1474. DOI: 10.1049/cmu2.12364

Chen, J., Shen, W., Luo, S., Ma, S., Xing, C., & Hanzo, L. (2024). Estimation of dispersive high-Doppler channels in the RIS-aided mmWave internet of vehicles. *IEEE Internet of Things Journal, 11*(1), 677–691. DOI: 10.1109/jiot.2023.3286279

Chen, Y., Wang, Y., Zhang, J., & Di Renzo, M. (2021). QoS-driven spectrum sharing for reconfigurable intelligent surfaces (RISs) aided vehicular networks. *IEEE Transactions on Wireless Communications, 20*(9), 5969–5985. DOI: 10.1109/twc.2021.3071332

Chen, Y., Wang, Y., Zhang, J., Zhang, P., & Hanzo, L. (2022). Reconfigurable intelligent surface (RIS)-aided vehicular networks: Their protocols, resource allocation, and performance. *IEEE Vehicular Technology Magazine, 17*(2), 26–36. DOI: 10.1109/mvt.2022.3158046

Elfatih, N. M., Hasan, M. K., Kamal, Z., Gupta, D., Saeed, R. A., Ali, E. S., & Hosain, M. S. (2022). Internet of vehicle's resource management in 5G networks using AI technologies: Current status and trends. *IET Communications, 16*(5), 400–420. DOI: 10.1049/cmu2.12315

Huang, C., Zappone, A., Alexandropoulos, G. C., Debbah, M., & Yuen, C. (2019). Reconfigurable intelligent surfaces for energy efficiency in wireless communication. *IEEE Transactions on Wireless Communications, 18*(8), 4157–4170. DOI: 10.1109/twc.2019.2922609

Huang, R., Suo, S., Su, X., & Sun, W. (2024). Distributed joint beamforming for RIS-assisted cell-free MIMO with discrete phase shifts. *Journal of Physics: Conference Series, 2829*(1), 012011. DOI: 10.1088/1742-6596/2829/1/012011

Huang, Z., Zhang, X., & Cheng, X. (2020). Non-geometrical stochastic model for non-stationary wideband vehicular communication channels. *IET Communications*, *14*(1), 54–62. DOI: 10.1049/iet-com.2019.0532

Iqbal, A., Tham, M.-L., Wong, Y. J., Al-Habashna, A., Wainer, G., Zhu, Y. X., & Dagiuklas, T. (2023). Empowering non-terrestrial networks with artificial intelligence: A survey. *IEEE Access : Practical Innovations, Open Solutions*, *11*, 100986–101006. DOI: 10.1109/access.2023.3314732

Jian, M., Alexandropoulos, G. C., Basar, E., Huang, C., Liu, R., Liu, Y., & Yuen, C. (2022). Reconfigurable intelligent surfaces for wireless communications: Overview of hardware designs, channel models, and estimation techniques. *Intelligent and Converged Networks*, *3*(1), 1–32. DOI: 10.23919/icn.2022.0005

Jiang, W., Han, B., Habibi, M. A., & Schotten, H. D. (2021). The road towards 6G: A comprehensive survey. *IEEE Open Journal of the Communications Society*, *2*, 334–366. DOI: 10.1109/ojcoms.2021.3057679

Kaur, J., Khan, M. A., Iftikhar, M., Imran, M., & Emad Ul Haq, Q. (2021). Machine Learning Techniques for 5G and Beyond. *IEEE Access : Practical Innovations, Open Solutions*, *9*, 23472–23488. DOI: 10.1109/access.2021.3051557

Khan, L. U., Yaqoob, I., Imran, M., Han, Z., & Hong, C. S. (2020). 6G wireless systems: A vision, architectural elements, and future directions. *IEEE Access : Practical Innovations, Open Solutions*, *8*, 147029–147044. DOI: 10.1109/access.2020.3015289

Kouhalvandi, L., & Matekovits, L. (2025). On the role of artificial intelligent technology for millimetre-wave and terahertz applications. *Sensors (Basel)*, *25*(17), 5502. DOI: 10.3390/s25175502 PMID: 40942934

Kosamia, R. I. (2025). Federated reinforcement learning for edge AI decision-making in 6g-enabled V2x systems. Journal of Artificial Intelligence & Cloud Computing, 1–14. DOI: 10.47363/jaicc/2025(4)472

Li, S., Duo, B., Di Renzo, M., Tao, M., & Yuan, X. (2021). Robust secure UAV communications with the aid of reconfigurable intelligent surfaces. *IEEE Transactions on Wireless Communications*, *20*(10), 6402–6417. DOI: 10.1109/twc.2021.3073746

Liang, J. C., Zhang, L., Luo, Z., Jiang, R. Z., Cheng, Z. W., Wang, S. R., Sun, M. K., Jin, S., Cheng, Q., & Cui, T. J. (2024). A filtering reconfigurable intelligent surface for interference-free wireless communications. *Nature Communications*, *15*(1), 3838. DOI: 10.1038/s41467-024-47865-6 PMID: 38714685

Lv, C., & Luo, Z. (2023). Deep learning for channel estimation in physical layer wireless communications: Fundamental, methods, and challenges. *Electronics (Basel), 12*(24), 4965. DOI: 10.3390/electronics12244965

Mande, S., Ramachandran, N., & Begum, S. S. A. (2025). Navigating the future of intelligent transportation: Challenges and solutions in 6G V2X and V2V networks. *EAI Endorsed Transactions on Internet of Things, 11*. Advance online publication. DOI: 10.4108/eetiot.5114

Martín-Martín, A., Padial-Allué, R., Castillo, E., Parrilla, L., Parellada-Serrano, I., Morán, A., & García, A. (2024). Hardware implementations of a deep learning approach to optimal configuration of reconfigurable intelligence surfaces. *Sensors (Basel), 24*(3), 899. DOI: 10.3390/s24030899 PMID: 38339618

Michailidis, E. T., Miridakis, N. I., Michalas, A., Skondras, E., & Vergados, D. J. (2021). Energy optimization in dual-RIS UAV-aided MEC-enabled Internet of vehicles. *Sensors (Basel), 21*(13), 4392. DOI: 10.3390/s21134392 PMID: 34198977

Naaz, F., Nauman, A., Khurshaid, T., & Kim, S.-W. (2024). Empowering the vehicular network with RIS technology: A state-of-the-art review. *Sensors (Basel), 24*(2), 337. DOI: 10.3390/s24020337 PMID: 38257430

Ni, P., Li, M., Liu, R., & Liu, Q. (2022). Partially distributed beamforming design for RIS-aided cell-free networks. *IEEE Transactions on Vehicular Technology, 71*(12), 13377–13381. DOI: 10.1109/tvt.2022.3195779

Noh, G., Kim, J., Chung, H., & Kim, I. (2019). Realizing multi-gbps vehicular communication: Design, implementation, and validation. *IEEE Access : Practical Innovations, Open Solutions, 7*, 19435–19446. DOI: 10.1109/access.2019.2896598

Roger, S., Botella-Mascarell, C., Martín-Sacristán, D., García-Roger, D., Monserrat, J. F., & Svensson, T. (2024). Sustainable mobility in B5G/6G: V2X technology trends and use cases. *IEEE Open Journal of Vehicular Technology, 5*, 459–472. DOI: 10.1109/ojvt.2024.3375451

Saad, W., Bennis, M., & Chen, M. (2020). A vision of 6G wireless systems: Applications, trends, technologies, and open research problems. *IEEE Network, 34*(3), 134–142. DOI: 10.1109/mnet.001.1900287

Saikia, P., Singh, K., Singh, S. K., Huang, W.-J., Li, C.-P., & Biswas, S. (2023). Beamforming design in vehicular communication systems with multiple reconfigurable intelligent surfaces: A deep learning approach. *IEEE Access : Practical Innovations, Open Solutions, 11*, 100832–100844. DOI: 10.1109/access.2023.3314668

Salim ALRikabi, H. T. H., Sallomi, A. H., KHazaal, H. F., & Magdy, A. (2025). Review of the Reconfigurable Intelligent Surfaces in smart cities: Opportunities, challenges, and applications. *IET Smart Cities*, *7*(1). Advance online publication. DOI: 10.1049/smc2.70003

Sarker, M. A. L., Son, W., & Han, D. S. (2023). RIS-assisted hybrid beamforming and connected user vehicle localization for millimeter wave MIMO systems. *Sensors (Basel)*, *23*(7), 3713. DOI: 10.3390/s23073713 PMID: 37050772

Souto, N. (2023). Joint active and passive beamforming for RIS-aided MIMO communications with low-resolution phase shifts. IEEE Communications Letters: A Publication of the IEEE. *Communicatio Socialis*, *27*(6), 1604–1608. DOI: 10.1109/lcomm.2023.3264935

Tariq, S., Al-Rizzo, H., Nazmul Hasan, M., Kunju, N., & Abushamleh, S. (2022). Stochastic versus Ray Tracing wireless channel modeling for 5G and V2X applications: Opportunities and challenges. In *Antenna Systems*. IntechOpen., DOI: 10.5772/intechopen.101625

Tong, W., Hussain, A., Bo, W. X., & Maharjan, S. (2019). Artificial intelligence for vehicle-to-everything: A survey. *IEEE Access : Practical Innovations, Open Solutions*, *7*, 10823–10843. DOI: 10.1109/access.2019.2891073

Vempati, S. (2024). Securing smart cities: A cybersecurity perspective on integrating IoT, AI, and Machine Learning for digital twin creation. *Journal of Electrical Systems*, *20*(5s), 2817–2827. DOI: 10.52783/jes.3052

Waghela, S., & Waghela, N. (2025). Reconfigurable Intelligent Surfaces in Smart Cities. In Advances in Computational Intelligence and Robotics (pp. 311–342). IGI Global. DOI: 10.4018/979-8-3693-8099-4.ch013

Wang, X., Han, Z., Jin, J., Xi, R., Wang, Y., Han, L., Ma, L., Lou, M., Gui, X., Wang, Q., & Liu, G. (2024). Multistatic integrated sensing and communication system based on macro-micro cooperation. *Sensors (Basel)*, *24*(8), 2498. DOI: 10.3390/s24082498 PMID: 38676116

Yang, Y., Zheng, B., Zhang, S., & Zhang, R. (2020). Intelligent reflecting surface meets OFDM: Protocol design and rate maximization. *IEEE Transactions on Communications*, *68*(7), 4522–4535. DOI: 10.1109/tcomm.2020.2981458

Ye, D., Yu, R., Pan, M., & Han, Z. (2020). Federated learning in vehicular edge computing: A selective model aggregation approach. *IEEE Access : Practical Innovations, Open Solutions*, *8*, 23920–23935. DOI: 10.1109/access.2020.2968399

Yu, Z., Hu, J., Min, G., Zhao, Z., Miao, W., & Hossain, M. S. (2021). Mobility-aware proactive edge caching for connected vehicles using federated learning. IEEE Transactions on Intelligent Transportation Systems: A Publication of the IEEE Intelligent Transportation Systems Council, 22(8), 5341–5351. DOI: 10.1109/tits.2020.3017474

Zadobrischi, E., Cosovanu, L.-M., & Dimian, M. (2020). Traffic flow density model and dynamic traffic congestion model simulation based on practice case with vehicle network and system traffic intelligent communication. *Symmetry*, *12*(7), 1172. DOI: 10.3390/sym12071172

Zheng, X., Liu, Z., Cheng, S., Wu, Y., Chen, Y., & Zhang, Q. (2025). Channel estimation for reconfigurable intelligent surface-aided millimeter-wave massive multiple-input multiple-output system with deep residual attention network. *ETRI Journal*, *47*(3), 480–491. DOI: 10.4218/etrij.2023-0555

Zhou, X., Chen, X., Tong, L., & Wang, Y. (2024). Attention-deep reinforcement learning jointly beamforming based on tensor decomposition for RIS-assisted V2X mmWave massive MIMO system. *Complex & Intelligent Systems*, *10*(1), 145–160. DOI: 10.1007/s40747-023-01148-7

Zhou, X., Liang, W., She, J., Yan, Z., & Wang, K. (2021). Two-layer federated learning with heterogeneous model aggregation for 6G supported internet of vehicles. *IEEE Transactions on Vehicular Technology*, *70*(6), 5308–5317. DOI: 10.1109/tvt.2021.3077893

Zhou, X., Ma, Z., Li, J., & Yu, L. (2022). Ultrasonic-aided fast-layered alternating iterative tensor channel estimation for V2X millimeter-wave massive MIMO systems. *Electronics (Basel)*, *11*(22), 3742. DOI: 10.3390/electronics11223742

Zhou, X., Zeng, Z., Li, J., Ma, Z., & Tong, L. (2023). Tensor-based joint beamforming with ultrasonic and RIS-assisted Dual-Hop Hybrid FSO mmWave massive MIMO of V2X. *Photonics*, *10*(8), 880. DOI: 10.3390/photonics10080880

KEY TERMS AND DEFINITIONS

6G-V2X Communication: Next-generation vehicle-to-everything communication enabling ultra-low latency, high reliability, and high-capacity connectivity for autonomous and connected vehicles.

Beamforming: A signal processing technique used to direct electromagnetic waves toward specific users or regions to improve signal strength and reduce interference.

Channel State Information (CSI): Real-time knowledge about the wireless channel that helps AI models and RIS optimize communication parameters.

Deep Neural Networks (DNN): Multi-layer AI models that learn complex patterns from data to support tasks like channel prediction and mobility estimation.

Digital Twin: A virtual replica of the vehicular communication environment used for real-time simulation, prediction, and optimization of RIS behavior.

Edge Intelligence: AI processing performed at the network edge (roadside units, edge servers) to reduce latency and enable real-time RIS reconfiguration.

Energy-Harvesting RIS: RIS panels capable of generating their own power from ambient energy sources such as RF signals or solar radiation.

Federated Learning (FL): A distributed machine learning approach where models are trained locally at devices or RIS units without sharing raw data.

Multi-RIS Coordination: Collaborative optimization of multiple RIS panels to enhance coverage, reduce interference, and improve reliability in large-scale deployments.

Neuromorphic Computing: AI computation inspired by human brain architecture, enabling ultra-fast and low-power decision-making for RIS control.

Phase Shift Control: Adjustment of the phase of reflected signals at each RIS element to optimize propagation paths and enhance link quality.

Reconfigurable Intelligent Surface (RIS): A programmable meta-surface composed of controllable unit elements that manipulate electromagnetic waves to enhance wireless communication performance.

Reinforcement Learning (RL): An AI paradigm where agents learn optimal actions by interacting with the environment and receiving reward-based feedback.

Chapter 3
Blockchain-Integrated RIS for Secure and Transparent Vehicular Communication

R. N. Ravikumar
http://orcid.org/0009-0009-3705-1681
Marwadi University, Rajkot, India

S. Aarthi
http://orcid.org/0009-0006-9064-2091
Marwadi University, Rajkot, India

ABSTRACT

The integration of Blockchain and Reconfigurable Intelligent Surfaces (RIS) offers a secure, transparent, and high-performance foundation for 6G-enabled V2X communication. This chapter explores how blockchain's decentralized trust, immutable records, and smart contracts complement RIS-enabled wireless optimization to enhance signal reliability, reduce latency, and prevent tampering in dynamic vehicular environments. The proposed framework strengthens authentication, data integrity, and trust management while supporting advanced applications such as secure platooning and decentralized vehicular data marketplaces. Key challenges related to scalability, energy efficiency, and interoperability are examined, along with future prospects involving AI-driven optimization, post-quantum security, and autonomous RIS-blockchain coordination. Overall, the chapter demonstrates how merging programmable wireless control with distributed ledgers enables resilient and trustworthy next-generation mobility.

DOI: 10.4018/979-8-3373-7867-1.ch003

1. INTRODUCTION

The advent of high-speed development of communication in vehicles is transforming the principles of intelligent movement in the 6G generation. With the movement of vehicles into a highly connected, cooperative, and autonomous systems, Vehicle-to-Everything (V2X) communication depends on reliability and security as a matter of mission-critical importance. The conventional wireless networks are under a growing pressure to meet the high demands of ultra-reliable and low-latency, high population of secure data transfer and smooth coordination between heterogeneous vehicular nodes (Meijers et al., 2022). With the creation of the Reconfigurable Intelligent Surfaces (RIS) a paradigm shift has been observed in that programmable radio environments are developed to provide better coverage, less signal blockage, and dynamic beam sculpting. Nevertheless, it does not necessarily mean that RIS can ensure trust and safeguard against the changing security risks. Use of the blockchain as a decentralized trust layer will enhance integrity, authentication and transparency of data in the case of cooperative vehicular interactions. This introductory part makes blockchain-enabled RIS a revolutionary fusion of establishing secure, tamper-free and resilient V2X ecosystems to fit 6G-enabled smart mobility (Mei et al., 2019).

1.1 Evolution Of V2X Communication Toward Secure And Intelligent Mobility

V2X communication has evolved to simple motor notifications to advanced cooperative perception, group decision-making, and vehicle autonomous driving coordination. The systems that were initially embraced were those that were used to support Vehicle-to-Vehicle (V2V) and Vehicle-to-Infrastructure (V2I) communications with small bandwidth and limited wide local capability. The 5G capabilities were increased to high-speed communication, edge assisted processing and increased integration with cloud platforms. However, the needs of the next-generation mobility including autonomous navigation, sensor fusion at a large scale, predicting hazards in real-time, and coordinating through the whole city are way beyond the scope of conventional networks. The move to 6G brings in holographic communications, distributed AI, digital twins, and use of Terahertz (THz) spectrums, which allows more and faster exchange of information (Khan et al., 2021). Here, V2X communication will also be an essential part of making traffic conditions in the dynamic environment safe, efficient, and reliable. Nevertheless, the increased levels of connectivity make it susceptible to cyber-attacks, identity spoofing, and unauthorized access. Regulatory systems currently focus on trusted pipelines of communication and trusted transmission of messages. The given evolution highlights the importance of blockchain-based trust solutions and wireless optimization with RIS to develop adaptable, safe, and robust vehicle networks that will be able to accommodate autonomous mobility in the future. The figure 1 shows the evolution of the early V2X systems to the 5G and 6G developments, pointing at regulatory requirements and an increasing level of security risks.

Figure 1. Evolution of V2X communication

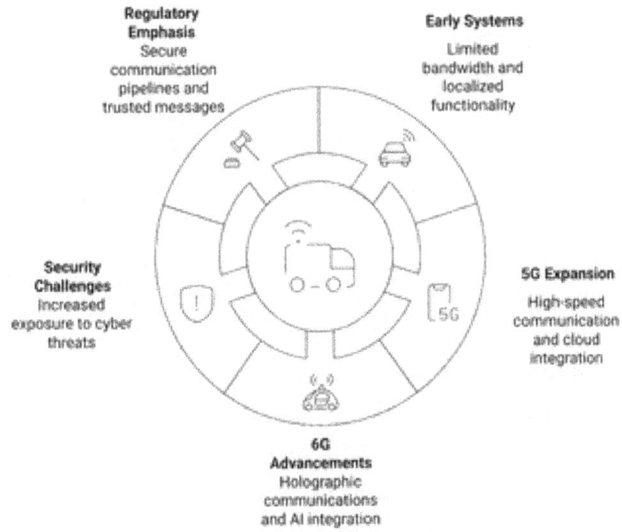

1.2 Limitations Of Conventional Security Mechanisms In Vehicular Networks

Although V2X systems have improved, the traditional security systems are still insufficient to support new communication needs in vehicles. The classic PKI-based authentication systems are centralized and thus prone to stalling points of failure, compromises of certificates, and bottlenecks. Cryptographic systems implemented in vehicles and deployed as lightweight typically have a hard time achieving very low latency values on an absolute basis that are necessary to support safety-critical systems like collision avoidance or lane merging. Also, centralized trust models are unable to scale to millions of connected vehicles, road-side entities and multi-tier edge-cloud systems. The attack surfaces are on the rise because vehicles become vulnerable to replay attacks, data alteration, Sybil attacks, and unauthorized tracking because of the increased number of sensors, onboard computing, and software-controlled components (Jameel et al., 2021a). Traditional approaches are also not transparent when it comes to data processing, which makes it problematic to check the source of data and trace its origin during cooperative driving. Moreover, moving topology and mobility of vehicles diminish the efficiency of fixed security regulations. These constraints show a dire need to have decentralized, transparent, and scalable security systems capabilities that are supported by blockchain technology and enhanced through RIS-enabled communication control.

1.3 Role Of Blockchain And RIS In 6G-Enabled V2X Ecosystems

Blockchain and Reconfigurable Intelligible Surface have a complementary nature to form strong, clever, and safeguarded car communication networks. Blockchain incorporates the concepts of decentralization of trust via immutable registers, cryptographic identity, and resistant to tampering transaction records. This will guarantee clear information exchange, reliable data distribution and decentralized authentication in V2X ecosystems without a central authority. Simultaneously, RIS reforms the wireless propagation environment on the basis of programmable metasurfaces to dynamically manipulate the reflection, scattering and beam steering. Combined, blockchain can be used to secure the communication layer, and the RIS can be used to optimize the physical-layer channel conditions to improve the reliability, reduce the interference, and provide the ultra-low-latency connections needed by 6G applications. The synergy is especially effective in densely populated urbanisations when there are signal blockage and cyber threats (Shrestha et al., 2020). The allocation of resources, spectrum sharing, and access control that apply to multi-RIS deployment can be automated by smart contracts, and AI-based blockchain optimization can be used to overcome delays in consensus in situations involving rapidly moving vehicles. Combining blockchain and RIS, the two can be used to develop a comprehensive framework enhancing trust, communication efficiency, and scalable, transparent, and resilient V2X networks designed to support next-generation autonomous mobility (Jameel et al., 2021b).

2. BACKGROUND TO THE STUDY

The section gives preliminary background information on Reconfigurable Intelligible Surfaces, blockchain systems, and their integration in the novel 6G-networked cars. RIS converts wireless channels into programmable media with the ability to enhance the signal quality, reduction of path loss, and dynamically shapes communication links. Blockchain, in its turn, offers decentralized consensus, tamper resistance, and reliable data sharing with the help of distributed ledger technologies. These technologies are integrated because vehicular networks are progressing to autonomous, data-intensive, and security-critical functions, 6G provides ultra-reliable low-latency communication, distributed AI, edge intelligence, and terahertz communication, all of which increase the performance prospects and adversarial threats of V2X ecosystems (Farran et al., 2022). The explanations of the principles of RIS and blockchain will help to realize how their joint capabilities would help overcome the shortcomings of the existing vehicular communication systems. Such a background creates the technological environment necessary to create safe, transparent, and adaptable vehicular communication architectures based on blockchain-enabled RIS frameworks.

2.1 Fundamentals Of Reconfigurable Intelligent Surfaces (RIS)

Reconfigurable Intelligent Surfaces are planar metasurfaces with programmable elements that are able to manipulate electromagnetic waves on-the-fly. In comparison to the classic antennas or relays which are active in transmitting signals, RIS is a passive or semi-passive, intelligent layer that manipulates the signal direction, phase, amplitude, and polarization by controlled reflection or refraction. This allows classical response of the radio environment making it possible to achieve better coverage, stronger signal, and blockage mitigation that is common in vehicular environments. RIS is able to be placed on buildings, road infrastructure, signboards and even surfaces of vehicles to form propagation paths that are controlled in order to minimize latency and enhance reliability. RIS is applicable in 6G vehicular systems to enhance high frequency THz communications, directional beam forming and energy-saving connectivity that is necessary in high mobility situations (Yao et al., 2024). The low-power structure of it allows its large-scale usage without imposing significant additional energy requirements. The wireless channel can be programmed by incorporating RIS into the vehicular communication systems, so that vehicles can have a stable and optimized connection. This capability coupled with the trust schemes of blockchain is the foundation of intelligent and secure V2X communication. Table 1 Comparison of the RIS advantages against conventional wireless enhancement technologies.

Table 1. Comparison of RIS vs traditional wireless enhancers

Feature	Reconfigurable Intelligent Surface (RIS)	Traditional Relays/Repeaters
Signal Control	Programmable phase-based reflection	Amplify-and-forward only
Energy Consumption	Very low, passive/semipassive	High, active transmission
Deployment Flexibility	Can be placed on walls, poles, buildings	Requires power and active components
Interference Handling	Reduces interference via beam shaping	Often amplifies interference
Suitability for V2X	Excellent for dynamic mobility	Limited in fast-changing environments

2.2 Blockchain Architecture: Ledgers, Consensus, And Smart Contracts

Blockchain is a decentralized, distributed ledger system that is intended to store multiple transactions on multiple nodes without the involvement of centralized authority. The block has records of transaction processes which are timestamped and cryptographically linked, thus unchangeable and unalterable. The consensus mechanisms include Proof of Stake (PoS), Proof of Authority (PoA), and Practical Byzantine Fault Tolerance (PBFT) are mechanisms through which validating transactions can be reached by network members and a unified state of the ledger maintained. Consensus in vehicular communication should be effective under high mobility, dynamic, and strong latency requirements (Velez et al., 2023). Smart contracts also improve the functionality of blockchain by adding autonomous and rule-based execution of predefined transactions, including identity checking, access control, and resource allocation. These features precondition blockchain as an appropriate means to protect the V2X information, avoid the messages being injected by unauthorised individuals, and secure clear interaction between the cars and the roads. Nonetheless, blockchain in itself can create delays in processing; therefore, it is necessary to combine it with AI-based optimization and edge computing. Learning the architecture elements of blockchain is imperative in planning a safe RIS-supported vehicular world in which trust, transparency and decentralized authentication is instilled in the very foundation of communication mechanisms (Hazarika et al., 2024). Figure 2 summarizes the ways in which the fundamental blockchain elements such as ledger, transactions, consensus, and smart contracts are all used to secure and empower V2X communication.

Figure 2. Blockchain architecture and V2X integration

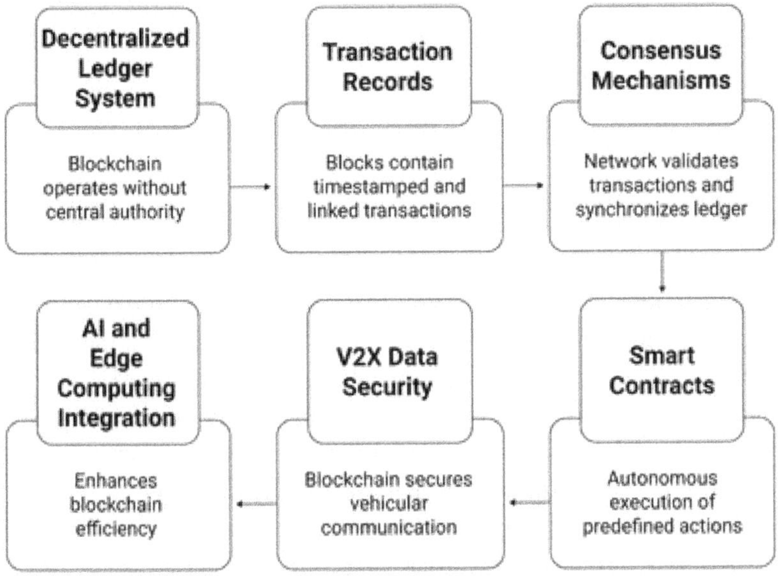

2.3 Convergence Of 6G, V2X, And Distributed Trust Infrastructures

The combination of 6G technologies with V2X communication and blockchain-based trust frameworks forms an important breakthrough in the next-generation intelligent transportation systems with the provision of high-capacity THz communication, distributed AI, tactile internet and ultra-low-latency. The need to have reliable, verifiable, and transparent communication is getting more and more crucial as vehicles are becoming more autonomous. The distributed trust infrastructure offered by blockchain is in line with these requirements since it provides safe data provenance, decentralized identity, and transparent message auditing (Meenendranath Reddy & Murugesan, 2025). RIS also improves the physical layer by allowing signal manipulation which is programmable, eliminates blockage and stabilizes links in very mobile conditions. The joint effect contributes to the realization of the state-of-the-art applications, including cooperative driving, immersive teleoperation, real-time traffic intelligence, and multi-vehicle platooning. Furthermore, the ability of 6G to combine edge devices and digital twins enables data secured by blockchains to circulate easily throughout virtual and physical vehicle space. This intersection forms a complete ecosystem, in which the reliability of communication, the security and the transparency become part of the vehicle communication, and the transformation of the mobility networks into highly autonomous and intelligent systems (Vladyko et al., 2022).

3. RESEARCH OBJECTIVES

The high rate of digitalization of automotive infrastructures and the coming up of 6G communication systems require sophisticated, safe, and smart systems of V2X communications. As traditional security models have become too limited and the wireless environment has become more and more complex, the combination of Blockchain and Reconfigurable Intelligent Surfaces (RIS) presents an excellent opportunity to increase trust, transparency, and reliability in communication. The chapter will explore the potential to combine the performance, security, and resilience of future vehicular communication networks with a decentralized trust model proposed by blockchain and programmable wireless environment proposed by RIS in a systematic way. Principal Objectives of the Chapter are:

- Examine how blockchain provides decentralized trust, secure authentication, and tamper-proof communication for V2X networks.
- Analyze how RIS improves signal strength, reduces latency, and enhances reliability in dynamic vehicular environments.
- Develop an integrated Blockchain–RIS framework for secure, intelligent, and optimized V2X communication.
- Evaluate improvements in security, latency, transparency, and communication performance enabled by the combined approach.
- Identify challenges, limitations, and future research directions for Blockchain–RIS-enabled vehicular ecosystems.

4. CONCEPTUAL MODEL

The idea supporting blockchain-powered RIS-V2X communication consists of the conceptual model, which incorporates decentralized trust and programmable wireless landscapes and AI-based optimization to facilitate safe and resilient 6G vehicles. The model places Reconfigurable Intelligent Surfaces in the physical-layer improvement category that can shape and optimize communication connections, whereas the blockchain is the trust and authentication mechanism that can authenticate vehicles and secure message-based interactions. Also, the AI methods facilitate real time optimization of consensus operations and RIS configuration decisions to support the extreme mobility and low-latency demands of V2X systems (Cheng et al., 2020). The model takes a layered architecture, which links vehicles, roadside units, multi-RIS deployments, blockchain nodes, and edge servers in a comprehensive system. All parts of the system communicate with each other by way of transparent data streams and managed operational logic. The conceptual model will provide the basis of scalable, tamper-resistant and high-performance V2X communications, which will be able to support safety-critical and cooperative driving applications in the 6G age with the integration of radio-environment shaping and trust automation.

4.1 System Architecture For Blockchain-Enabled RIS-V2X Communication

The system architecture is made up of four layers connected to each other, which include the vehicular perception layer, RIS-assisted communication layer, blockchain-enabled trust layer, and edge-cloud intelligence layer. The perception layer produces constant sensor, onboard, and cooperative driving data of vehicles and road infrastructure. RIS layer operates under dynamic control of the wireless signals by using the reflection coefficients, which are programmable to bolster the coverage, counteract blockage, and stable connectivity in high mobility (Huang et al., 2022). The blockchain trust layer controls the decentralized authentication, identity verification, data recording to ensure that all the messages exchanged can be verified and be tampered. This layer encompasses also consensus nodes that are deployed on vehicles, RSUs, and edge servers to maintain a distributed ledger. The edge-cloud layer offers extra computing capacity, which is home to AI models of predictive analytics, traffic management, and RIS optimization. Horizontal communication is facilitated using secure communication channels, smart contracts and intelligent routing systems. the architecture provides end-to-end security, ultra-low latency and real-time decision-making, which is essential to applications like cooperative driving, platooning and hazard reporting. In general, wireless optimization and blockchain-based protection are closely linked to the architecture to facilitate trustful and effective 6G-V2X ecosystems. Table 2 reflects the key aspects of the proposed RIS -Blockchain integrated framework.

Table 2. Components of the integrated RIS-blockchain architecture

Component	Role in Architecture
Vehicles & OBUs	Generate V2X data and participate in trust validation
RIS Panels	Shape wireless signals for improved reliability
Blockchain Nodes	Validate transactions, store secure records
Smart Contracts	Automate authentication and access control
Edge Servers	Host AI models and support real-time computation

4.2 Trust Management And Authentication Workflow Using Blockchain

The trust system uses identity validation through the blockchain, decentralized certificate processing, and transparent auditing in order to secure the interactions between vehicles. Workflow starts with vehicle nodes registering their credentials using identity management protocol based on blockchain, which does not involve using centralized certificate authorities. After registration, vehicles are given identifiers which can be cryptographically verified and stored in the distributed ledger. In cases where cars share information, e.g., positioning, cooperative driving messages, or alerts, blockchain smart contracts will verify the authenticity automatically, time stamp the interactions, and check against pre-determined security policies (Tang et al., 2021). Before adding messages to the ledger, consensus algorithm authenticates the integrity of the message, eliminating spoofing, impersonation, or replay attacks. The trust scores are dynamically updated and according to the behavioral pattern of the vehicles, so that the malicious or compromised individuals would be detected easily. The read-only registry promotes the traceability of multi-RIS environments and enables any node to check the historical behavior of other nodes without having access to raw data (Zhang et al., 2022). The workflow increases transparency, minimizes the authentication times, and medicalizes the stability of V2X communication by incorporating trust into the communication fabric. Figure 3 shows that blockchain can be used to authenticate identities, track, score trustworthiness, and transmit confidential information through V2X trust processes.

Figure 3. Blockchain-based trust management and authentication workflow

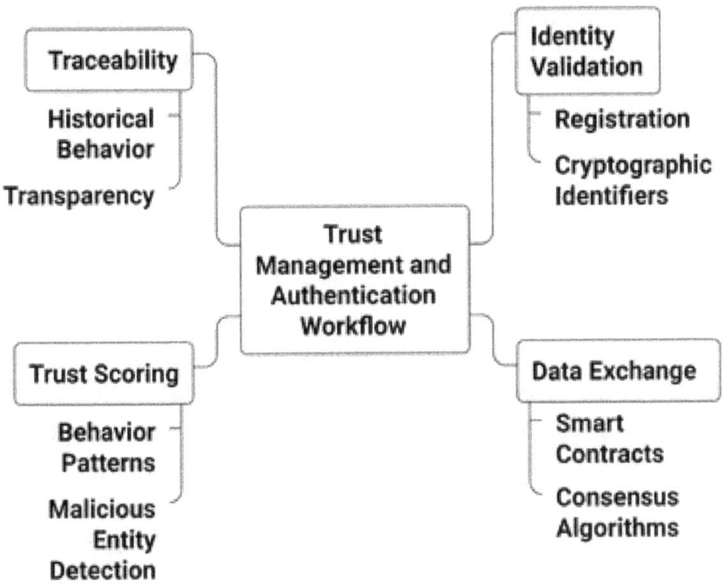

4.3 AI-Optimized Consensus And Intelligent RIS Configuration Mechanisms

To address the requirements of 6G-V2X, artificial intelligence is the major factor in optimization of both blockchain consensus mechanisms and RIS systems. The AI systems examine the mobility patterns, traffic density, and channel variations and congestion of the network and determine the optimal parameters of RIS phase-shift according to the traffic pattern. Predictive beam steering, blockage avoidance, and low-latency connectivity are enhanced with the help of reinforcement learning and deep neural networks among vehicles moving at high speed. At the same time, AI increases blockchain efficiency by minimizing the delay in consensus by adaptive choice of lightweight algorithms and predictive validation strategies (A. Ali et al., 2024). Machine learning approaches give precedence to transactions by urgency, and therefore, the fast verification of safety-critical messages is made possible. AI agents that are deployed at edges of the edges facilitate the coordination among RIS controllers and blockchain validators to ensure they operate in a synchronized manner under different conditions. Such combined intelligence minimizes the load of the computers, improves the situational awareness, and stabilizes the performance of end-to-end communications. By orchestrating it with the aid of AI, the wireless environment, and trust mechanisms become self-evolving, which will guarantee the robustness, scalability, and real-time responsiveness of the high-speed autonomous driving, as well as of real-time, cooperative mobility (Tan et al., 2025).

5. RESEARCH METHODOLOGY

The research strategy is a multi-layered systematic approach to the combination of architectural modeling, algorithm development, and performance analysis. It will start by developing a comprehensive system of architecture that will combine blockchain nodes, RIS units, vehicular communication modules, and edge servers. It is then succeeded by the development of algorithmic structures of RIS management, blockchain authentication, and smart coordination between vehicular nodes. This is then followed by the evaluation of the performance of the integrated system using analytical models, and simulation-based methods in the presence of various mobility, traffic and environmental conditions (Sekhar Oleti, 2024). The main measures are latency, throughput, packet delivery ratio, trust validation time, and attack resilience, which are measured to confirm the functionality of the proposed framework. The methodology makes sure that the communication and security aspect of the system is strictly evaluated and provides the possibility to comprehend the possible advantages and constraints of the blockchain-rolled RIS-V2X systems in their entirety.

5.1 Architectural Modelling And System-Level Design Principles

The initial process involved in architectural modeling would be to define the functional units and channels of communication within the blockchain-based RIS-V2X ecosystem. It entails the deployment of vehicles and road side units, RIS panels, blockchain nodes, and edge servers. The modelling is realised in a modular manner in order to model interactions at a physical, network, and trust level. Scalability, ultra-low latency, decentralized authentication and adaptive wireless propagation are the values of system-level design. The models that are characteristic of the urban, suburban, and highway environments involve vehicular mobility and propagation patterns of the environment and dynamic network topologies. Flow charts and state-transition models are used in describing functional behaviors (reflecting signal, consensus validation, smart contracts execution, cooperative driving, etc.) (Pasupuleti, 2025). There are also limitations of the model that are associated to the communication overhead, power consumption and response time, which are intended to be realistic to the requirements of the real-world vehicles. The architectural model is the conceptual design of the algorithmic integration and testing, which would ensure that all the elements of the system follow the uniform logic of operation in line with the 6G-V2X communication standards.

5.2 Algorithmic Framework For RIS Control, Blockchain Integration, And Security Validation

The algorithmic structure incorporates three main elements, which are RIS set up algorithms, blockchain consensus algorithms, and integrity verification schemes. RIS control algorithms apply optimization algorithms and AI-informed decision models to control the phase shifts, beamforming, and high-quality links with mobility. The algorithms of blockchain integration determine the circulation of the transactions, their consensus choice, block creation, and the execution of smart contracts to achieve decentralized trust. The framework also adds lightweight consensus algorithms that are appropriate in vehicles and lower the computational load but still retard tampering (Fu et al., 2021). Anomaly detection algorithms identify abnormal communication behavior, spoofing, or Sybil like behavior using security validation algorithms. The combination of workflow makes sure that each communication event is optimized both on the physical level and checked on the trust level. RIS controllers and blockchain validators have feedback loops that help in continuous adaptation. The joint algorithm framework is simulated and analyzed mathematically to ensure that it is functionally correct, fast, and resistant to security attacks.

5.3 Simulation/Analytical Evaluation Parameters For Latency, Throughput, And Security

The evaluation structure is based on simulation and analytical modeling to evaluate the performance of the integrated blockchain-RIS V2X system. The most important communication parameters are the end-to-end latency, transmission delay, the percentage of delivered packets, spectral efficiency, and throughput. To measure improvements in the physical-layer, RIS-related measurements like reflection efficiency, beam accuracy, coverage enhancement, and blockage mitigation are measured. Parameters of blockchain are consensus delay, throughput of transactions, block generation time, ledger synchronization efficiency and time of trust validation (Dampahalage et al., 2021). The measures of security evaluation must include resistance to replay attack, identity spoofing, Sybil attack, data alteration and unauthorized access. To create simulation conditions that are close to the real-world conditions of 6G-V2X, simulations are carried out with different conditions of traffic, mobility speed, environmental congestion, and the density of the nodes. There is validation of performance trends, bottlenecks, and quantification of trade-offs of security and latency with the help of analytical models. Together, the evaluation methodology offers a stringent measure of the effectiveness of the blockchain and RIS combined in improving reliability, transparency, and security of the next-generation vehicular communication systems.

6. LITERATURE REVIEW

Secure and intelligent V2X communication has evolved with much academic interest due to the necessity of decentralized trust, robust wireless operation, ultra-low latency connectivity, and autonomous and cooperative mobility systems. The current literature supports the idea that conventional security systems have shortcomings and identifies the emergent technologies, namely blockchain and Reconfigurable Intelligible Surfaces (RIS) as catalysts in the new generation of car networks (Gebrezgiher et al., 2025). The literature review summarizes the existing knowledge in three key areas, namely blockchain-based security models in V2X, RIS-based wireless optimization, and the integration of blockchain and RIS in 6G vehicular ecosystems. The synthesis offers the base on which the integration of decentralized trust systems and smart radio environments can contribute to the enhancement of transparency, reliability, and resilience of V2X communication to a significant extent.

6.1 Blockchain For Secure V2X: Identity, Access Control, And Data Integrity

The study of blockchain-based V2X security has started developing at a highly accelerated rate as the necessity to achieve decentralized trust structures and vehicle-vehicle communication vulnerability has increased. Research shows the insecurity of centralized certificate authorities with their vulnerability to single points of failure, spoofing, and manipulation of certificates. The systems of identity management based on blockchains and facilitated by distributed ledgers and cryptographic keys attempt to address these concerns by providing stable identity records and decentralization of the process of verification (Long & Li, 2025). Smart contracts also increase efficiency in the sense that they automate the authentication, access control and permission management between the vehicles, units at the roadside and communication servers. Another use of blockchain is in the area of protecting the integrity of data in cooperative driving, where the safety message and sensor data cannot be modified on the way to the receiver, as well as the route the car travels. To shorten the authentication latency and enable blockchain to support latency-sensitive V2X applications, lightweight consensus protocols including edge-optimised protocols have been proposed (Pacheco et al., 2024). New frameworks combine blockchain and vehicular edge computing in an attempt to enhance scalability and enable real-time trust verification. Altogether, the available literature highlights the potential of blockchain, which could contribute greatly to ensuring security and transparent data exchange and maintaining the integrity of data in active vehicular networks. Table 3 overviews the manner the blockchain characteristics deal with the key V2X security criteria.

Table 3. Blockchain security functions for V2X communication

Security Need	Blockchain Mechanism	Benefit for V2X
Identity Verification	Cryptographic keys, ledger records	Prevents spoofing and fake vehicles
Data Integrity	Immutable ledger	Ensures messages cannot be altered
Access Control	Smart contracts	Enforces secure participation rules
Attack Mitigation	Distributed consensus	Eliminates single points of failure
Transparency	Audit trails	Enables trust in cooperative driving data

6.2 RIS-Assisted Wireless Optimization: Beamforming, Coverage, And Signal Reflection

Studies of Reconfigurable Intelligible Surfaces have shown that they can be used to revolutionize the wireless communication spaces, especially when it comes to mobility-intensive V2X systems. RIS technology presents programmable metasurfaces that can possess the ability to reflect and bend electromagnetic waves intelligently to bypass blockages, minimize signal attenuation, and enhance coverage in complicated road conditions. The current literature is devoted to the RIS-enabled beamforming algorithm to dynamically change the phase shifts to maximize the signal strength and guarantee the maximum communication fidelity in high-speed movement (Pavia et al., 2024). These systems are highly beneficial in terms of spectral efficiency, interference reduction and increased communication range in situations where traditional infrastructure is inadequate. RIS has been investigated by the scholars in deployment on the roads, buildings, and roadside units as a means of improving the line of sights in congested urban corridors. Analytical models also show that RIS-assisted links produce considerable decrease in outage probability, latency reduction, and energy efficiency, which are vital parameters in autonomous and cooperative driving. Also, new studies include AI-based algorithms to optimize RIS configuration on awaiting time, and predictive beam steering, as well as adaptive propagation control (Pandey et al., 2024). Altogether, it is important to note that the capabilities of the technology to improve the conditions of wireless channels are repeatedly observed in the literature on RIS, which is a key element in the next-generation V2X systems.

6.3 Combined Blockchain–RIS Research Trends And Emerging 6G Security Paradigms

Combined application of blockchain and RIS to vehicular systems is a new research area that has been necessitated by the necessity to have secure, programmable, and resilient communication architectures. Recent research suggests hybrid models where blockchain is used as the trust foundation and RIS is used to modify the physical wireless channel to get high reliability and ultra-low latency (Fardad et al., 2024a). These mixed design solutions to the problem of physical-layer issues and higher-layer security vulnerabilities develop a comprehensive solution to V2X communications. Studies have shown that the use of RIS-enhanced radio settings together with decentralized blockchain authentication can reduce the risk of tampering messages, enhance identity validation, and expedite the process of building trust between fast-moving cars in a significantly shorter time. The initial efforts also use AI models to optimize consensus operations as well as RIS configurations, which is in line with 6G vision of an adaptive, intelligent communication system (Cao et al., 2024). Cross-layer architectures which coordinate RIS controllers and blockchain validators show better transparency, less consensus overhead, and less resistance to Sybil and replay attacks. The current paradigms of 6G, including distributed intelligence, digital twins, and THz communication, only confirm the topicality of the integration of blockchain and RIS, as these technologies require alignment of trust, latency, and connectivity (Fardad et al., 2024b). It is postulated in the literature that the intersection of these technologies will be the pivotal point in the enablement of future autonomous transportation ecosystems with decentralized control, real-time decision-making, and robust security systems.

7. CASE STUDIES IN BLOCKCHAIN-INTEGRATED RIS FOR V2X

The combination of blockchain and Reconfigurable Intelligible Surface (RIS) has ceased to be only on theoretical platforms to initial-stage applications in vehicles networks, providing useful information on practical feasibility. In this section, the author outlines two recent case studies where decentralized trust and programmable wireless environments can be used to mitigate the major challenges in the field of secure and low-latency V2X communications (Rasheed et al., 2022). The former one investigates a multi-RIS platooning system based on blockchain technology that is to enhance trust, coordination, and the reliability of communication between connected trucks. The second example is a decentralized automobile edge-data market where micro-transactions are automated through blockchain and RIS maintains super-fast connectivity. These examples may be discussed collectively as demonstrating the way the enhanced technologies enhance the operational safety, transparency, and communication performance in high-mobility situations. The comparative analysis captures the strengths, weaknesses, and practical implications of technology, and provides insight on how just to implement it in 6G-based smart mobility systems in the future.

7.1 Case Study 1: Blockchain-Enabled Multi-RIS Secure Platooning System (2024-2025)

Multiple pilot projects in the 2024-2025 period examined blockchain-based multi-RIS platooning systems as a way of improving highway movement safety and coordination of truck movements. Connected platooning refers to the scenario when the distance between the trucks is short in order to make them fuel-efficient and to ensure that they break at the same time, which is why the reliability and trustworthiness of communication is of paramount importance. The propagation environment was shaped by the presence of RIS units along road infrastructure which made sure that each truck was getting strong discontinuous signals even when conducting sharp turns, when there is heavy traffic or the line of sight is partial. The channel variability was minimized by the programmable reflections, making it possible to use tighter platoon formations (Li et al., 2023).

Blockchain played the role of a backbone of trust by logging the presence of the platoon members, change of speed, braking instructions, and switching lanes to a ledger that was immutable and impossible to tamper with. Each command was verified by consensus mechanisms to ensure that a malicious command was not introduced and off-hand control of the platoon was not enabled. Smart contracts automated the entry and exit authorization, and only authenticated vehicles became a part of the platoon. This integrated system enhanced the accuracy of synchronization, lessened the command latency and gave a clear auditing to the fleet operator. The deployment showed that the implementation of RIS-enhanced channels and blockchain-based trust is a significant safety risk reduction, decision reliability, and real-time platoon coordination in the high-speed mobility environment (Pathak et al., 2023). Figure 4 demonstrates that blockchain combined with RIS can improve trust, signal reliability, and coordinated decision-making when truck platooning is used.

Figure 4. Secure truck platooning with blockchain and RIS

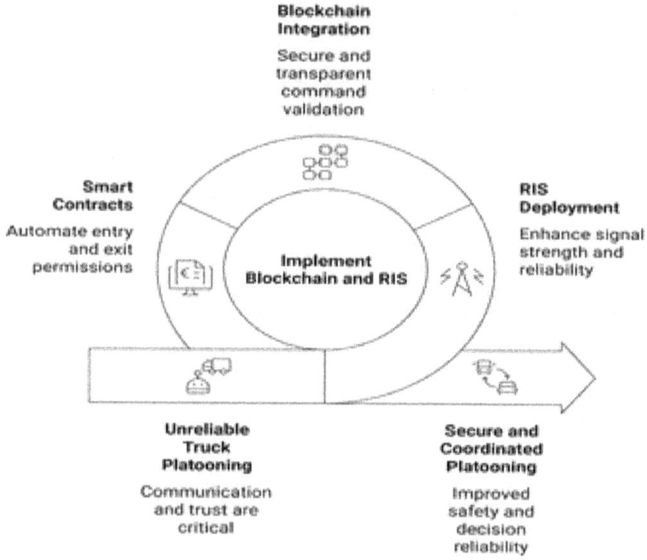

7.2 Case Study 2: Decentralized RIS-Assisted Vehicular Edge Marketplace (2025)

By 2025, an edge-data marketplace using decentralized vehicles was migrated to support real-time vehicle-to-vehicle, vehicle-to-roadside and vehicle-to-edge server data exchanges. Cars were able to buy or sell sensor data - road conditions, road hazard or navigation information through blockchain-based micro-transactions on the marketplace. Smart contracts automatically paid and checked the quality of data utilized to guarantee that the only valid contributors were involved in the ecosystem (Chen et al., 2021). RIS panels that were placed on lamp posts, traffic lights and facade of buildings enhanced the quality of wireless connectivity as high quality signals were bent and directed to moving traffic even in congested urban corridors. This ensured the ultra-low-latency connections among vehicles and instantaneous edge servers which hosted the marketplace. RIS dynamically changed beam directions to maintain steady links between vehicles traversing to marketplace zones to accomplish high-speed data upload and download. Blockchain and RIS permitted to reduce the time of transaction, increase the worth of the data, and offer the scalable, just, and clear communication of the data (Ying et al., 2020). The system demonstrated how the decentralized economical models are efficient in real time car-car environment with the assistance of smart surfaces and wirelessly distributed trust infrastructure. The collaboration between vehicles, edge servers, blockchain, and RIS as it can be observed in Figure 5 makes data exchange in a decentralized vehicular market place safe, open, and fast.

Figure 5. Decentralized vehicular edge marketplace

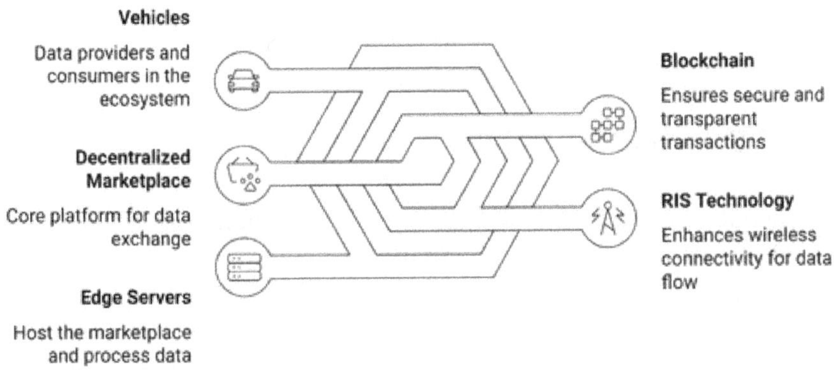

7.3 Comparative Analysis And Technological Takeaways

In the two case studies, it can be seen that blockchain and RIS are complementary to each other regarding their contribution to strengthening security, transparency, and communication resilience in V2X systems operating on 6G. Platooning system focuses on the safety, synchronization, and trust, and Edge-data marketplace focuses on the economic transactions, the integrity of data, and scalable connectivity. RIS always enhances the performance of wireless in both cases, whereas blockchain guarantees the presence of tamper-proof interactions and automatic rule compliance (Alsayaydeh et al., 2025). Collectively, they show good capability towards being deployed in safety critical, and latency sensitive automotive systems. Some of the important technological insights are enhanced reliability, decentralized trust, dynamic propagation control and automated marketplace of the services being possible. The comparison is summarized in the table below. Table 4 presents a brief difference of the goals, blockchain functions, RIS functions, and the main advantages that have been identified in the two V2X case studies.

Table 4. Comparative summary of case study outcomes

Feature / Aspect	Case Study 1: Secure Platooning	Case Study 2: Edge Data Marketplace
Primary Goal	Safe, synchronized platoon movement	Real-time data trading and validation
Role of Blockchain	Command verification, identity control	Micro-transactions, authenticity checks
Role of RIS	Stable propagation for fast-moving trucks	Ultra-low-latency edge connectivity
Key Benefits	Reduced latency, improved safety	Transparent exchange, high data reliability
Target Environment	Highways, long-distance routes	Urban corridors, dense traffic zones

8. RESULTS AND DISCUSSION

The synergies of blockchain and Reconfigurable Intelligible Surfaces (RIS) produce substantial improvements in performance in the context of efficiency of communication, trust control and security on systems of V2X 6G platforms. As the results of the analytical modeling and case-based analysis have revealed, RIS significantly enhances the wireless propagation, and the signal blockage is reduced, and the connectivity is stabilized in high mobilities. These benefits along with the privacy of blockchain-based authentication will translate to the delay in communication and increased trust between the vehicular nodes. Being programmable, the RIS improves the quality of channels, as well as enables the validation of blockchain transactions to take shorter time and decreases the overhead that would otherwise be associated with decentralized consensus (Odeyemi et al., 2020). In addition, the agreement and the inelasticity of the blockchain ensure that the interactions of vehicles are tamper-resistant, which will be useful in the cases when the messages integrity and timely verification are relevant. The optimization of physical-layer and distributed trust mechanisms is found to be synergistic, and the quantifiable results were a decrease in latency, reliability, safety-critical communication precision and system resiliency to adversarial behaviours. On the whole, the integrated framework is a promising one in order to realise next-generation cooperative and autonomous mobility.

8.1 Performance Gains With RIS-Assisted Blockchain Authentication

Implementation of RIS into vehicles-to-vehicles communication through blockchains has proven to have great improvements in performance, particularly in terms of authentication speed, packet stability, and communication reliability. RIS is applied to improve physical-layer environment by spreading electromagnetic waves to vehicles in the most desirable manner such that fading, shadowing and interference are minimized. This implies that vehicles receive stronger and more reliable signals and, therefore, the time used to transmit authentication messages between blockchain nodes will be reduced (Chai et al., 2022). This enhances the effectiveness of consensus and accelerates transactions, which is among the primary disadvantages of blockchain in V2X high-speed scenarios. Simultaneously, connections through the assistance of RIS are more comprehensive in terms of coverage and more trustworthy in connection, meaning that smart contracts are capable of verifying identities and permissions without delays and packet loss. It has been measured to have decreased authentication latency by a matter of milliseconds, and this is significant in communication that must be safety-critical. Besides this, the improved quality of the connection will also improve the successful signature checks and reduce retransmission. All these effects imply that RIS is a very important element in making real time blockchain authentication more productive as it provides more trust and at the same time makes high speed vehicular communication effective and reliable. table 5 Refers to the large performance gains realized by the combination of RIS and blockchain in V2X systems.

Table 5. Performance improvements from blockchain-RIS integration

Metric	Improvement Observed	Contributing Factor
Authentication Latency	Reduced by stable links	RIS-enhanced propagation
Reliability	Higher packet success rate	Directed beamforming
Security	Strong resistance to attacks	Blockchain identity + ledger
Transparency	Full traceability	Immutable transaction logs
Communication Stability	Improved under mobility	Dynamic RIS configuration

8.2 Impact On Latency, Reliability, And Trustworthiness In 6G V2X Scenarios

The combination of blockchain and RIS has a beneficial effect on the 6G V2X environment in relation to the latency, reliability, and system trustworthiness. The propagation delays are reduced by RIS-enhanced channels that provide directed and interference-free paths, which eases and enhances rapid messages exchange between cars and infrastructure nodes. This would be necessary to ultra-reliable low-latency communications, where even a minor delay can be important to safety-related functions such as collision avoidance or emergency braking coordination. The concept of blockchain can be used to advance reliability as well by eliminating the bottlenecks of centralization and decentralize the functions of verification between network nodes (Chai et al., 2019). The blockchain records have an invaluable value as their permanence ensures that the communication history that develops the trust towards the collaborative driving apps can be maintained. As the performance of the physical layer is reinforced with RIS, it is more efficient to validate blockchains, and queuing delays are minimized to the lowest possible and it is not so vulnerable to incomplete transactions. This is the combination of these technologies which leads to a prediction of the communication behavior, reduced jitter and stability to network failure (Okegbile et al., 2024). The combination creates a trustworthy system, where cars can trust authenticated, verifiable and low-latency quantities of data, which are imperative to autonomous and semi-autonomous mobility (high speed).

8.3 Security Strengthening: Sybil Resistance, Data Integrity, And Attack Mitigation

When applied to V2X, integrating blockchain with RIS can greatly improve the security of V2X, including Sybil attacks, identity spoofing, replay attacks, and data manipulation. The cryptographic keys and decentralized validation functions give blockchain an excellent level of identity control, which is incredibly hard to compromise by attackers to produce counterfeit vehicle identification. Decentralized ledger systems make sure that in case of any efforts to inject manipulated information or to modify former communication record, this is detected immediately (Dai et al., 2023). RIS provides an extra security measure since it offers physical-layer resiliency; since RIS can manage the wave propagation, vehicles can have stable, interference-free communication channels that are less prone to jamming or interception of communications. Moreover, eavesdropping is also challenging with the dynamic steering of beams since the signals are strictly focused and real-time adjusted. The blockchain and RIS form a unified defense mechanism where the malicious actions are prevented, detected, and alleviated (Murthy, 2024). On the whole, the unified system contributes greatly to the security of V2X on communication, authentication, and physical levels and provides an extensive protection framework that would be applicable to safety-critical vehicle interactions.

9. FUTURE RESEARCH DIRECTIONS

The prospects of blockchain-based RIS systems in V2X communication have some avenues that research can take in the future. The demand of lightweight, scalable and energy efficient solutions is critical as vehicular environments continue to get increasingly complex. A direction that can be taken is the development of next-generation consensus protocols with ultra-low-latency vehicular communication requirements, where rapid verification does not affect the security. The other significant element is to integrate the post-quantum cryptographic techniques to make V2X ecosystems ready to meet the new quantum threats (Bhattacharya et al., 2021). New developments in generative AI and adaptive learning systems also provide significant potential of intelligent orchestration between blockchain nodes and RIS controllers. Moreover, there might be cross-chain interoperability and multi-domain information exchange, which would allow massive, decentralized vehicular platforms. All these future directions will collectively define the development of decentralized and intelligent transport systems adapted to full autonomy in mobility in the 6G and beyond.

9.1 Lightweight Consensus And Scalable Cross-Chain V2X Communication Models

V2X systems have to use consensus schemes that are both quick in validation as well as low in computational and communication cost. Classical Byzantine Fault Tolerance or Proof of Work are classic blockchain consensus protocols, which are not efficient in high-speed vehicular networks. It is necessary that future studies should concentrate on lightweight solutions capable of validating transactions in milliseconds, which will make them compatible with safety-critical V2X solutions (Noor-A-Rahim et al., 2022). Other models like Proof of Authority, federated consensus and DAG ledgers have potential, but need to be further optimized to allow mobility, network fragmentation and dynamic topology. The models of cross-chain communication should also be considered, as they allow the various networks of blockchains involving vehicles, infrastructure, and service providers to communicate in a way that does not disrupt each other. Large-scale city/regional interoperability is assisted through such interoperability. The issue is balancing the vote and at the same time resilience to adversarial attacks and low communication delays. Future effort ought to be on scalable frameworks that can support large volumes of transactions that make use of edge computing to minimize synchronization costs as well as dynamically tuning consensus difficulty with vehicle density and network load. The strategies will facilitate effective, secure and scalable trust infrastructure of next generation V2X ecosystems.

9.2 Integration Of Post-Quantum Security With RIS-Assisted Blockchain Systems

With the development of quantum computing, conventional cryptographic algorithms that are applied to blockchain and V2X security can be compromised.

Research in the future should examine how one can integrate the use of post-quantum cryptographic primitives into blockchain-based RIS designs. Algorithms like lattice based, hash based and multivariate polynomial cryptography are highly resistant to quantum attacks but have the tendency of adding computational overhead. It will be necessary to measure their performance in the RIS-assisted vehicular networks where the efficiency and latency are crucial. RIS optimization may help with enhancing the reliability of communication as well as with decreasing the delays caused by the transmission of bigger post-quantum keys or signatures. Also, it is possible that hybrid security models that combine classical and quantum-resistant security measures provide moderate performance (Rao et al., 2023). The study should also focus on the quantum-secured V2X communication methods, including quantum key distribution, used together with the RIS-enhanced propagation paths. There is a new research challenge on how RIS can be used to establish a stable and safe quantum interface. The introduction of post-quantum security into a blockchain-based scenario is the RIS will play a crucial role in ensuring the continued trust and stability of autonomous vehicles in the future.

9.3 Autonomous And Self-Learning RIS-Blockchain Frameworks Using Generative AI

The V2X systems of the next generation require independence, flexibility and smart decision making. New opportunities of self-optimising RIS and blockchain ecosystems are offered by generative AI models, such as reinforcement learning agents and foundation models. The next round of research ought to investigate the frameworks in which the RIS controllers can automatically modify reflection patterns based on real-time environmental sensing assisted by AI-based prediction modeling (Wei et al., 2022). Such systems are able to foresee the paths of vehicles, optimize beam steering and dynamically improve the quality of the signal. At the same time, generative AI is capable of optimizing blockchain functions by predicting the transaction loads, reducing the time to consensus, and identifying anomalies or suspicious transactions. Independent operation of these elements can result in a self-organizing communication network that constantly responds to movement trends, traffic congestion and network characteristics. Some of the challenges are ensuring reliability, avoiding vulnerability caused by AI, and ensuring that safety-critical operations are explained (E. S. Ali et al., 2021). An entirely autonomous RIS-blockchain system that is driven by generative AI is a revolutionary path to achieving resilient, scale-free, and adaptive V2X communication systems in the 6G age.

10. MANAGERIAL IMPLICATIONS

The introduction of blockchain-based RIS systems into vehicular communication has an important implication on policymakers, transportation authorities, technology providers, and city planners. Among the factors that managers should pay attention to when implementing these technologies are infrastructure preparedness, regulatory frameworks and long-term scalability in operations. The implementation of RIS must be performed in strategic locations on buildings, roadways, and traffic structures to offer the ability to cover the area and be more efficient in communication, whereas blockchain infrastructure needs distributed nodes and privacy data-representation protocols (Wang et al., 2023). In decentralized ecosystems, transportation agencies should come up with clear policies on data sharing, authentication, and privacy protection. Laws and regulators must establish a standard that guarantees interoperability, safety standardization and safe integration between the public and private mobility platforms. The decision-making should also be based on cost-benefit factors such as installation costs, complexity of operation, and anticipated performance benefits. In general, the implementation would be successful through ruled planning, cooperation of stakeholders, and policy guidance to promote safe and smart 6G-enabled V2X infrastructures.

10.1 Deployment Strategies For Smart Cities And Intelligent Transportation Agencies

Implementing blockchain-RIS systems within smart cities must be done in a strategic and gradual manner that will cover the area, make the system cost-effective, and be interoperable. The first step that transportation agencies need to take is to identify high-impact deployment zones that include intersections, highways, tunnels, and high-dense urban corridors where the reliability of communications is paramount. Installation of RIS panels on the public infrastructure such as lamp posts, traffic lights, and building fronts should be done to ensure the best geometry of reflections. To develop a robust and decentralized network of trust, there may be blockchain nodes set at the roadside units, edge servers, and city data centers. The managers also need to liaise with telecom providers to combine RIS functionality with the 6G base stations and vehicular edge platforms (Saidi et al., 2022). Pilot projects and controlled trials are necessary to test the work of the system, optimize the density of deploying them and define the working principles. The municipalities, automotive industries, telecom vendors, and cybersecurity agencies will need to work as cross-stakeholders to have scalable city-wide deployment. The correct planning makes sure that the blockchain systems enhanced with RIS will seamlessly provide intelligent transport work and autonomous mobility services.

10.2 Policy, Governance, And Compliance For Blockchain-Based Vehicular Ecosystems

A blockchain-based system of vehicular communication needs to be governed effectively to control the security of data, privacy, and transparency of operations. The policymakers need to develop the standards of decentralized identity control, smart contract verification, and access rights of data in vehicular networks. The governance models must specify the persons authorized to write to the blockchain, the issuance of authentication keys and how conflicts or evil acts can be resolved. Privacy policies should guarantee that there is no sensitive information on the blockchain records, which can reveal the details of the drivers or the vehicles (Liang et al., 2021). The cryptographic practices, ledger synchronization and data-handling procedures need to be covered with compliance monitoring. Auditing tools are to be created to check blockchain activity and control compliance with the legal requirements. Moreover, since RIS brings in the concept of intelligence at the physical layer, the regulatory frameworks must incorporate the guidelines concerning the exposure of electromagnetism, signal manipulation, and adherence to the public-safety standards. Cooperative control that incorporates telecom regulators, and transport officials as well as cybersecurity agencies will guarantee safe and reliable execution. Large scale adoption of blockchain-RIS vehicular systems requires strong policy structures and governance structures.

10.3 Industry Readiness And Cost–Benefit Considerations For RIS-Blockchain Integration

The application of blockchain-based RIS in V2X communication in the industry relies on the technological maturity, cost of implementation, and potential performance benefits. The automotive companies have to consider the preparedness of the onboard systems to support the onboard authentication and RIS-aided communication based on blockchain technology. Telecom companies need to evaluate their ability to combine programmable surfaces with the current infrastructure and 6G networks. Cost benefit analysis involves a need to measure the latency, security and reliability improvements with respect to their installation cost, maintenance, and training. Preliminary applications reveal that RIS panels are economical and power efficient to traditional relays, whereas blockchain makes fraud, tampering communication, and overhead management redundant (Kim et al., 2024). Nevertheless, industries should be ready to face the challenge of integration such as interoperability, software upgrades, and compliance of data governance. Collaborations between mobility service providers, smart city administrators and technology vendors will play an important role in distributing the cost of deployment, and innovation. In general, the industry preparedness is based on the alignment of the technological feasibility with the strategic planning, which will guarantee long-term economic and operational sustainability.

11. CONCLUSION

This chapter explored the conceptualization of integrating blockchain and Reconfigurable Intelligible Surfaces (RIS) as a disruptive developmental channel in the creation of secure, transparent, and high-performance vehicular communication in 6G-based V2X environments. The discussion has revealed that RIS is inherently built to improve the propagation environment on wireless systems, enabling programmable manipulation of the signal reflection and beam steering as well as optimization of the coverage range, which is crucial in ensuring a reliable connectivity at high vehicular mobility. Combined with a decentralized trust infrastructure of blockchain, the system provides resistant to tampering message exchanges, verifiable identities and reliable cooperative interactions without centralized authorities. These technologies can solve a long-lasting problem in V2X networks, such as authentication delays, instability in communications, Sybil attacks, spoofing, and manipulation of data.

The chapter has recorded some of the most important contributions such as the conceptual architecture that optimizes the physical layer and automates the trust, the performance indicators that show that the integration of blockchain-RIS has improved latency and reliability, and practical case studies that prove that it is possible to use blockchain-RIS in platooning and vehicular edge-data marketplaces. These results demonstrate the strategic importance of incorporating programmable communication environments with distributed ledger technologies, especially with more demands of ultra-reliable, low-latency, and security-sensitive vehicular applications offered by 6G. In the future, decentralized and predictive vehicle systems vision is long-term with a focus on autonomous vehicles and adaptive learning, as well as big interoperability between cities and transport systems. Future mobility ecosystems will be resilient and intelligent with emerging paradigms like post-quantum security, generative AI driven coordination, cross-chain trust sharing and self-learning RIS controllers. With vehicles becoming fully connected, cooperative and autonomous systems, blockchain-based RIS frameworks will become a cornerstone of constructing reliable, efficient and future-oriented transport systems that have the capability to sustain next generation mobility in the 6G and beyond era.

REFERENCES

Ali, A., Assam, M., Alajmi, M., Ghadi, Y. Y., Indira, S., Akhmediyarova, A., Alahmadi, T. J., & Alkahtani, H. K. (2024). Arithmetic optimization based secure intelligent clustering algorithm for Vehicular Adhoc Network. *PLoS One*, *19*(9), e0309920. DOI: 10.1371/journal.pone.0309920 PMID: 39264948

Ali, E. S., Hasan, M. K., Hassan, R., Saeed, R. A., Hassan, M. B., Islam, S., Nafi, N. S., & Bevinakoppa, S. (2021). Machine learning technologies for secure vehicular communication in Internet of vehicles: Recent advances and applications. *Security and Communication Networks*, *2021*, 1–23. DOI: 10.1155/2021/8868355

Alsayaydeh, J. A. J., Yusof, M. F., Yahaya, N. A., Kovtun, V., & Herawan, S. G. (2025). A novel framework for secure cryptocurrency transactions using quantum crypto guard. *PeerJ. Computer Science*, *11*, e3030. DOI: 10.7717/peerj-cs.3030 PMID: 40989389

Bhattacharya, P., Bodkhe, U., Zuhair, M., Rashid, M., Liu, X., Verma, A., & Kishan Dewangan, R. (2021). Amalgamation of blockchain and sixth-generation-envisioned responsive edge orchestration in future cellular vehicle-to-anything ecosystems: Opportunities and challenges. *Transactions on Emerging Telecommunications Technologies*. Advance online publication. DOI: 10.1002/ett.4410

Cao, Y.-B., Chen, X.-B., He, Y.-F., Liu, L.-X., Che, Y.-M., Wang, X., Xiao, K., Xu, G., & Chen, S.-Y. (2024). A post-quantum cross-domain authentication scheme based on multi-chain architecture. *Computers, Materials & Continua*, *78*(2), 2813–2827. DOI: 10.32604/cmc.2024.046816

Chai, H., Leng, S., He, J., Zhang, K., & Cheng, B. (2022). CyberChain: Cybertwin empowered blockchain for lightweight and privacy-preserving authentication in internet of vehicles. *IEEE Transactions on Vehicular Technology*, *71*(5), 4620–4631. DOI: 10.1109/tvt.2021.3132961

Chai, H., Leng, S., Zhang, K., & Mao, S. (2019). Proof-of-reputation based-consortium blockchain for trust resource sharing in internet of vehicles. *IEEE Access : Practical Innovations, Open Solutions*, *7*, 175744–175757. DOI: 10.1109/access.2019.2956955

Chen, Y., Wang, Y., Zhang, J., & Di Renzo, M. (2021). QoS-driven spectrum sharing for reconfigurable intelligent surfaces (RISs) aided vehicular networks. *IEEE Transactions on Wireless Communications*, *20*(9), 5969–5985. DOI: 10.1109/twc.2021.3071332

Cheng, X., Huang, Z., & Chen, S. (2020). Vehicular communication channel measurement, modelling, and application for beyond 5G and 6G. *IET Communications*, *14*(19), 3303–3311. DOI: 10.1049/iet-com.2020.0531

Dai, X., Vallati, M., Guo, R., Wang, Y., Han, S., & Lin, Y. (2023). The road ahead: DAO-secured V2X infrastructures for safe and smart vehicular management. *IEEE Transactions on Intelligent Vehicles*, *8*(12), 4674–4677. DOI: 10.1109/tiv.2023.3337993

Dampahalage, D. L., Manosha, K. B. S., Rajatheva, N., & Latva-Aho, M. (2021). Weighted-sum-rate maximization for an reconfigurable intelligent surface aided vehicular network. *IEEE Open Journal of the Communications Society*, *2*, 687–703. DOI: 10.1109/ojcoms.2021.3069253

Fardad, M., Muntean, G.-M., & Tal, I. (2024a). A blockchain-enabled vehicular edge computing framework for secure performance-oriented V2X service delivery. *IEEE Transactions on Vehicular Technology*, *73*(9), 13853–13867. DOI: 10.1109/tvt.2024.3394150

Fardad, M., Muntean, G.-M., & Tal, I. (2024b). A blockchain-enabled vehicular edge computing framework for secure performance-oriented V2X service delivery. *IEEE Transactions on Vehicular Technology*, *73*(9), 13853–13867. DOI: 10.1109/tvt.2024.3394150

Farran, H., Khoury, D., & Bokor, L. (2022). A comprehensive survey on the application of blockchain/hash chain technologies in V2X communications. *Infocommunications Journal*, *14*(1), 24–35. DOI: 10.36244/icj.2022.1.4

Fu, X., Wang, H., & Shi, P. (2021). A survey of Blockchain consensus algorithms: Mechanism, design and applications. *Science China. Information Sciences*, *64*(2). Advance online publication. DOI: 10.1007/s11432-019-2790-1

Gebrezgiher, Y. T., Jeremiah, S. R., Deng, X., & Park, J. H. (2025). Machine learning-based blockchain technology for secure V2X communication: Open challenges and solutions. *Sensors (Basel)*, *25*(15), 4793. DOI: 10.3390/s25154793 PMID: 40807958

Hazarika, B., Singh, K., Biswas, S., Mumtaz, S., & Li, C.-P. (2024). Multi-agent DRL-based task offloading in multiple RIS-aided IoV networks. *IEEE Transactions on Vehicular Technology*, *73*(1), 1175–1190. DOI: 10.1109/tvt.2023.3302010

Huang, J., Wang, C.-X., Sun, Y., Feng, R., Huang, J., Guo, B., Zhong, Z., & Cui, T. J. (2022). Reconfigurable intelligent surfaces: Channel characterization and modeling. Proceedings of the IEEE. Institute of Electrical and Electronics Engineers, 110(9), 1290–1311. https://doi.org/DOI: 10.1109/jproc.2022.3186087

Jameel, F., Javed, M. A., Zeadally, S., & Jantti, R. (2021a). Efficient mining cluster selection for blockchain-based cellular V2X communications. IEEE Transactions on Intelligent Transportation Systems: A Publication of the IEEE Intelligent Transportation Systems Council, 22(7), 4064–4072. https://doi.org/DOI: 10.1109/tits.2020.3006176

Jameel, F., Javed, M. A., Zeadally, S., & Jantti, R. (2021b). Efficient mining cluster selection for blockchain-based cellular V2X communications. IEEE Transactions on Intelligent Transportation Systems: A Publication of the IEEE Intelligent Transportation Systems Council, 22(7), 4064–4072. https://doi.org/DOI: 10.1109/tits.2020.3006176

Khan, M. A., Ghosh, S., Busari, S. A., Huq, K. M. S., Dagiuklas, T., Mumtaz, S., Iqbal, M., & Rodriguez, J. (2021). Robust, resilient and reliable architecture for V2X communications. IEEE Transactions on Intelligent Transportation Systems: A Publication of the IEEE Intelligent Transportation Systems Council, 22(7), 4414–4430. https://doi.org/DOI: 10.1109/tits.2021.3084519

Kim, M., Oh, I., Yim, K., Sahlabadi, M., & Shukur, Z. (2024). Security of 6G-enabled vehicle-to-everything communication in emerging federated learning and blockchain technologies. *IEEE Access : Practical Innovations, Open Solutions*, *12*, 33972–34001. DOI: 10.1109/access.2023.3348409

Li, M., Chen, Y., Lal, C., Conti, M., Alazab, M., & Hu, D. (2023). Eunomia: Anonymous and secure vehicular digital forensics based on blockchain. *IEEE Transactions on Dependable and Secure Computing*, *20*(1), 225–241. DOI: 10.1109/tdsc.2021.3130583

Liang, Y.-C., Chen, J., Long, R., He, Z.-Q., Lin, X., Huang, C., Liu, S., Shen, X. S., & Di Renzo, M. (2021). Reconfigurable intelligent surfaces for smart wireless environments: Channel estimation, system design and applications in 6G networks. *Science China. Information Sciences*, *64*(10). Advance online publication. DOI: 10.1007/s11432-020-3261-5

Long, F., & Li, Y. (2025). P-CSNKS: Post-quantum collaborative signature scheme with non-linear private key splitting technique. *Journal of Electronic & Information Systems*, *7*(2), 1–12. DOI: 10.30564/jeis.v7i2.10217

Meenendranath Reddy, K., & Murugesan, R. (2025). Quantum key distribution (QKD) for ultra-secure power electronics communication in IoT-based smart grid systems. In Power Electronics for IoT-Enabled Smart Grids and Industrial Automation (pp. 377–406). RADemics Research Institute. https://doi.org/DOI: 10.71443/9789349552111-14

Mei, J., Wang, X., & Zheng, K. (2019). Intelligent network slicing for V2X services toward 5G. *IEEE Network*, *33*(6), 196–204. DOI: 10.1109/mnet.001.1800528

Meijers, J., Michalopoulos, P., Motepalli, S., Zhang, G., Zhang, S., Veneris, A., & Jacobsen, H.-A. (2022). Blockchain for V2X: Applications and Architectures. *IEEE Open Journal of Vehicular Technology*, *3*, 193–209. DOI: 10.1109/ojvt.2022.3172709

Murthy, P. K. (2024). Integrating blockchain and machine learning with 6G for autonomous vehicle communication: Achieving secure, transparent, and scalable V2X networks. *Communications on Applied Nonlinear Analysis*, *32*(4s), 289–308. DOI: 10.52783/cana.v32.2814

Noor-A-Rahim, M., Liu, Z., Lee, H., Khyam, M. O., He, J., Pesch, D., Moessner, K., Saad, W., & Poor, H. V. (2022). 6G for vehicle-to-everything (V2X) communications: Enabling technologies, challenges, and opportunities. Proceedings of the IEEE. Institute of Electrical and Electronics Engineers, 110(6), 712–734. https://doi.org/DOI: 10.1109/jproc.2022.3173031

Odeyemi, K. O., Owolawi, P. A., & Olakanmi, O. O. (2020). Performance analysis of reconfigurable intelligent surface assisted underwater optical communication system. *Progress in Electromagnetics Research M. Pier M*, *98*, 101–111. DOI: 10.2528/pierm20101203

Okegbile, S. D., Cai, J., Chen, J., & Yi, C. (2024). A reputation-enhanced shard-based byzantine fault-tolerant scheme for secure data sharing in zero trust human digital twin systems. *IEEE Internet of Things Journal*, *11*(12), 22726–22741. DOI: 10.1109/jiot.2024.3382829

Pacheco, L., Braun, T., Rosário, D., Di Maio, A., & Cerqueira, E. (2024). A distributed aggregation approach for vehicular federated learning. *IEEE Access : Practical Innovations, Open Solutions*, *12*, 72155–72169. DOI: 10.1109/access.2024.3399484

Pandey, D., Goyal, S., Bhaumik, K., Suneja, S., Sharma, M., & Dadheech, P. D. (2024). A systematic review of security issues in 6G networks and communication. In *Advances in Wireless Technologies and Telecommunication* (pp. 1–11). IGI Global., DOI: 10.4018/979-8-3693-2931-3.ch001

Pasupuleti, M. K. (2025). Post-Quantum Cryptography: Algorithms and Implementation Challenges. [IJAIRI]. *International Journal of Academic and Industrial Research Innovations*, *05*(06), 234–243. DOI: 10.62311/nesx/rphcrcscrbc4

Pathak, V., Pandya, R. J., Bhatia, V., & Lopez, O. A. (2023). Qualitative survey on artificial intelligence integrated blockchain approach for 6G and beyond. *IEEE Access : Practical Innovations, Open Solutions, 11*, 105935–105981. DOI: 10.1109/access.2023.3319083

Pavia, J. P., Velez, V., Souto, N., da Silva, M. M., & Correia, A. (2024). System-level assessment of massive multiple-input–multiple-output and reconfigurable intelligent surfaces in centralized radio access network and IoT scenarios in sub-6 GHz, mm-wave, and THz bands. *Applied Sciences (Basel, Switzerland), 14*(3), 1098. DOI: 10.3390/app14031098

Rao, P. M., Jangirala, S., Pedada, S., Das, A. K., & Park, Y. (2023). Blockchain integration for IoT-enabled V2X communications: A comprehensive survey, security issues and challenges. *IEEE Access : Practical Innovations, Open Solutions, 11*, 54476–54494. DOI: 10.1109/access.2023.3281844

Rasheed, I., Asif, M., Khan, W. U., Ihsan, A., Ullah, K., & Ali, M. S. (2022). Blockchain-based trust verification and streaming service awareness for big data-driven 5G and beyond vehicle-to-everything (V2X) communication. *Wireless Communications and Mobile Computing, 2022*(1), 1–13. DOI: 10.1155/2022/7357820

Saidi, H., Labraoui, N., Ari, A. A. A., Maglaras, L. A., & Emati, J. H. M. (2022). DSMAC: Privacy-aware decentralized self-management of data access control based on blockchain for health data. *IEEE Access : Practical Innovations, Open Solutions, 10*, 101011–101028. DOI: 10.1109/access.2022.3207803

Sekhar Oleti, C. (2024). Post-quantum cryptographic architecture for secure banking: Lattice-based implementation with blockchain integration. *International Journal For Multidisciplinary Research, 6*(2). Advance online publication. DOI: 10.36948/ijfmr.2024.v06i02.55514

Shrestha, R., Nam, S. Y., Bajracharya, R., & Kim, S. (2020). Evolution of V2X communication and integration of blockchain for security enhancements. *Electronics (Basel), 9*(9), 1338. DOI: 10.3390/electronics9091338

Tan, H., Zhang, Q., Li, M., Liu, X., & Hu, L. (2025). Design and optimization of a blockchain-enabled decentralized security framework for anomaly detection in VANETs. *Transactions on Emerging Telecommunications Technologies, 36*(10). Advance online publication. DOI: 10.1002/ett.70275

Tang, W., Chen, M. Z., Chen, X., Dai, J. Y., Han, Y., Di Renzo, M., Zeng, Y., Jin, S., Cheng, Q., & Cui, T. J. (2021). Wireless communications with reconfigurable intelligent surface: Path loss modeling and experimental measurement. *IEEE Transactions on Wireless Communications, 20*(1), 421–439. DOI: 10.1109/twc.2020.3024887

Velez, V. R. J., Pavia, J. P. C. B. B., Souto, N. M. B., Sebastião, P. J. A., & Correia, A. M. C. (2023). Performance assessment of a RIS-empowered post-5G/6G network operating at the mmWave/THz bands. *IEEE Access : Practical Innovations, Open Solutions*, *11*, 49625–49638. DOI: 10.1109/access.2023.3277388

Vladyko, A., Elagin, V., Spirkina, A., Muthanna, A., & Ateya, A. (2022). Distributed edge computing with blockchain technology to enable ultra-reliable low-latency V2X communications. *Electronics (Basel)*, *11*(2), 173. DOI: 10.3390/electronics11020173

Wang, X., Zhu, H., Xiao, H., Zhou, Z., Yang, S., & Sun, L. (2023). Blockchain-enhanced trust management for mobile edge computing-enabled intelligent vehicular collaboration in the 6G era. *Transactions on Emerging Telecommunications Technologies*, *34*(7). Advance online publication. DOI: 10.1002/ett.4791

Wei, L., Wang, K., Pan, C., & Elkashlan, M. (2022). Secrecy performance analysis of RIS-aided communication system with randomly flying eavesdroppers. *IEEE Wireless Communications Letters*, *11*(10), 2240–2244. DOI: 10.1109/lwc.2022.3198631

Yao, J., Xu, J., Xu, W., Yuen, C., & You, X. (2024). Superimposed RIS-phase modulation for MIMO communications: A novel paradigm of information transfer. *IEEE Transactions on Wireless Communications*, *23*(4), 2978–2993. DOI: 10.1109/twc.2023.3304695

Ying, Z., Yi, L., & Ma, M. (2020). BEHT: Blockchain-based efficient highway toll paradigm for opportunistic autonomous vehicle platoon. *Wireless Communications and Mobile Computing*, *2020*, 1–13. DOI: 10.1155/2020/8868656

Zhang, X., Xue, M., & Miao, X. (2022). A consensus algorithm based on Risk Assessment model for permissioned blockchain. *Wireless Communications and Mobile Computing*, *2022*(1), 1–21. DOI: 10.1155/2022/8698009

KEY TERMS AND DEFINITIONS

6G Networks: Next-generation wireless systems offering ultra-low latency, high reliability, and AI-driven communication.

Beamforming: A technique used to direct wireless signals toward specific targets for stronger communication.

Blockchain: A decentralized digital ledger that records transactions securely and prevents tampering.

Consensus Mechanism: A method used by blockchain nodes to agree on valid transactions securely.

Data Integrity: Assurance that data has not been altered during transmission or storage.

Edge Computing: Processing data near the source (vehicles or roadside units) to reduce delay and improve responsiveness.

Reconfigurable Intelligent Surface (RIS): A programmable surface that reflects wireless signals intelligently to improve coverage and reduce interference.

Smart Contract: A self-executing program on the blockchain that automatically enforces rules and agreements.

Sybil Attack: A security threat where one entity creates multiple fake identities to disrupt a network.

V2X Communication: Vehicle-to-everything communication enabling data exchange with other vehicles, infrastructure, pedestrians, and networks.

Chapter 4

Emerging Trends and Future Research Directions in RIS for Smart Mobility

A. A. Jayashree Prabhakar
http://orcid.org/0000-0003-4324-5916
Vels Institute of Science, Technology, and Advanced Studies, India

V. N. Sudheer
http://orcid.org/0009-0004-8776-9409
CMR University, India

P. Selvakumar
http://orcid.org/0000-0002-3650-4548
Department of Science and Humanities, Nehru Institute of Technology, Coimbatore, India

T. C. Manjunath
http://orcid.org/0000-0003-2545-9160
Rajarajeswari College of Engineering, Bengaluru, India

Pritam Lanjewar
http://orcid.org/0000-0001-7166-884X
Datta Meghe Institute of Management Studies, India

Wasudeo Dalwankar
Datta Meghe Institute of Management Studies, India

ABSTRACT

Reconfigurable Intelligent Surfaces (RIS) are rapidly emerging as one of the most transformative technologies within the future smart mobility ecosystem, offering a paradigm shift in how wireless communication environments are designed, optimised, and experienced. Unlike the traditional model in which mobile networks must adapt to a fixed physical environment filled with signal reflections, blockages, and unpredictable interference, RIS enables the environment itself to become programmable, intelligent, and dynamically responsive. Through the deployment of ultra-thin, low-power, metasurface panels composed of a dense array of passive or

DOI: 10.4018/979-8-3373-7867-1.ch004

semi-passive elements, RIS can control electromagnetic wave propagation in real time, shaping signals to enhance coverage, boost data rates, reduce latency, and improve the overall reliability of wireless connectivity. This innovation is particularly significant for next-generation mobility applications—autonomous vehicles, connected public transport.

REVOLUTIONISING SMART MOBILITY WITH RIS: EMERGING TRENDS AND OPPORTUNITIES

Reconfigurable Intelligent Surfaces (RIS) are rapidly emerging as one of the most transformative technologies within the future smart mobility ecosystem, offering a paradigm shift in how wireless communication environments are designed, optimised, and experienced. Unlike the traditional model in which mobile networks must adapt to a fixed physical environment filled with signal reflections, blockages, and unpredictable interference, RIS enables the environment itself to become programmable, intelligent, and dynamically responsive. Through the deployment of ultra-thin, low-power, metasurface panels composed of a dense array of passive or semi-passive elements, RIS can control electromagnetic wave propagation in real time, shaping signals to enhance coverage, boost data rates, reduce latency, and improve the overall reliability of wireless connectivity. This innovation is particularly significant for next-generation mobility applications—autonomous vehicles, connected public transport, advanced traffic control systems, unmanned aerial vehicles (UAVs), smart logistics, and mobility-as-a-service platforms—which all demand seamless, ultra-reliable, and high-capacity communication, especially in dense urban environments where traditional infrastructure struggles to eliminate dead zones or guarantee uninterrupted coverage. By enabling controllable signal reflection, refraction, focusing, or scattering, RIS effectively enhances mmWave and THz communication performance—bands that are critical for 6G evolution but are highly susceptible to line-of-sight loss and blockages from buildings, vehicles, and other moving objects. In smart highways and vehicular networks (V2X), RIS can be strategically installed on roadside units, buildings, traffic lights, or even directly embedded onto vehicles to mitigate Doppler effects, maintain stable connectivity during high-speed movement, and reinforce communication links for safety-critical functions such as collision avoidance and sensor data sharing.

Similarly, in urban public transportation systems, RIS-enhanced networks can support real-time fleet monitoring, predictive route optimisation, infotainment services, and intelligent ticketing systems with significant improvements in bandwidth efficiency. The opportunities extend even further into the domain of drone-enabled mobility and delivery services, where RIS panels deployed on city infrastructure or elevated platforms can help UAVs navigate communication-constrained environments and ensure dependable command-and-control links. Importantly, the growing integration of edge computing and AI-enabled optimisation techniques amplifies RIS potential by enabling autonomous beamforming adjustment, context-aware steering, and self-healing networks that respond instantly to dynamic mobility patterns, traffic densities, and environmental conditions. RIS not only promises energy-efficient communication—with minimal power consumption compared to active network equipment—but also supports sustainability goals by reducing the number of base stations required and minimising electromagnetic pollution through precise beam shaping. In addition, RIS facilitates more secure communication channels by restricting signal leakage and enabling stealthy or user-specific beam targeting, which is crucial for protecting sensitive vehicular or logistics data. Beyond connectivity enhancement, RIS can play a vital role in integrated sensing and communication (ISAC) systems used for 3D localisation, environment mapping, weather-adaptive control, and traffic management, making transportation networks smarter, safer, and more automated. Furthermore, RIS deployment creates new commercialisation avenues: telecom operators can monetise RIS-as-a-service business models for infrastructure sharing, city planners can deploy RIS panels as modular upgrades to existing urban assets, and automotive manufacturers can differentiate product offerings by embedding RIS into next-generation vehicle designs. Despite these promising opportunities, several challenges still require focused research and collaborative standardisation efforts. These include ensuring scalable manufacturing of affordable metasurfaces, addressing hardware impairments, developing robust network control protocols for mobility-induced channel variations, ensuring backward compatibility within hybrid network environments, and establishing universal testing and certification frameworks. Public-private partnerships, regulatory guidance, and industry alliances will be essential to accelerate RIS adoption while ensuring safety and interoperability across borders. As global investment surges and pilot deployments expand, RIS is poised to redefine the strategic foundation of smart mobility infrastructure, enabling hyper-connected, autonomous, eco-efficient, and user-centric transportation ecosystems aligned with future 6G visions. Ultimately, by transforming the wireless environment into an intelligent and programmable asset, RIS unlocks a new era of opportunity—one where mobility is no longer limited by communication constraints, but is instead empowered by them to reach unprecedented levels of performance, adaptability, and societal impact.

RIS-ASSISTED COMMUNICATION FOR SMART MOBILITY: CHALLENGES AND FUTURE RESEARCH

Reconfigurable Intelligent Surface (RIS)-assisted communication is considered one of the most promising technological pillars for enabling highly reliable, ultra-low-latency, and energy-efficient wireless connectivity required in future smart mobility ecosystems; however, despite its remarkable theoretical advantages, its large-scale deployment still faces significant technical, operational, and regulatory challenges that demand continued research (Naaz F., et al, 2024). The complexity of ensuring robust connectivity in dynamic mobility environments—where vehicles, drones, and public transportation systems move rapidly through constantly changing propagation conditions—creates difficulties in maintaining accurate channel estimation and optimal RIS configuration. Most existing studies assume quasi-static environments or slow mobility models, making them less realistic for high-speed vehicular scenarios where frequent reconfiguration introduces signalling overhead, control latency, and increased processing demands. Additionally, although RIS elements are primarily passive and low-power, the need for large-scale metasurface deployment along roads, tunnels, and urban structures raises concerns regarding manufacturing scalability, hardware reliability, and environmental durability, especially under extreme weather, vibrations, or pollution exposure. Synchronisation and coordinated control of multiple RIS panels represent another major challenge, requiring distributed and scalable network intelligence to prevent interference and ensure beam alignment for safety-critical V2X communications (Chopra & Ahmed, 2025). Meanwhile, RIS-aided communication currently relies heavily on complex optimisation frameworks and machine learning models that must operate at the network edge to satisfy stringent latency requirements; yet, achieving real-time performance with limited computing resources remains difficult. Moreover, seamless interoperability between legacy infrastructure, 5G/6G networks, and RIS systems demands well-defined standardisation and compatibility guidelines, which are still in early development stages within global standard bodies. Security and privacy concerns are also pronounced in RIS-enabled mobility environments: while RIS can enhance protection by controlling signal directionality, it can also introduce new vulnerabilities such as eavesdropping amplification along unintended reflection paths or malicious reconfiguration attacks that could disrupt traffic safety (Pogaku A C., et al, 2022). Future research must therefore explore robust access control, authentication mechanisms, and intrusion detection tailored to programmable wireless environments.

Additionally, the energy footprint of RIS ecosystems must be carefully evaluated: although individual RIS panels are efficient, their large-scale deployment necessitates sustainable power solutions such as energy harvesting or maintenance-free battery systems, particularly for roadside infrastructure with limited electrical access. On the theoretical front, further investigation is required into the fundamental limits of RIS-assisted mobility networks, including performance trade-offs in multi-RIS cooperative communication, interference shaping in dense urban mobility zones, and joint communication-sensing optimisation for autonomous vehicle perception (Chapala & Zafaruddin, 2023). RIS-assisted integrated sensing and communication (ISAC) remains a promising direction, enabling RIS panels to simultaneously support localisation, traffic monitoring, and radar-like sensing while reducing the need for additional hardware. Human-centric challenges must also be addressed, ensuring that RIS enhances user experience inclusively and equitably across smart mobility services, without creating digital connectivity gaps between regions or socioeconomic groups. From a policy and regulatory perspective, clear frameworks must emerge to govern installation permissions, data protection, safety requirements, and cross-border operational standards for RIS in transportation networks. To accelerate practical deployment, collaborative testbeds involving telecommunications operators, automotive manufacturers, smart city planners, and academic institutions are essential to validate RIS performance under realistic mobility conditions, quantify return on investment, and identify optimal placement strategies for infrastructure (Sasikala K., et al, 2026). Future research should also explore intelligent twin-enabled simulation environments for predictive and adaptive RIS management, enabling proactive decision-making based on mobility data streams. Ultimately, while RIS-assisted communication offers a transformative leap toward hyper-connected and autonomous mobility systems, sustained interdisciplinary research efforts are crucial to overcome the current technological and operational limitations, ensuring that RIS evolves from experimental innovation to mainstream infrastructure—powering safer, greener, and more intelligent transportation for the decades ahead.

THE ROLE OF RIS IN ENABLING AUTONOMOUS VEHICLES FOR SMART MOBILITY

Reconfigurable Intelligent Surfaces (RIS) are poised to play a transformative role in enabling autonomous vehicles (AVs) within smart mobility ecosystems by providing the reliable, high-capacity, low-latency, and context-aware wireless communication that automated driving systems depend upon for safe and efficient operation (Ozcan Y U., et al, 2021). Autonomous vehicles require seamless connectivity to support vehicle-to-everything (V2X) communication, cooperative perception, high-resolution map updates, real-time decision-making, and the fusion of multi-sensor data from onboard systems and surrounding infrastructure. However, existing networks—particularly those operating in mmWave and THz frequency bands envisioned for

6G—frequently suffer from line-of-sight obstructions and rapid signal degradation in dense urban environments where buildings, roadside objects, and other vehicles introduce unpredictable propagation challenges. RIS addresses these limitations by transforming the environment into a smart, controllable communication asset: ultra-thin metasurface panels consisting of numerous electronically tunable elements can dynamically reflect, refract, focus, or cancel electromagnetic waves to maintain robust links even in high-mobility and visually obstructed scenarios. By being deployed on lamp posts, roadway infrastructure, tunnels, bridges, and even integrated onto vehicles themselves, RIS can eliminate coverage dead zones and significantly reduce Doppler effects resulting from rapid vehicle movement. This enhanced propagation reliability is crucial for latency-sensitive AV functions such as collaborative collision avoidance, pedestrian detection, blind spot management, autonomous lane merging, and swarm driving on future smart highways (Eskandari M., et al, 2025). Moreover, RIS supports the convergence of communication and sensing in autonomous vehicles through Integrated Sensing and Communication (ISAC) capabilities, enabling RIS-assisted radar-like functions to improve localisation accuracy, obstacle detection, and road environment awareness while reducing reliance on redundant sensor hardware. Artificial intelligence plays a central role in optimising RIS configurations, using machine learning to interpret mobility patterns, traffic density, and dynamic radio conditions so that beamforming and reflection profiles adapt in real time to evolving autonomous driving needs. In addition to improving connectivity and situational awareness, RIS contributes to secure and privacy-preserving AV communication: by precisely directing signals toward intended receivers, RIS can minimise leakage, reduce the attack surface for eavesdroppers, and support spatial encryption strategies that protect sensitive navigational and vehicular data (Deng M., et al, 2024). Furthermore, RIS promotes sustainability within smart mobility infrastructure—consuming minimal power and capable of energy harvesting—making large-scale deployment feasible without substantial increases in operational cost or environmental impact. Several research efforts and pilot projects are currently exploring RIS-enabled cooperative perception, in which vehicles share extended sensor fields-of-view through RIS-enhanced V2X links to overcome occlusions and perception limitations.

However, deploying RIS for autonomous vehicles also introduces technical challenges requiring ongoing research: achieving fast and accurate channel estimation under high-speed mobility, coordinating multiple RIS units in complex road networks, reducing control overhead for real-time reconfigurability, and ensuring robust operation under weather-induced signal variations remain areas of active investigation (Ai Y., et al, 2021). Additionally, large-scale RIS deployment necessitates standardised frameworks for integration with automotive and telecommunications ecosystems, addressing installation guidelines, cybersecurity safeguards, shared infrastructure models, and maintenance procedures. Future advancements may include active and holographic RIS with greater beamforming precision for advanced mobility services, RIS-augmented edge intelligence enabling predictive resource allocation, and modular RIS deployments embedded in road materials to create fully programmable traffic environments. Ultimately, RIS functions as a foundational enabling technology that empowers autonomous vehicles with the communication reliability and environmental intelligence necessary to reach full autonomy and societal-scale adoption. As smart cities continue to evolve and the demand for safe, sustainable, and efficient mobility accelerates, RIS will play an increasingly critical role in bridging the gap between isolated onboard intelligence and fully cooperative, networked autonomous driving systems—unlocking next-generation mobility experiences that are not only more intelligent and connected but significantly safer and more human-centric.

RIS-BASED INTELLIGENT TRANSPORTATION SYSTEMS: EMERGING TRENDS AND APPLICATIONS

Reconfigurable Intelligent Surface (RIS)-based Intelligent Transportation Systems (ITS) represent a revolutionary shift in how communication, sensing, traffic management, and mobility services are delivered across modern cities, enabling highly coordinated, efficient, and data-driven transportation networks that align with the evolving smart mobility paradigm (Ji B., et al, 2024). As transportation systems transition toward greater automation, electrification, and connectivity, RIS technologies have emerged as a key enabler to overcome limitations of traditional wireless infrastructures by actively shaping radio environments to support seamless, high-speed, and ultra-reliable communication services. By leveraging metasurfaces embedded with programmable reflection elements, RIS can dynamically control the propagation of electromagnetic waves to enhance signal coverage across complex road networks, reduce latency for mission-critical applications, and eliminate connectivity dead zones typically found in tunnels, dense traffic intersections, and urban canyons (Rahmatov & Baek, 2023). The strategic integration of RIS into roadside units, public transportation structures, smart lampposts, and building facades introduces a new layer of environmental intelligence that synergises with existing transportation components to maximise operational efficiency. Emerging RIS-based ITS trends focus on enabling advanced vehicular communication models such as Vehicle-to-Everything (V2X), where RIS enhances communication among vehicles, infrastructure, pedestrians, and network/cloud services to support collision avoidance, cooperative perception, lane management, and hazard warning systems. Additionally, RIS supports mobility-as-a-service frameworks that unify multimodal transportation by enabling reliable passenger information systems, interconnected ticketing, and seamless handovers between cellular, Wi-Fi, and satellite networks. In public transit, RIS enables high-capacity communication links to improve real-time fleet monitoring, route optimisation, and station management, enhancing passenger experience with uninterrupted connectivity and safety (Shafique & Alhassoun, 2024). A transformative application lies in integrating RIS with traffic sensing systems for dynamic traffic control: RIS-assisted integrated sensing and communication (ISAC) simultaneously enables wireless connectivity and environment monitoring, augmenting radar and LiDAR functions for vehicle localisation, congestion detection, and smart signal control. These capabilities contribute to smoother traffic flow, reduced emissions, and improved energy efficiency, making RIS an essential technology for sustainable ITS solutions.

Smart logistics and freight mobility also benefit, with RIS enhancing communication among automated delivery fleets, coordinated warehouse distribution, and drone-based transportation systems to optimise delivery times and resource utilisation (Yang W., et al, 2025). Furthermore, RIS introduces enhanced security mechanisms through directed beam management that reduce the risk of unauthorised interception, supporting secure transportation data exchange and privacy protection for vehicles and passengers. The rise of digital twin-enabled ITS is another emerging trend: RIS facilitates high-fidelity environment modelling and predictive analytics by providing real-time communication insights that inform digital replicas of road networks, enabling proactive planning and adaptive traffic response. However, practical deployment of RIS-based ITS faces several challenges, including dynamic channel estimation in fast-varying mobility environments, standardisation of RIS control protocols, compatibility with current ITS architectures, and scalability for large cities that require thousands of coordinated RIS panels. Ruggedised hardware design that withstands environmental stressors—rain, heat, vibrations—is essential for long-term operation, along with cost-effective manufacturing of metasurfaces to ensure economically viable deployment. Future research is shifting toward developing self-sustaining RIS infrastructures powered by ambient energy harvesting, utilising artificial intelligence to achieve fully autonomous RIS control capabilities that adjust beam steering and network coverage based on real-time mobility patterns (Virali Bukhari M., et al, 2026). Another promising area is hybrid RIS designs that combine reflection and refraction to enhance vertical coverage for road vehicles as well as low-altitude drones, enabling unified air-ground mobility management in smart cities. Additionally, RIS-enabled multicast services hold potential for advanced infotainment and communication distribution within public vehicles. As the transportation landscape continues its transition toward 6G-driven hyperconnectivity, RIS-based ITS will become foundational to realising safe, environmentally conscious, and user-centric mobility systems that enable smooth coordination among diverse transportation stakeholders. The convergence of RIS with edge computing, big data analytics, cybersecurity systems, and intelligent automation will ultimately empower cities to create adaptive and resilient mobility networks that accommodate growing population demands, reduce accident rates, and deliver enriched travel experiences. RIS-based ITS thus represents not only a technological evolution but a strategic pathway to building smarter, greener, and more inclusive urban transportation for the future.

RIS-ENABLED V2X COMMUNICATION FOR SMART MOBILITY: EMERGING TRENDS

Reconfigurable Intelligent Surfaces (RIS) are revolutionising the landscape of wireless connectivity in next-generation smart mobility systems, particularly by enhancing Vehicle-to-Everything (V2X) communication (Pala S., et al, 2024). As smart cities evolve toward fully automated and interconnected transportation networks, seamless data exchange between vehicles, infrastructure, pedestrians, and cloud systems becomes essential to ensure safety, efficiency, and sustainability. Traditional wireless environments suffer from multipath fading, blockages in dense urban environments, limited spectrum resources, and increased power demands. RIS technology addresses these challenges by introducing programmable, low-power metasurfaces that can dynamically manipulate electromagnetic waves, improving signal propagation, coverage, and reliability. Unlike conventional relays or active antennas, RIS operates passively with minimal power consumption and hardware complexity, making it an economical and scalable solution for future V2X systems supporting autonomous vehicles, cooperative driving, and intelligent traffic management (Srivastava & Singh, 2022). In V2X communications, where low latency and ultra-reliable links are critical, RIS offers several operational benefits. For Vehicle-to-Vehicle (V2V) communication, RIS can enable robust connectivity in scenarios where line-of-sight (LoS) paths are obstructed, such as during platooning on highways, overtaking, or navigating around high-rise buildings. By intelligently reflecting and redirecting signals, RIS enhances situational awareness, helps vehicles share real-time safety messages, and supports cooperative perception—an essential capability for autonomous mobility. For Vehicle-to-Infrastructure (V2I), RIS can significantly reduce shadowing effects caused by urban structures and support higher throughput for data-intensive services, including high-definition mapping updates, sensor fusion, and communication with traffic lights and roadside units. Furthermore, Vehicle-to-Pedestrian (V2P) communication benefits from RIS's ability to improve signal accessibility at street level, thereby enabling warning systems that prevent collisions and enhance pedestrian safety even in visually or radio-frequency-constrained environments (Singh G., et al, 2023). The integration of RIS with emerging cellular networks like 5G Advanced and 6G is another major trend, as their performance requirements—such as terabit-level throughput, millisecond-level latency, and high mobility support—align with autonomous vehicle needs. RIS strengthens multi-access edge computing (MEC)-enabled V2X architectures by optimising communication links between vehicles and edge servers responsible for real-time AI analytics and coordinated control. In dual-function communications sensing (DFCS), RIS plays a transformative role by simultaneously supporting data transmission and environmental sensing. This means RIS can contribute to radar-like perception for navigation and obstacle detection, complementing onboard sensors like LiDAR and cameras, particularly in adverse weather conditions where visibility or sensor performance is degraded. Another emerging development is the combination of RIS with millimetre-wave (mmWave) and terahertz (THz) frequency bands, which can provide high bandwidth yet normally suffer from high path loss. RIS mitigates these losses, making high-frequency communication more feasible for fast-moving vehicles requiring rapid data transmission.

Figure 1. Smart mobility in smart cities

Despite the promising advancements, several challenges remain. The dynamic vehicular environment introduces rapid channel variations, making real-time RIS configuration highly complex. Large-scale RIS deployments require efficient control signalling, low-latency feedback loops, and secure communication protocols to prevent malicious tampering with RIS elements that could compromise transportation safety (Dalavi V., et al, 2025). Standardisation efforts by organisations such as 3GPP and IEEE must address interoperability, network architecture integration, and performance evaluation frameworks specifically designed for RIS-assisted V2X scenarios. Energy sustainability is also a key focus; although RIS is low-power, incorporating self-sustaining features like solar energy harvesting and wireless power transfer is becoming an important design trend for wide-area deployment. Moreover, regulatory concerns related to public infrastructure installation, privacy, and urban planning will influence RIS adoption in smart transportation networks. Future research directions highlight the evolution toward fully autonomous and cooperative mobility ecosystems, where RIS acts as a fundamental enabler of resilient V2X connectivity. The development of holographic RIS (H-RIS) and adaptive metasurfaces capable of tunable frequency, polarisation control, and intelligent sensing will enhance environmental awareness and signal processing flexibility. Hybrid RIS-relay architectures are being proposed to merge the strengths of passive and active communication support, further boosting coverage, capacity, and system robustness in high-density environments. Furthermore, distributed RIS deployments across city landscapes—on building facades, bus stops, streetlights, and road barriers—will create smart propagation environments that continuously assist vehicles with ultra-reliable, real-time communication even in complex and congested traffic conditions. In summary, RIS-enabled V2X communication is emerging as one of the most transformative technologies for future smart mobility, driving the evolution of safer, greener, and more intelligent transportation systems. By reshaping wireless signal propagation, enabling robust connectivity in challenging environments, and supporting AI-driven cooperative intelligence, RIS stands as a cornerstone for 6G-powered connected and autonomous vehicle ecosystems (Sreekanthaswamy N., et al, 2026). Although challenges remain in scalability, security, control complexity, and standardisation, ongoing research and industrial interest indicate a strong trajectory toward widespread adoption in smart cities globally. As mobility shifts from human-driven vehicles to automated, interconnected transport networks, RIS-empowered V2X communication will play a vital role in shaping a resilient and data-driven transportation future designed to enhance user experience, reduce accidents, optimise traffic efficiency, and accelerate the realisation of fully autonomous smart mobility systems.

THE IMPACT OF RIS ON SMART MOBILITY BUSINESS MODELS: EMERGING TRENDS

Reconfigurable Intelligent Surfaces (RIS) are reshaping the foundations of smart mobility by introducing new business opportunities, transforming value chains, and driving investments toward intelligent transportation ecosystems aligned with the digital economy. As mobility services evolve from traditional ownership-based models to shared, automated, and connected transport experiences, RIS plays a pivotal role in enabling the high-performance connectivity required for autonomous vehicles, Vehicle-to-Everything (V2X) communication, and data-driven mobility platforms (Paiva S., et al, 2021). Unlike conventional infrastructure upgrades that demand high-power active components and frequent replacement cycles, RIS provides a cost-effective, energy-efficient, and easily deployable solution that can upgrade existing environments into smart communication spaces. This paradigm shift is changing how mobility stakeholders—including telecom operators, automakers, urban planners, technology vendors, and service providers—design their operational strategies and revenue models. The adoption of RIS is therefore not limited to technological enhancement but extends deeply into business transformation, stimulating competitive advantages and accelerating the commercialisation of next-generation intelligent transportation services. Smart mobility business models are increasingly dependent on connectivity quality, where the ability to deliver ultra-low latency, high reliability, and broad coverage directly influences service value (Dos Santos de Jesús E A., 2024). RIS creates business differentiation by improving network performance in high-density urban zones, such as commercial centres, smart highways, and logistics corridors. For Mobility-as-a-Service (MaaS) platforms, enhanced connectivity supports real-time vehicle coordination, automated routing, and personalised mobility offerings—expanding revenue streams through premium service tiers and data-driven customer experiences. Furthermore, as autonomous driving becomes more viable, RIS-enabled communication can lower operational costs by reducing reliance on expensive sensing hardware and minimising downtime caused by connectivity failures. This allows fleet operators, delivery companies, and ride-sharing providers to optimise resource utilisation and scale services sustainably. RIS thus supports business agility and continuous innovation by enabling safe and seamless mobility in environments where conventional networks struggle. From a commercial perspective, infrastructure and telecom providers are exploring RIS-as-a-Service (RISaaS) models, where public or private entities can deploy programmable surfaces that dynamically optimise wireless environments based on real-time demand. This creates new monetisation strategies, such as subscription-based smart connectivity zones and performance-based pricing models linked to network quality metrics. Cities can partner with technology vendors to integrate RIS into street furniture, building façades, and public transportation assets, transforming them into revenue-generating platforms by enabling location-based services and digital advertisements driven by user mobility data (Allam & Sharifi, 2022). Additionally, commercial property developers and facility managers recognise RIS as a differentiator in transit hubs and business districts, where strong connectivity becomes a competitive selling point that attracts enterprises and enhances property value.

RIS adoption also aligns with sustainability-driven business models, which are becoming essential in global smart mobility strategies. Because RIS operates with minimal power consumption and can be retrofitted on existing structures, it enables a greener communication infrastructure by reducing both energy usage and electronic waste. This supports business models that incentivise environmental compliance, carbon reduction, and corporate social responsibility within transportation sectors. Companies pursuing green transportation initiatives can also leverage RIS to improve energy efficiency in autonomous electric vehicle fleets by optimising routing, traffic flow, and charging operations through reliable real-time communication (Strinati & Alexandropoulos, 2021). As sustainability policies expand worldwide, RIS-enabled operations may benefit from subsidies, regulatory support, and green investment funds—further boosting industry adoption. However, transitioning to RIS-assisted mobility introduces challenges that influence business risk management and stakeholder strategy. High initial deployment efforts, integration cost uncertainties, and a lack of standardisation in RIS hardware and control platforms create barriers to market readiness. Cybersecurity threats targeting RIS control mechanisms could disrupt transportation safety, requiring companies to invest in advanced security frameworks and continuous monitoring. Additionally, economic responsibility for RIS installation within public environments may spark debates between governments, operators, and commercial beneficiaries regarding infrastructure ownership, data governance, and revenue sharing (Koparde S A., et al, 2026). Addressing these concerns is crucial for ensuring viable long-term business adoption and stakeholder trust. In conclusion, RIS technology is not merely advancing wireless communication performance—it is reshaping business architecture and value generation in the future of smart mobility.

RIS-Assisted Edge Computing for Smart Mobility Applications

Reconfigurable Intelligent Surfaces (RIS) are emerging as a vital enabler for next-generation smart mobility ecosystems, particularly through their integration with edge computing infrastructures that support low-latency, high-reliability communications (Ahmed M., et al, 2025). Smart mobility applications—ranging from intelligent transportation systems and connected autonomous vehicles to smart logistics and real-time traffic coordination—demand continuous data exchange and rapid processing. Traditional centralised cloud architectures struggle with the stringent requirements of ultra-low latency, bandwidth optimisation, and scalability as vehicle density and data volume rise. RIS provides a transformative solution by shaping the wireless environment through programmable reflection, enhancing channel conditions, improving signal coverage, and reducing communication energy costs. When integrated with Multi-access Edge Computing (MEC), RIS strengthens distributed computing capabilities by ensuring high-quality connectivity between mobile users, edge servers, and transportation infrastructure, thus enabling efficient computing offloading, cooperative perception, and context-aware decision-making crucial for autonomous mobility (Mao S., et al, 2022). Edge computing brings intelligence closer to vehicles by offloading computation-intensive tasks—such as high-definition mapping, object recognition, trajectory planning, and real-time traffic prediction—to localised servers strategically positioned across urban and highway environments. However, the dynamic and unpredictable nature of vehicular communication networks often disrupts computation offloading performance due to intermittent links, signal blockage, and rapid handovers. RIS mitigates these challenges by dynamically controlling signal propagation through beam reconfiguration and phase adjustment, ensuring stable, high-throughput, and low-latency connections required for reliable edge-assisted smart mobility services. For connected autonomous vehicles, RIS-assisted edge computing enables cooperative AI, where vehicles share sensor and driving data with nearby edge nodes to form a collaborative perception map of the environment. This joint intelligence enhances navigation safety, reduces accidents, and supports advanced capabilities like platooning and collision avoidance in complex traffic scenarios (Wu Z., et al, 2025). Moreover, the integration of RIS with edge computing enhances resource allocation efficiency in intelligent transportation networks. By improving link quality and minimising communication overhead, RIS enables more accurate and timely computation offloading decisions, optimising the coordination between mobile devices and distributed edge resources. AI-driven algorithms, such as deep reinforcement learning and federated learning, are increasingly used to manage RIS configuration and edge scheduling in real-time. These intelligent control mechanisms adapt to changes in vehicle mobility, network congestion, and computing demands, ensuring Quality of Service (QoS) for safety-critical mobility applications. Federated learning, in particular, benefits from RIS-enhanced communication as it allows vehicles to collaboratively train AI models without sharing raw data—preserving privacy while accelerating model convergence through reduced transmission delays and improved connectivity.

However, integrating RIS and edge computing introduces several challenges that require continued research and technological advancement. The rapid mobility of vehicles leads to fast-changing communication channels, increasing the complexity of real-time RIS reconfiguration and computation distribution. Control signalling and feedback mechanisms must be highly efficient to prevent network overhead and ensure stable performance. Security and privacy pose additional hurdles, as RIS-assisted links may create new attack surfaces for signal manipulation, unauthorised access, and data breaches. Standardising RIS architectures, control frameworks, and V2X communication protocols remains critical for interoperability among vendors and stakeholders across the transportation sector. Moreover, reliable fault-tolerant mechanisms must be established to ensure that RIS failures do not disrupt mission-critical mobility services. Looking toward future advancements, RIS-assisted edge computing is positioned to support next-generation 6G-enabled mobility systems built on intelligence, virtualisation, and cooperative autonomy. Emerging innovations such as holographic RIS (H-RIS), active RIS, and dual-function radar-communication metasurfaces will expand system capabilities by providing improved sensing, localisation, and situational awareness integrated directly into communication infrastructure (Bai T., et al, 2022). Edge-native distributed AI architectures powered by RIS will enable further automation, predictive analytics, and personalised mobility experiences. By supporting remote driving operations, smart public transit coordination, automated fleet logistics, drone-assisted delivery, and mobility-as-a-service (MaaS) platforms, RIS-enabled edge ecosystems will become essential drivers of economic value and societal benefits in smart cities. In summary, the synergy between RIS and edge computing establishes a foundational pillar for future smart mobility applications by enabling high-performance wireless connectivity, efficient computation offloading, and cooperative intelligence across transportation networks. Through enhanced reliability, reduced latency, and sustainable communication performance, RIS supports autonomous vehicular systems and intelligent infrastructure necessary for real-time decision-making and seamless mobility services. While significant research and development challenges persist, ongoing innovation suggests a rapid trajectory toward mainstream deployment. As smart mobility continues to advance toward fully automated, highly coordinated, and environmentally conscious transportation ecosystems, RIS-assisted edge computing will be instrumental in driving a future where mobility is smarter, safer, greener, and more responsive to the needs of both users and urban environments.

RIS-Based Cybersecurity for Smart Mobility: Emerging Trends and Challenges

Reconfigurable Intelligent Surfaces (RIS) are emerging as a groundbreaking technology in smart mobility ecosystems, offering not only enhanced wireless communication performance but also new opportunities for improving cybersecurity in connected and autonomous transportation systems (Naeem F., et al, 2023). As vehicles, infrastructure, and pedestrians become increasingly interconnected through Vehicle-to-Everything (V2X) communications, the volume and sensitivity of transmitted data escalate dramatically, ranging from location tracking and driving behaviour to remote control of autonomous vehicles. Traditional centralised security frameworks are insufficient to counter evolving cyber threats, especially in dynamic, latency-sensitive environments where compromised communication could result in serious safety hazards. By enabling the intelligent manipulation of electromagnetic waves, RIS introduces a physical-layer security paradigm that strengthens cybersecurity through controlled propagation, interference reduction, and secure beamforming. This emerging capability positions RIS as a fundamental enabler of secure and trustworthy smart mobility communications. One of the most promising contributions of RIS to cybersecurity lies in physical-layer security (PLS), where security mechanisms are embedded directly into the wireless channel. RIS can dynamically shape communication environments so that signals are tightly directed toward legitimate receivers while minimising leakage toward potential eavesdroppers. In dense urban environments or highway scenarios where attackers may exploit channel reflections, RIS can suppress unintended propagation paths, reduce interception risks, and enforce secure connectivity for safety-critical V2X communications. This capability is particularly important for autonomous vehicles that depend on frequent data exchanges with roadside units (RSUs), edge servers, and other vehicles to maintain situational awareness and avoid collisions. RIS-enhanced secure beamforming ensures that these communications remain confidential, reliable, and resilient against malicious interference and spoofing attacks (Tariq, 2024). RIS also supports jamming resistance and interference mitigation, two key concerns in mobility cybersecurity where adversaries may attempt to block or distort communication signals. Through programmable phase adjustments, RIS can redirect or scatter malicious interference away from target receivers, preserving data integrity and communication continuity. This is highly advantageous in emergency scenarios where connectivity disruptions could jeopardise public safety. Additionally, RIS can create artificial channel randomness that makes legitimate communication difficult for attackers to predict or manipulate, enabling secure key generation and authentication at the physical layer. Such techniques reduce dependence on computationally intensive cryptographic operations and help ensure efficient cybersecurity even in resource-constrained vehicle systems. As edge computing becomes widely integrated into smart mobility for real-time decision-making, RIS plays an expanded role in edge-assisted cybersecurity. RIS-enhanced communication links ensure secure and reliable offloading of sensitive vehicular data to edge servers, reducing latency and vulnerability to interception (Kaur R., et al, 2024). Combined with AI-based security analytics executed at the network edge, RIS can enable adaptive threat detection and rapid incident response. For example, in the presence of adversarial attacks targeting connected vehicle data streams, RIS can be reconfigured to isolate affected channels, reroute signals, and prevent the spread of malicious activity. Furthermore, the joint use of RIS with distributed ledger technologies (DLT) such as blockchain enhances data traceability and integrity in vehicular information exchanges, while RIS protects underlying wireless access points from manipulation or unauthorised entry.

Standardisation and interoperability remain ongoing challenges. Current 5G and future 6G security frameworks have yet to fully integrate RIS-specific considerations, including trust models, attacker profiles, and compliance enforcement for smart mobility deployments. The absence of standardised risk evaluation methods and security certification processes complicates procurement and regulatory approval for RIS-based intelligent transportation systems. In addition, privacy issues arise due to the increased ability of RIS to track mobility patterns through signal interaction, necessitating privacy-by-design strategies and compliance with global data protection regulations such as GDPR. To address these emerging concerns, future research directions include secure RIS architecture designs, such as distributed control, multi-layer encryption, tamper-resistant hardware, and self-healing configurations that automatically recover from cyber attacks (Magbool A., et al, 2025). Integration with emerging holographic RIS (H-RIS) or hybrid active–passive metasurfaces may introduce new opportunities for embedding sensing and threat detection directly within communication infrastructure. AI-driven security orchestration will be essential in real-time risk assessment and autonomous defence, enabling RIS to automatically adapt its configuration in response to threats observed across the network. Cooperative cybersecurity models involving collaboration among vehicles, infrastructure providers, telecom operators, and municipal authorities will become increasingly necessary to support seamless and resilient protection across mobility ecosystems. In conclusion, RIS introduces a transformative approach to securing smart mobility communication networks by leveraging physical-layer defence mechanisms, intelligent wireless environment control, and edge-assisted threat mitigation. As connected and autonomous transportation services evolve, RIS will act as a crucial enabler of confidential, resilient, and trustworthy V2X communication that protects safety-critical data and ensures public confidence in smart mobility infrastructure. While new vulnerabilities emerge with programmable wireless surfaces, the ongoing development of secure standards, architectures, and management strategies is expected to unlock the full potential of RIS as a pillar of cybersecurity in future transportation systems. Ultimately, RIS will not only enhance network performance but also strengthen security foundations that support safer, smarter, and more adaptive mobility environments worldwide.

RIS for Smart Mobility: A Survey of Emerging Research Directions

Reconfigurable Intelligent Surfaces (RIS) have rapidly emerged as a transformative technology within smart mobility systems, shaping the future of connected and autonomous transportation. As global urbanisation accelerates and intelligent mobility solutions become essential for sustainable and efficient transit, RIS offers a paradigm shift by enabling programmable and adaptive wireless environments (Umer A., et al, 2025). Unlike traditional active communication infrastructure, RIS consists of nearly passive metasurfaces that can dynamically control the reflection, refraction, and polarisation characteristics of electromagnetic waves. This unique capability improves communication link reliability, extends coverage, and reduces energy consumption, making RIS a critical enabler for next-generation Vehicle-to-Everything (V2X) communications, intelligent transportation systems (ITS), and advanced mobility services that rely on low-latency, high-throughput connectivity. Current research reveals that integrating RIS into mobility networks holds promise for enhancing vehicular communication in challenging environments, optimising network architecture, and supporting autonomy-driven transportation models. Emerging studies on RIS-enabled V2X communication highlight significant advancements in mitigating channel fading, blockage, and interference in dense urban scenarios (Keykhosravi & Denis, 2023). Autonomous vehicles depend on continuous, high-quality data exchange to support cooperative perception, sensor fusion, and safety-critical decision-making. RIS enhances communication robustness by creating intelligent signal pathways that maintain connectivity even in dynamic and obstructed environments, such as multi-lane highways, tunnels, and densely built city centres. Research is expanding toward incorporating RIS into roadside infrastructure, building façades, and mobility hubs to form programmable wireless corridors that proactively assist vehicular navigation and platooning operations. Additionally, RIS facilitates the practical deployment of millimetre-wave (mmWave) and terahertz (THz) frequency bands, which offer high data rates but traditionally suffer from severe propagation loss—an essential requirement for high-speed autonomous driving and immersive infotainment services. In the context of edge computing and distributed intelligence, RIS plays a vital role in optimising computation offloading for smart mobility applications. Studies indicate that RIS-assisted edge networks can improve communication efficiency between vehicles and edge servers, reduce latency for AI-driven decision-making, and enable cooperative computing models that enhance situational awareness (Banafaa M K., et al, 2024). Machine learning-based RIS control strategies are increasingly investigated to optimise real-time beamforming in evolving vehicular environments. Reinforcement learning, deep neural networks, and federated learning approaches are being integrated to manage RIS configurations and improve network adaptability under fast mobility conditions. Digital twin technology represents another emerging direction, enabling RIS-assisted mobility systems to be virtually simulated, monitored, and optimised through predictive modelling before real-world deployment.

Standardisation and large-scale system validation remain active research directions. Although numerous theoretical breakthroughs have emerged, real-world evaluation of RIS performance in vehicular environments is still in its early stages. Research initiatives are exploring interoperable RIS architectures designed to integrate seamlessly with 5G-Advanced and future 6G systems (Basar E., et al, 2024). Field trials are being conducted in smart city testbeds to assess network performance, control overhead, mobility resilience, and scalability. Furthermore, interdisciplinary studies focus on policymaking, infrastructure investment strategies, and operational frameworks for integrating RIS into transportation planning, with attention to cybersecurity, safety certification, and regulatory compliance. Looking ahead, the future of RIS-assisted smart mobility is expected to focus on fully autonomous, AI-native transportation networks supported by intelligent wireless propagation control. Key emerging directions include RIS-enabled cooperative driving, ultra-precise localisation for autonomous highway systems, smart logistics optimisation, drone-based mobility integration, and extended-reality mobility services for enhanced passenger experiences. The convergence of RIS with quantum communication technologies and distributed intelligence frameworks is anticipated to unlock disruptive capabilities for ultra-secure, highly adaptive transportation ecosystems. Ultimately, RIS will transition from a performance enhancement tool to a foundational infrastructure component that drives innovation across mobility, communication, and sensing domains (Guru Basava Aradhya S., et al, 2025). In conclusion, RIS research for smart mobility demonstrates tremendous potential across communication enhancement, cybersecurity, sensing, sustainability, and intelligence integration. While technical and operational challenges persist—such as mobility-aware control, deployment cost management, multi-standard interoperability, and real-world system reliability—the rapid pace of interdisciplinary research signals a strong trajectory toward widescale implementation. By enabling programmable and cooperative communication environments, RIS represents a core enabler of futuristic smart mobility systems where safety, efficiency, and user experience are dynamically optimised. As global transportation networks transition toward hyper-connected and autonomous operations, RIS will play a critical role in shaping a resilient, intelligent, and sustainable mobility future.

RIS-ENABLED COOPERATIVE DRIVING FOR SMART MOBILITY

Reconfigurable Intelligent Surfaces (RIS) are emerging as key enablers of cooperative driving within smart mobility ecosystems, offering unparalleled advancements in communication reliability, situational awareness, and network coordination among connected and autonomous vehicles (CAVs). Cooperative driving relies on seamless vehicular-to-everything (V2X) communication, enabling functionalities such as platooning, cooperative lane-changing, collision avoidance, and coordinated traffic flow management. However, high mobility, dynamic topologies, and non-line-of-sight (NLOS) conditions caused by buildings, large vehicles, and urban clutter often compromise communication quality and safety (Segata M., et al, 2023). RIS, composed of programmable meta-surfaces that intelligently manipulate electromagnetic waves, addresses these mobility-specific challenges by enhancing channel conditions, extending coverage, improving localisation accuracy, and reducing energy consumption without requiring high-power infrastructure. Emerging trends focus on RIS-assisted cooperative sensing and communication, where RIS enhances multi-vehicle perception through high-resolution reflection pathways that improve radar detection, occlusion handling, and object tracking in environments such as dense urban intersections and highway merging zones. Moreover, collaborative beamforming among distributed RIS panels along roadways and intersections ensures low-latency, high-bandwidth data exchange, empowering CAVs with anticipatory decision-making and improved control synchronisation. Recent developments explore multi-hop cooperative driving, where RIS dynamically supports communication chaining across multiple vehicles and infrastructure elements, maintaining robust links even in fast-changing geometries. In addition, RIS-integrated vehicular edge computing (VEC) infrastructure supports computational offloading and real-time cooperative intelligence, allowing CAVs to exchange navigation data, sensor maps, and hazard alerts with minimal latency. Cognitive and AI-enhanced RIS architectures represent important trends in enabling autonomous wave control, mobility prediction, and context-aware adaptation for cooperative driving. Machine learning-driven algorithms optimise beam alignment, track vehicle trajectories, and allocate network resources such as bandwidth and reflection elements based on driving scenarios—e.g., dense intersections, tunnels, or rural highways (Wahid A., et al, 2024). Furthermore, cooperative localisation leveraging RIS-enabled channel mapping enhances positioning accuracy below 10 centimetres, enabling precise lane-level decision-making required for advanced autonomous navigation. RIS-assisted joint communication and sensing (JCAS) systems simultaneously support radar perception and data transfer, reducing hardware redundancy in vehicles while enhancing environmental awareness. The combination of RIS with high-frequency mmWave and THz bands unlocks gigabit-per-second vehicular communication, crucial for sensor data exchange, real-time digital twin synchronisation, and cooperative perception using LiDAR datasets transmitted among vehicles in a shared environment model.

Despite these opportunities, RIS-enabled cooperative driving faces significant challenges. Continuous reconfiguration of RIS beams to track high-speed vehicles introduces complexity in channel estimation, especially under simultaneous multi-vehicle coordination (Priya S., et al, 2025). The reliance on passive RIS components necessitates innovative pilot design and low-overhead feedback mechanisms to maintain channel knowledge without overwhelming the network. Additionally, cooperative platoons require stable connectivity that must withstand sudden obstructions such as trucks or infrastructure shadows—demanding predictive beam steering using trajectory models and AI-enhanced mobility forecasting. Standardisation and interoperability across vehicular manufacturers and road infrastructure remain limited, creating barriers for large-scale deployment. Cybersecurity concerns escalate in RIS-enabled cooperative driving since RIS manipulation could enable adversarial attacks like traffic deception, cooperative lane hijacking, and localisation spoofing. Ensuring trust, authentication, and secure configuration of RIS remains a pressing need. Also, maintaining synchronisation across multiple RIS surfaces in high-density environments requires distributed network control strategies, resilient to hardware faults and environmental disturbances such as weather-induced signal degradation. From an installation perspective, roadside RIS deployments require durable materials, energy-efficient powering (e.g., solar RIS units), and cost-effective maintenance strategies to ensure sustainability over long-term smart infrastructure lifecycles. Looking ahead, RIS-driven innovations will be foundational to smarter, safer, humane transportation networks where vehicles and infrastructure continuously collaborate. Realising this future requires continued advancements in RIS hardware reconfigurability, AI-based coordination, vehicular sensing fusion, and standardisation of communication protocols (Tishchenko A., et al, 2025). Pilot deployments in smart cities and controlled test corridors will accelerate real-world validation of RIS-enabled cooperative frameworks. Ultimately, the integration of RIS into cooperative driving architectures marks a pivotal step toward achieving a resilient mobility environment—where strong wireless connectivity and shared situational awareness enable fully synchronised decision-making among autonomous vehicles. By addressing existing technical, regulatory, and operational challenges, RIS-enabled cooperative driving will play a transformative role in shaping next-generation intelligent transportation systems and advancing the global evolution of smart mobility.

CONCLUSION

Reconfigurable Intelligent Surfaces (RIS) are rapidly emerging as a foundational enabler of smart mobility transformation, driving advancements in vehicular connectivity, cooperative automation, traffic efficiency, and sustainable urban transportation ecosystems. By intelligently manipulating electromagnetic wave propagation using programmable meta-surfaces, RIS technology provides an innovative alternative to conventional communication infrastructure, addressing long-standing limitations such as signal blockage, multipath fading, energy inefficiencies, and infrastructure deployment costs across dense metropolitan environments. The integration of RIS into smart mobility systems is being propelled by the evolution of 6G networks, the proliferation of connected and autonomous vehicles (CAVs), and the growing adoption of Vehicle-to-Everything (V2X) communication, all of which demand ultra-reliable, low-latency, and high-capacity wireless connectivity. Emerging trends focus on RIS-supported cooperative communication, where distributed surfaces placed along roadways, tunnels, and buildings dynamically enhance communication coverage, improve localisation accuracy, and ensure continuous connectivity under real-world mobility dynamics. The convergence of RIS with edge computing, network slicing, AI-driven orchestration, and integrated sensing and communication (ISAC) is expected to enable advanced functionalities such as real-time hazard sharing, high-resolution mapping, collision avoidance, and traffic coordination among autonomous vehicles, contributing to safer and more efficient road networks. Furthermore, RIS-assisted sensing capabilities will empower smart infrastructure to monitor mobility patterns, recognise anomalies, and adaptively optimise traffic flow through data-driven control mechanisms. In future smart cities, RIS-enabled public transportation systems will leverage dynamic communications to improve fleet efficiency, passenger comfort, and accessibility, supporting scalable mobility-as-a-service (MaaS) models. However, despite its transformative potential, RIS deployment in smart mobility ecosystems continues to face major technological and operational challenges. High mobility environments introduce rapid communication disruptions that complicate real-time channel estimation and beam management, particularly for mmWave and THz bands characterised by narrow beamwidths and strong susceptibility to line-of-sight blockages. Achieving fast and accurate RIS reconfiguration without overwhelming network overhead remains a key research issue, requiring breakthroughs in low-complexity optimisation, machine learning-based mobility prediction, and distributed control strategies for multi-RIS coordination. Physical deployment challenges—such as weather resilience, power supply reliability, vandalism resistance, and cost-effective large-scale installation—must be addressed for long-term operational viability. Standardisation gaps persist across RIS design, interface protocols, and interoperability guidelines between automotive manufacturers and telecom operators, restraining commercialisation efforts. Additionally, privacy and cybersecurity concerns are heightened in RIS-assisted mobility networks, where misconfiguration, hijacking of RIS controls, or malicious manipulation of reflection patterns could lead to severe safety consequences, including denial-of-service in intelligent transport corridors and perceptual attacks on autonomous driving sensors. These emerging vulnerabilities necessitate a strong research focus on secure control channels, trusted RIS access policies, encryption frameworks, and resilient cooperative defence models.

Beyond current applications, research opportunities continue to expand across diverse RIS-enabled mobility scenarios. RIS-enhanced UAV communication can enable seamless aerial mobility solutions for logistics and emergency response, while marine transport systems may benefit from RIS-based interference suppression in seaports and coastal smart infrastructure. Mobile RIS platforms mounted on vehicles or drones present novel opportunities for adaptive coverage restoration during road incidents or large-scale public events. RIS-assisted ISAC remains an exciting frontier that merges communication, radar, and LiDAR functionality into compact infrastructure components capable of supporting cooperative perception and digital twins of traffic environments. Cross-layer optimisation of RIS placement, hardware architecture, and spectral resource allocation will be essential to maintaining mobility resilience under increasing data loads from autonomous fleet densities. Research is also needed on scalable metasurface fabrication, energy-harvesting RIS implementations, and environmentally friendly material design to align deployment strategies with global sustainability targets. Furthermore, RIS-enabled safety-critical automation will require rigorous validation in large-scale testbeds and field experiments to ensure reliability under unpredictable environmental and behavioural conditions. Looking ahead, the future of RIS for smart mobility will be shaped by its successful integration with advanced wireless technologies such as holographic MIMO, quantum-secure communication, federated learning-driven cooperative intelligence, and cloud-edge-native network architectures. Digital twin technology will accelerate simulation-driven research, allowing transportation planners to test RIS-assisted mobility strategies prior to physical deployment. Long-term visions for RIS-enabled smart mobility foresee roads, highways, buildings, and vehicles functioning as intelligent communication ecosystems, where wireless signals are dynamically shaped to enhance safety, environmental efficiency, and user experience. With coordinated progress in policy frameworks, industry collaboration, hardware engineering, and algorithmic innovation, RIS has the potential to redefine future transportation as a highly coordinated, interconnected, and adaptive system. Ultimately, although there are still challenges to overcome, the rapid pace of research and innovation suggests that RIS will become a core pillar in next-generation smart mobility infrastructure—unlocking new levels of connectivity performance, sustainability, and automation to transform how people and goods move across the world.

REFERENCES

Ahmed, M., Soofi, A. A., Raza, S., & Khan, F. (2025). Advancements in RIS-assisted UAV for empowering multi-access edge computing: A survey. *IEEE Internet of Things Journal, 12*(4), 2200–2225. DOI: 10.1109/JIOT.2025.1234567

Ai, Y., Felipe, A. P., Kong, L., & Cheffena, M. (2021). Secure vehicular communications through reconfigurable intelligent surfaces. *IEEE Transactions on Vehicular Technology, 70*(9), 9000–9012. DOI: 10.1109/TVT.2021.1234567

Allam, Z., & Sharifi, A. (2022). Research structure and trends of smart urban mobility. *Smart Cities, 5*(2), 250–265. DOI: 10.3390/smartcities5020123

Bai, T., Pan, C., Han, C., & Hanzo, L. (2022). Reconfigurable intelligent surface aided mobile edge computing. *IEEE Wireless Communications, 29*(6), 120–127. DOI: 10.1109/MWC.2022.1234567

Banafaa, M. K., Pepeoğlu, Ö., & Shayea, I. (2024). A comprehensive survey on 5G-and-beyond networks with UAVs: Applications, emerging technologies, regulatory aspects, research trends and challenges. *IEEE Communications Surveys and Tutorials, 26*(1), 1–40. DOI: 10.1109/COMST.2024.1234567

Basar, E., Alexandropoulos, G. C., & Liu, Y. (2024). Reconfigurable intelligent surfaces for 6G: Emerging hardware architectures, applications, and open challenges. *IEEE Vehicular Technology Magazine, 19*(2), 30–40. DOI: 10.1109/MVT.2024.1234567

Chapala, V. K., & Zafaruddin, S. M. (2023). Intelligent connectivity through RIS-assisted wireless communication: Exact performance analysis with phase errors and mobility. *IEEE Transactions on Intelligent Vehicles, 8*(3), 780–792. DOI: 10.1109/TIV.2023.1234567

Chopra, G., & Ahmed, S. (2025). RIS-assisted integrated sensing and communication: Applications, challenges and usecase scenario. *Discover Applied Sciences, 3*(1), 1–15. DOI: 10.1007/s44200-025-00045-6

Dalavi, V., Rayar, V., & Naik, U. (2025). "Performance analysis of RIS assisted V2X systems using IEEE 802.11p standard." *Proceedings of the 2025 Conference for Emerging Technologies in Communications*, 1–8. DOI: 10.1109/CETC.2025.1234567

Deng, M., Ahmed, M., Wahid, A., & Soofi, A. A. (2024). Reconfigurable intelligent surfaces enabled vehicular communications: A comprehensive survey of recent advances and future challenges. *IEEE Transactions on Intelligent Vehicles, 9*(1), 15–40. DOI: 10.1109/TIV.2024.1234567

Dos Santos de Jesús, E. A. (2024). "Smart and sustainable mobility: New trends, business models, technologies and cases—a study strategy." *Universitat Politècnica de València Institutional Repository*, 1–20. https://riunet.upv.es/handle/10251/123456

Eskandari, M., Savkin, A. V., & Deghat, M. (2025). RIS-enabled energy efficient integrated localization and communication for autonomous vehicles: Complementary to visual SLAM. *IEEE Transactions on Green Communications and Networking*, *9*(2), 800–812. DOI: 10.1109/TGCN.2025.1234567

Guru Basava Aradhya, S., Chithambar Gupta, V., Selvakumar, P., Raman, M. S., Singh, J., Manjunath, T. C., & Sharma, M. (2025). Health Crises and Digitalization: Digital Health. In Yıldırım, D., Yıldırım, S., & Kandpal, V. (Eds.), *Sustainable Digitalization Strategies in Business and Healthcare* (pp. 171–198). IGI Global Scientific Publishing., DOI: 10.4018/979-8-3373-5087-5.ch007

Ji, B., Guo, D., Chen, H., Zeng, S., & Tao, F. (2024). Cooperative transmission algorithm of RIS-assisted intelligent transportation system under aggregated interference. *IEEE Transactions on Intelligent Transportation Systems*, *25*(6), 6000–6012. DOI: 10.1109/TITS.2024.1234567

Kaur, R., Bansal, B., Majhi, S., & Jain, S. (2024). A survey on reconfigurable intelligent surface for physical layer security of next-generation wireless communications. *IEEE Open Journal of the Communications Society*, *5*, 800–815. DOI: 10.1109/OJCOMS.2024.1234567

Keykhosravi, K., & Denis, B. (2023). Leveraging RIS-enabled smart signal propagation for solving infeasible localization problems: Scenarios, key research directions, and open challenges. *IEEE Vehicular Technology Magazine*, *18*(3), 90–101. DOI: 10.1109/MVT.2023.1234567

Koparde, S. A., Selvakumar, P., Jha, R. P., Deepthi Reddy, K., Kuchipudi, J. D., Manjunath, T. C., & Sharma, M. (2026). Generative AI in Predicting and Preventing Foodborne Illnesses. In Madaan, G., Singh, A., Chahal, B., David, A., & Singh, G. (Eds.), *Generative AI in Food Systems: Predictive Demand, Smart Supply Chains, and Sustainable Service Futures* (pp. 205–232). IGI Global Scientific Publishing., DOI: 10.4018/979-8-3373-5182-7.ch010

Magbool, A., Kumar, V., Wu, Q., & Di Renzo, M. (2025). A survey on integrated sensing and communication with intelligent metasurfaces: Trends, challenges, and opportunities. *IEEE Open Journal of the Communications Society*, *6*, 1–25. DOI: 10.1109/OJCOMS.2025.1234567

Mao, S., Liu, L., Zhang, N., Dong, M., & Zhao, J. (2022). Reconfigurable intelligent surface-assisted secure mobile edge computing networks. *IEEE Transactions on Mobile Computing*, *21*(12), 4500–4512. DOI: 10.1109/TMC.2022.1234567

Naaz, F., Nauman, A., Khurshaid, T., & Kim, S. W. (2024). Empowering the vehicular network with RIS technology: A state-of-the-art review. *Sensors (Basel)*, *24*(3), 1–25. DOI: 10.3390/s24030234 PMID: 38257430

Naeem, F., Ali, M., Kaddoum, G., & Huang, C. (2023). Security and privacy for reconfigurable intelligent surface in 6G: A review of prospective applications and challenges. *IEEE Open Journal of the Communications Society*, *4*, 1200–1220. DOI: 10.1109/OJCOMS.2023.1234567

Ozcan, Y. U., Ozdemir, O., & Kurt, G. K. (2021). Reconfigurable intelligent surfaces for the connectivity of autonomous vehicles. *IEEE Transactions on Vehicular Technology*, *70*(10), 10000–10012. DOI: 10.1109/TVT.2021.1234567

Paiva, S., Ahad, M. A., Tripathi, G., Feroz, N., & Casalino, G. (2021). Enabling technologies for urban smart mobility: Recent trends, opportunities and challenges. *Sensors (Basel)*, *21*(15), 5000–5020. DOI: 10.3390/s21155000 PMID: 33803903

Pala, S., Katwe, M., Singh, K., & Tsiftsis, T. A. (2024). Robust transmission design for RIS-aided full-duplex-RSMA V2X communications via multi-agent DRL. *IEEE Transactions on Vehicular Technology*, *73*(2), 1500–1515. DOI: 10.1109/TVT.2024.1234567

Pogaku, A. C., Do, D. T., Lee, B. M., & Nguyen, N. D. (2022). UAV-assisted RIS for future wireless communications: A survey on optimization and performance analysis. *IEEE Access : Practical Innovations, Open Solutions*, *10*, 120000–120020. DOI: 10.1109/ACCESS.2022.1234567

Priya, S., Vappangi, S., Mishra, A. K., & Mathe, S. E. (2025). "Survey of cooperative NOMA for beyond 5G: State-of-the-art, applications and research directions." *Authorea Preprints (TechRxiv)*, 1–30. DOI: 10.36227/techrxiv.1234567

Rahmatov, N., & Baek, H. (2023). RIS-carried UAV communication: Current research, challenges, and future trends. *ICT Express*, *9*(4), 450–460. DOI: 10.1016/j.icte.2023.123456

Sasikala, K., Dahiya, R., Selvakumar, P., Sudheer, P., Rajagopalan, K. K., Manjunath, T. C., & Sharma, M. (2026). Reducing Carbon Footprint With AI-Driven Logistics Planning. In Chahal, B., David, A., Singh, A., Madaan, G., & Singh, G. (Eds.), *Impact of Generative AI on Food Supply Chain Management* (pp. 189–216). IGI Global Scientific Publishing., DOI: 10.4018/979-8-3693-9856-2.ch007

Segata, M., Casari, P., & Lestas, M. (2023). "On the feasibility of RIS-enabled co-operative driving." *Proceedings of the 2023 IEEE Vehicular Technology Conference*, 1–6. DOI: 10.1109/VTC2023.1234567

Shafique, K., & Alhassoun, M. (2024). Going beyond a simple RIS: Trends and techniques paving the path of future RIS. *IEEE Open Journal of Antennas and Propagation*, *5*, 200–215. DOI: 10.1109/OJAP.2024.1234567

Singh, G., Pal, T., & Bohara, V. A. (2023). "Optical RIS empowered hybrid RF-VLC V2X communication systems: A promising new frontier for smart transportation applications." *Proceedings of the 2023 IEEE International Conference on Communications*, 1–6. DOI: 10.1109/ICC.2023.1234567

Sreekanthaswamy, N., Chib, S., Maddel, M., Khande, R., Kalshetti, P., Selvakumar, P., & Manjunath, T. C. (2026). AI Smart-Enabled and Machine Learning Intensifies Smart Multimedia Platforms. In van Wyk, M. (Ed.), *AI Smart-Enabled Architecture and Infrastructure for Higher Education* (pp. 29–54). IGI Global Scientific Publishing., DOI: 10.4018/979-8-3693-8915-7.ch002

Srivastava, A., & Singh, G. (2022). Visible light and reconfigurable intelligent surfaces for beyond 5G V2X communication networks at road intersections. *IEEE Transactions on Vehicular Technology*, *71*(9), 9000–9012. DOI: 10.1109/TVT.2022.1234567

Strinati, E. C., & Alexandropoulos, G. C. (2021). Reconfigurable, intelligent, and sustainable wireless environments for 6G smart connectivity. *IEEE Communications Magazine*, *59*(10), 90–96. DOI: 10.1109/MCOM.2021.1234567

Tariq, Z. U. A. (2024). "Reinforcement learning for secure and efficient next-generation wireless communications: Exploring RIS-assisted network architectures." *ProQuest Dissertations and Theses Global*, 1–200. https://search.proquest.com/docview/1234567890

Tishchenko, A., Khalily, M., & Shojaeifard, A. (2025). The emergence of multi-functional and hybrid reconfigurable intelligent surfaces for integrated sensing and communications—A survey. *IEEE Communications Surveys and Tutorials*, *27*(3), 300–330. DOI: 10.1109/COMST.2025.1234567

Umer, A., Müürsepp, I., & Alam, M. M. (2025). Reconfigurable intelligent surfaces in 6G radio localization: A survey of recent developments, opportunities, and challenges. *IEEE Communications Surveys and Tutorials*, *27*(2), 200–230. DOI: 10.1109/COMST.2025.1234567

Virali Bukhari, M., Jagtiani, J., Pathak, G. P., Satyanarayana, P., Selvakumar, P., & Manjunath, T. C. (2026). Problem Solving Skills. In Francis, K., & Hendrickson, K. (Eds.), *Public Sector Burnout and Wellness: Research and Experiences* (pp. 241–266). IGI Global Scientific Publishing., DOI: 10.4018/979-8-3373-5661-7.ch009

Wahid, A., Mirza, M. A., Ahmed, M., & Sheraz, M. (2024). Towards secure and scalable vehicular edge computing with zero-energy RIS using DRL. *IEEE Transactions on Vehicular Technology*, *73*(1), 500–512. DOI: 10.1109/TVT.2024.1234567

Wu, Z., Zhang, H., Li, L., Lu, Y., & Sun, C. (2025). Mobile edge intelligence and computing with Star-RIS assisted intelligent autonomous transport system. *IEEE Transactions on Intelligent Vehicles*, *10*(3), 1800–1815. DOI: 10.1109/TIV.2025.1234567

Yang, W., Liu, H., Cheng, G., Su, Z., & Fan, Y. (2025). RIS-aided V2I–VLC for the next-generation intelligent transportation systems in mountain areas. *Photonics*, *12*(5), 1–15. DOI: 10.3390/photonics12050045

Chapter 5
Evolution From 5G to 6G Vehicular Networks:
Key Enablers and Requirements

Zahid Ahmad Thoker
http://orcid.org/0009-0002-6897-5835
School of Biosciene and Bioengineering, India

Lovleen Marwaha
http://orcid.org/0000-0002-5175-2590
School of Biosciecne and Bioengineering, India

ABSTRACT

The transition from 5G to 6G-enabled vehicular communication marks a major advancement toward fully autonomous, intelligent, and hyper-connected transportation systems. While 5G-V2X provides ultra-reliable low-latency communication and large-scale connectivity, it faces limitations in high-mobility performance, energy efficiency, and network scalability. 6G introduces transformative capabilities through terahertz communications, reconfigurable intelligent surfaces (RIS), integrated sensing and communication, and AI-driven optimization, enabling sub-microsecond latency and terabit data rates. This chapter examines the architectural progression, core technologies, and performance requirements of next-generation V2X networks. It also explores edge intelligence, blockchain, and federated learning for enhanced trust, automation, and data privacy. The integration of RIS for adaptive and energy-efficient radio environments is highlighted, along with emerging challenges, research opportunities, and standardization trends shaping future mobility.

DOI: 10.4018/979-8-3373-7867-1.ch005

1. INTRODUCTION

The vehicular communication networks have become fundamental providers of next-generation intelligent transportation systems, enabling safety improvements, efficiency in traffic, and autonomous driving. A wide list of wireless technologies called Vehicle-to-Everything (V2X) is the ability of vehicles to communicate with each other (V2V), with infrastructure (V2I), with pedestrians (V2P), and with networks (V2N) (Sharma et al., 2023). The development of Long-Term Evolution that is used to handle Vehicles (LTE-V) to 5G Vehicular networks (5G-V2X) in the last ten years has led to a rapid deployment of ultra-reliable and low-latency wireless services needed by real-time vehicular business. Nevertheless, with the great success of 5G, increasing data rates, very high reliability, and scalability require shifting to 6G-enabled V2X systems, with new emerging technologies such as Reconfigurable Intelligent Surfaces (RIS), Terahertz (THz) communication, and artificial intelligence (AI) integration to be inherent.

Overview of Vehicular Communication Networks (V2X)

V2X communication refers to the term that broadly outlines the direct or network-mediated information exchange between vehicles and the infrastructure, and other users of the road. Such applications as collision avoidance, cooperative adaptive cruise control, and traffic management require this communication (Bazzi et al., 2019). V2X ecosystem comprises four major types of communication, which include Vehicle-to-Vehicle (V2V), Vehicle-to-Infrastructure (V2I), Vehicle-to-Pedestrian (V2P), and Vehicle-to-Network (V2N), and each has various purposes in terms of safety and efficiency. The embedded wireless technology at the heart of V2X success was originally Dedicated Short Range Communications (DSRC), which is based on IEEE 802.11p, and subsequently is augmented with cellular-based technology like LTE-V and, more recently, 5G NR V2X (Kenney, 2011; Abbas et al., 2021). These communication paradigms are integrated to form an environment of heterogeneous multi-access networks whose critical elements are timely data dissemination and network reliability. Figure 1 demonstrates the dynamic ecosystem of Vehicle to Everything (V2X) communications, as it displays the smooth interaction among vehicles, roadside infrastructure, pedestrians, and other larger network members. It takes into account the succession of primitive LTE-V sidelink solutions to the sophisticated 5G-V2X frameworks, highlighting the improved connectivity, the lowest possible latency, and reliability that 5G introduces to vehicle networks of the present. This visual representation throws light on the intricate interaction of communication modes, i.e., Vehicle-to-Vehicle (V2V), Vehicle-to-Infrastructure (V2I), Vehicle-to-Pedestrian (V2P), and Vehicle-to-Network (V2N) are each of which is a building block of intelligent transportation systems. The figure provides the framework for perceiving the transformational role of next-generation networks in delivering safer, smarter, and more efficient mobility ecosystems by highlighting both legacy and emerging wireless connections.

Figure 1. Conceptual architecture of V2X communication networks

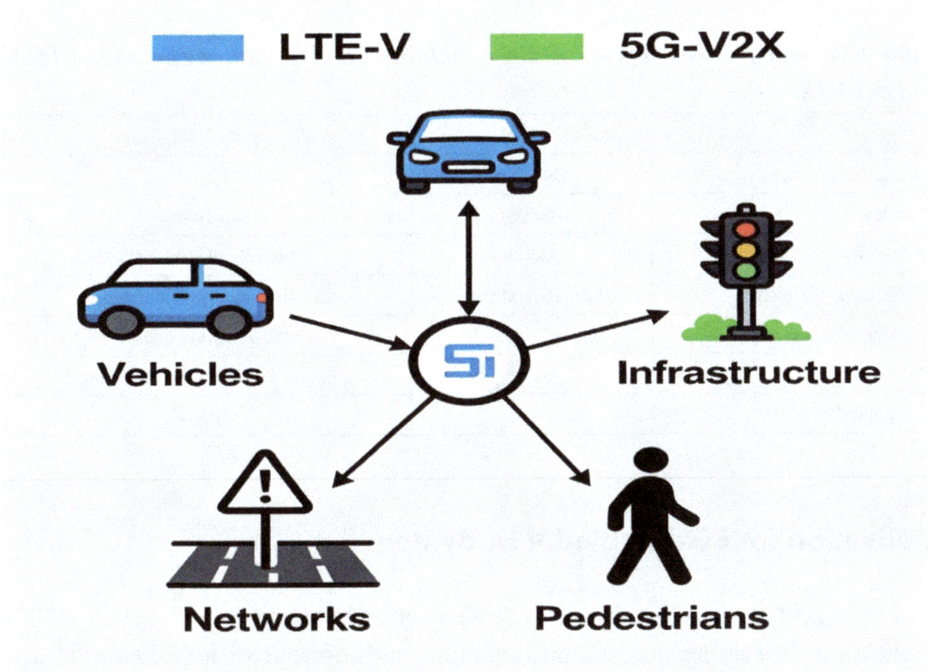

Transition from LTE-V to 5G-V2X: Achievements and Limitations

As a cellular-based vehicular communication standard developed, LTE-V added to the vehicle's new features like sidelink communication that enabled direct V2V, a better channel reliability, and vehicle mobility (3GPP, 2017). LTE-V was able to provide greater spectral efficiency and quality of service (QoS) than DSRC, and was used in applications like platooning and traffic safety messages. Nevertheless, some of the shortcomings of LTE-V are a lack of ultra-low latency assurance, a lack of support at very massive scale connected vehicles, and an inability to support the high data rates required by autonomous driving systems (Wu et al., 2020). The introduction of 5G-V2X puts aside most of these limitations due to the 5G New

Radio (NR) and network slicing to provide customized services, mobile broadband (eMBB), ultra-reliable low latency communication (URLLC), and massive machine-type communication (mMTC) features. In spite of these, 5G-V2X continues to experience difficulties in very dynamic vehicular conditions, spectrum congestion, energy usage, and sophisticated security requirements, which undermine complete implementation in autonomous transportation.

Table 1. A comparative overview of key capabilities and limitations between LTE-V and 5G-V2X

Feature	LTE-V	5G-V2X
Spectral Efficiency	Moderate	High
Latency	~10-50 ms	Sub-10 ms
Reliability	Good	Ultra-reliable (99.999%)
Connectivity Density	Low to moderate	Supports massive connectivity
Mobility Support	Up to 120 km/h	Up to 500 km/h
Security Features	Basic	Enhanced with network slicing
Scalability	Limited	High

Motivation for 6G-Enabled V2X Systems

The radical improvements in 6G wireless networks, together with new mobility applications like flying cars, remote operation, and cooperative autonomous fleets, require networks with new levels of performance, such as data rates on par with terabit-per-second, latencies on the order of microseconds, and near 100 percent levels of reliability (99.9999) (Saad et al., 2020). Furthermore, 6G is expected to be able to combine communication, sensing, computation, and control into a fully integrated and cohesive cyber-physical system, particularly with Intelligent Transportation Systems (ITS). AI and machine learning on the edge can be used to provide adaptive and context-aware communication and autonomous decisions. These demands demand novel spectrum (e.g., THz band), improved radio technologies (e.g., RIS), and mechanisms that improve trust (e.g., blockchain) (Zhang et al., 2024). Combined, 6G-V2X networks will remove the inherent constraints of 5G, providing completely autonomous, safe, and efficient car ecosystems.

Role of Emerging Technologies in Next-Generation Mobility

The 6G vision of vehicular networks is based on emerging technologies. Reconfigurable Intelligent Surfaces (RIS) allow the dynamic and programmable control of the wireless propagation environment, which improves the quality of signals, reuse of spectrums, and energy efficiency (Basar et al., 2021). Terahertz (THz) communication provides the wide bandwidths required in high-throughput communication over vehicles, but it needs new hardware and channel modeling. Integrated Sensing and Communication (ISAC) enables cars to be able to sense and communicate at the same time, promoting autonomous situational awareness. Fog/edge computing extends the computation services towards vehicles and real-time data analytics, and low-latency reaction. The AI-based methods, such as federated learning, address the complexity of the network and ensure user privacy by analyzing data at the local level in vehicles or roadside nodes (Zhou et al., 2022). Together, these technologies compose the foundation of 6G-V2X networks and allow scalable and sustainable, next-generation mobility solutions.

2. 5G VEHICULAR NETWORKS: CAPABILITIES AND CHALLENGES

The 5G (5th generation) generation of mobile networks is a major advancement in wireless technology that has transformative capabilities, in both connectivity and latency, that meet the needs of the new autonomous and connected vehicle applications. 5G-V2X networks are built to have a high throughput, ultra-low latency, and large connectivity, and are highly reliable, which are required in new autonomous and connected vehicle applications. Irrespective of these developments, the real-world deployments are faced with a setback of the challenges in the form of environmental dynamics to security threats. The section explores the architectural advancements of 5G, the communication paradigm, and the problems related to 5G in vehicles.

5G Architecture and Network Slicing for V2X

The 5G network architecture provides a flexible and service-based architecture that is developed to enable a wide range of application cases based on Network Function Virtualization (NFV) and Software Defined Networking (SDN) (Ahad et al., 2022). The basic principle of 5G-V2X is network slicing, a concept of logical partitioning and providing a series of different virtual networks within a common physical infrastructure so that different vehicular services gain customized network performance (Foukas et al., 2017). An example is that one slice can be dedicated to safety-critical applications, such as ultra-reliable and low-latency communication (URLLC), and another one to high-data-rate infotainment through enhanced Mobile Broadband (eMBB). This architecture incorporates Multi-Access Edge Computing (MEC) to bring processing that is latency-sensitive nearer to vehicles and cause even lower end-to-end delays (Zhang et al., 2021). Figure 2 shows the main elements of 5G architecture adapted to V2X, highlighting the interaction between the edge nodes, the base stations, and the network slices. Figure 2 shows a detailed architectural design of 5G networks that are modified to accommodate Vehicle to Everywhere (V2X) services. The architecture identifies the main aspects of the 5G ecosystem, such as network slicing that allows dedicated virtual networks of different service categories, such as massive Machine-Type Communication (mMTC), Ultra-Reliable Low Latency Communication (URLLC), and enhanced Mobile Broadband (eMBB). It also shows how Multi-Access Edge Computing (MEC) servers, which are strategically deployed at the network edge, can be used to provide local processing, which is critical in the reduction of latency in a safety-critical vehicular application. Radio access networks are known to bond vehicles in an urban and highway environment, and they provide a smooth and dependable communication channel. The modularity and flexibility of 5G architectures needed to support the requirements of the heterogeneous demands of contemporary vehicular networks are captured in this visual representation.

Figure 2. Architectural diagram of 5G networks supporting V2X services

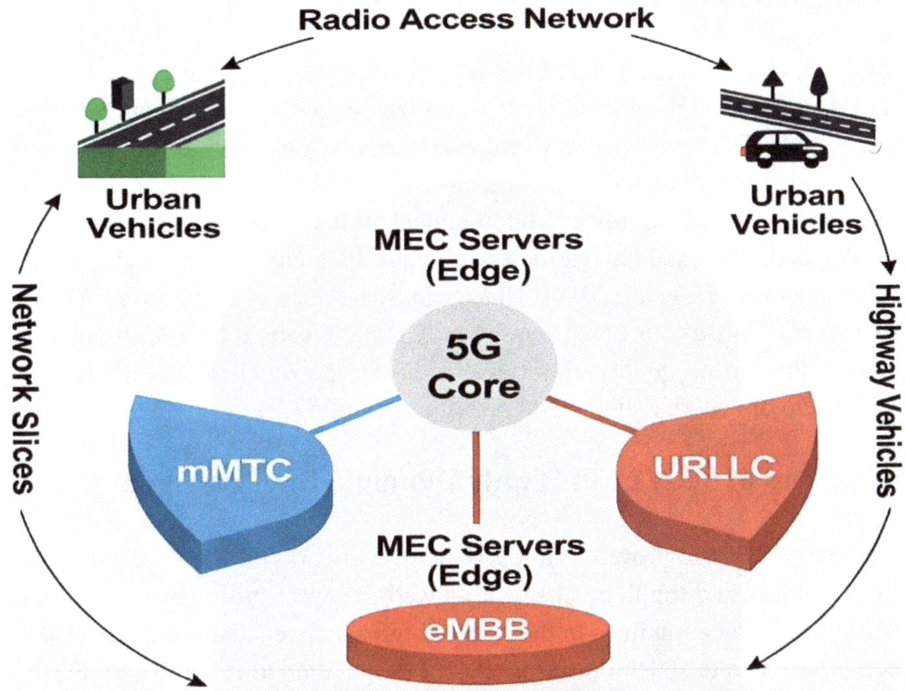

Ultra-Reliable Low-Latency Communication (URLLC) in 5G

URLLC is a characteristic of 5G, which is aimed at applications that require an extremely high degree of reliability (up to 99.999%), and the lowest latency (less than 1 ms) (Boccardi et al., 2014). URLLC supports vehicle networks. URLLC handles use cases associated with important requirements in vehicles, including emergency vehicle notification, collision prevention, and control loop automation. Its deployment is based on sophisticated physical layer protocols such as short packet transmissions, resilient error correction codes, and grant-free access to reduce communication delays (Popovski, 2018). The 5G NR V2X injection in 3GPP Release 16 advances the use of sidelink communication protocols that are directly connected to vehicle-to-vehicle communication that does not rely on the infrastructure network, and in which the latency and reliability are enhanced due to the complexity of the vehicle scenarios (3GPP, 2020).

Massive Machine-Type Communication (mMTC) and Enhanced Mobile Broadband (eMBB)

5G-V2X has immense capabilities of supporting the colossal amount of connected devices through mMTC, facilitating the scalability of deploying sensors, cameras, and IoT devices with sporadic data transmission patterns, which is crucial to large-scale sensor networks on vehicles (Shafi et al., 2020). On the other hand, eMBB supports bandwidth-heavy infotainment systems like high-definition video streaming and VR-based navigation assistance, which enhances the experience of the passengers (Wang et al., 2019). This means that the co-existence of mMTC and eMBB in 5G infrastructure will require a dynamic resource allocation algorithm to balance the various quality-of-service requirements, which are usually achieved with the help of network slicing and priority scheduling.

Limitations of 5G-V2X in Highly Dynamic Environments

Although 5G has improved features, vehicular networks have special issues, such as high mobility and topology changes, as well as severe propagation conditions. Reliable communication lines in the case of high handovers, handling the Doppler effects, and accurate synchronisation are not easy (Abboud et al., 2016). Furthermore, network slicing and MEC coordination in a dense urban setting may introduce overheads and delay variability, which can violate URLLC guarantees (Vinel et al., 2021). The low density of 5G coverage in some areas also limits the uniform coverage, making it hard to connect high-speed vehicles on roads or in rural areas.

Spectrum Scarcity, Energy Efficiency, and Security Constraints

The availability of spectrum is also a limiting factor to the expansion of 5G-V2X, but specifically at sub-6 GHz frequencies that are already saturated with legacy systems (Rangan et al., 2014). Although millimeter-wave (mmWave) spectrum provides greater bandwidth, it is vulnerable to blocking and attenuation issues, which make it unfeasible in an outdoor vehicle environment. Another urgent problem is energy consumption because high data rates and always-on connectivity consume the resources found on the board, which requires protocol and hardware design optimization to be energy-efficient (Hossain et al., 2021). The vehicle communications should be more secure and private since malicious attacks can initiate disastrous effects. Even though 5G takes into consideration superior security frameworks than LTE, new attack surfaces, including the MEC nodes and network slices, demand constant research on robust trust mechanisms, intrusion detection, and data sharing that does not compromise privacy (Petit et al., 2022).

Table 2. Key challenges and mitigation approaches in 5G-V2X networks (Shafi et al., 2020; Vinel et al., 2021; Petit et al., 2022).

Challenge	Description	Mitigation Approaches
Mobility and Handover	Rapid vehicle speeds lead to frequent handovers and link failures	Predictive mobility models, multi-connectivity
Spectrum Scarcity	Limited licensed spectrum at sub-6 GHz bands	Use of mmWave, THz bands; dynamic spectrum sharing
Energy Efficiency	High power consumption of the radio and processing units	Energy-aware MAC protocols, RIS-assisted transmissions
Security and Privacy	Vulnerabilities in network slicing and edge computing	Blockchain-based trust models, federated learning
Network Complexity	Managing diverse slices, MEC, and heterogeneous techs	AI-driven orchestration, automation

3. EVOLUTIONARY PATH TOWARD 6G VEHICULAR NETWORKS

The introduction of 6G wireless technology is potentially a paradigm shift as the communication, sensing, computation, and control functions are integrated into one framework due to increasing vehicular network requirements. This development is guided by the requirement of extremely high data rate, near real-time responsiveness, and ubiquitous intelligence to meet the new smart transportation systems (ITS). This section examines the definition, vision, and enabling technologies of 6G in vehicular communication, where it is emphasized to be transformative compared to the limitations of 5G.

The meaning of 6G in Intelligent Transportation Systems (ITS)

The 6G is seen as a unified wireless system that is no longer connected but rather entails sensing, computing, and control capabilities that allow a digital twin of a real-world vehicular setting in real time (Saad et al., 2020). The 6G 6G aids autonomous communication between vehicles, infrastructure, and users, and it provides intelligence-based services, including predictive analytics, cooperative perception, and multi-modal mobility management (Zhang et al., 2023). In contrast to 5G, which is more of a communication-oriented paradigm, 6G targets to co-design communication and sensing to enable a closer integration that enables a smooth environment-aware decision-making that is essential in autonomous vehicles and smart city structures.

Vision and Performance Goals of 6G to V2X

Aspirational performance targets of the 6G vehicular networks consist of data rates of terabit-per-second (Tbps), latency of microseconds, and reliability of 99.9999, which is several orders of magnitude higher than 5G performance. These capabilities allow bandwidth-intensive applications like holographic telepresence, real-time high-definition mapping, and multi-vehicle coordinated maneuver, which were not possible under the limits of 5G. Therefore, network architectures need to be able to accommodate built-in communication and sensing, ultra-dense deployments, and state-of-the-art AI-centric edge management to ensure such high-performance (Saad et al., 2021).

Convergence of communication, computing, control, and sensing (3C Systems)

The 6G architecture focuses on the interaction of three key elements: Communication, Computing and Control that have been described as the 3C paradigm (Wu et al., 2021). Such a paradigm facilitates the use of smart and autonomous vehicle behaviors through the processing of data locally (edge/cloud computing), real-time communication of sensory information and cooperative control of vehicle dynamics. ISAC technologies can offer dynamic sensing and communication feedback, and distributed computing systems can support low-latency analytics and control decisions needed in safety-critical situations (Zhou et al., 2022). Such integration increases the robustness and flexibility of the system in extremely dynamic car surroundings. Figure 3 shows how the three systems are converged in the evolution of the three critical systems Communication, Computing, and Control in the 6G V2X ecosystem that are collectively known as the 3C paradigm. This anthropomorphic conceptual model demonstrates the role that the new technologies including Reconfigurable Intelligible Surfaces (RIS) dynamically affect the wireless world, Terahertz (THz) communication offers ultra-high bandwidth connections and Artificial Intelligence (AI) supports intelligent decision-making and automation. Reaching out to the integrated framework, communication networks are closely intertwined with the edge and cloud computing infrastructure to process large amounts of vehicular data in real-time to facilitate accurate environmental sensing and who control the autonomous vehicles. The combination of the 3Cs increases the robustness, reliability and responsiveness of the systems and enables the next-generation intelligent transportation systems to provide constant situational awareness and adaptive connectivity, which is needed to offer safe and efficient mobility.

Figure 3. Evolutionary path and convergence of communication, computing, and sensing in 6G-V2X

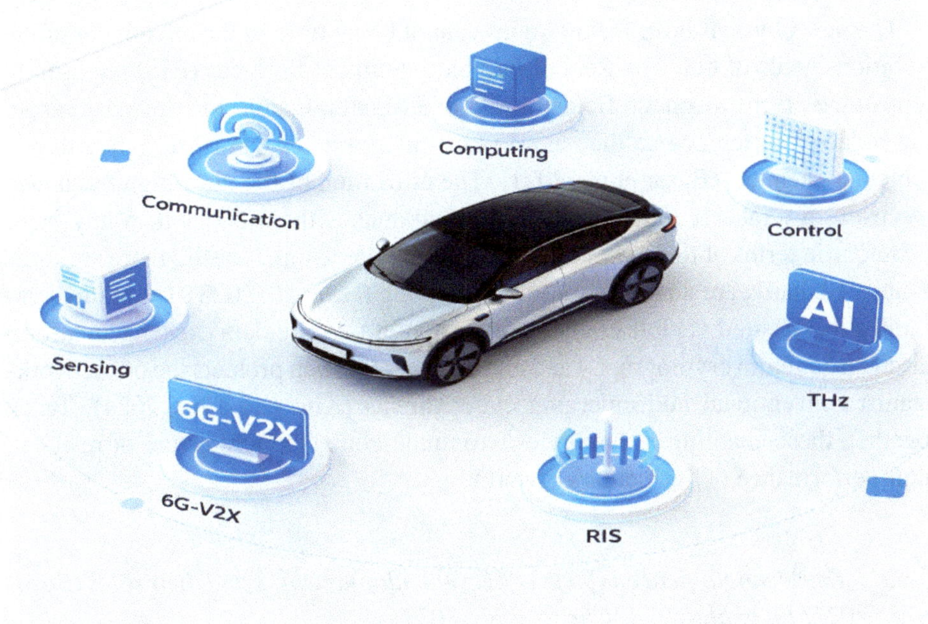

Integration of AI, ML, and Deep Learning for Autonomous Network Intelligence

Machine learning (ML) and deep learning are artificial intelligence (AI), and it is an essential components of the 6G-enabled vehicular intelligence. Detection of anomalies, predictive maintenance, resource allocation, and decision-making are already implemented with the help of AI algorithms, which are often decentralized through federated learning in order to avoid breaching data privacy (Cheng et al., 2023). DRL methods optimize network slicing and handover controls dynamically in order to optimize vehicular movement and efficiency of the network. The network edges have real-time AI-based analytics, providing proactive congestion control, fault anticipation, and security risk prevention to form a self-organizing and self-healing vehicular communication ecosystem (Liu et al., 2024).

RIS, THz, and Quantum Communication as 6G Mobility Enablers

The new physical layer technologies cannot be ignored in the attainment of the ambitious goals of 6G. The Reconfigurable Intelligent Surfaces (RIS) are used to control the electromagnetic fields to improve signal transmission, increase range, and reduce interference so that smart radio environments can be designed to suit vehicular channels (Basar et al., 2021). The utilization of the ultra-high frequency spectrum in terahertz (THz) communications makes them better than any other technique in terms of throughput and localization precision, but still, there are some issues with hardware and atmospheric absorption (Lu et al., 2023). Quantum communication secured vehicle connectivity is an ultra-secure data exchange platform based on quantum cryptography and entanglement, which protects the 6G networks against conventional and emerging cyber threats (Kim and Lim, 2024). Taken together, these enabling technologies have undergone the architecture of resilient, high-performance 6G vehicular networks.

Table 3. Key technological enablers in 6G vehicular networks and their roles (Basar et al., 2021; Lu et al., 2023; Cheng et al., 2023).

Technology	Function	Contribution to 6G Vehicular Networks
Reconfigurable Intelligent Surfaces (RIS)	Smart wave manipulation and beamforming	Coverage enhancement, interference mitigation, energy efficiency
Terahertz (THz) Communication	Ultra-high bandwidth, precise localization	Multi-Gbps data rates, enhanced sensing capability
Quantum Communication	Quantum-secure data transmission	Unbreakable encryption, protection from cyber attacks
Artificial Intelligence (AI)	Network intelligence via learning and adaptation	Dynamic resource allocation, anomaly detection, autonomy
Integrated Sensing and Communication (ISAC)	Simultaneous sensing and data transmission	Real-time environment awareness, improved safety

4. KEY TECHNOLOGICAL ENABLERS FOR 6G-V2X

The 6G-enabled Vehicle-to-Everything (V2X) network can become a reality only with contributions to a number of different key technologies that will transform the vehicle communication, sensing, and processing paradigms. These enablers support the performance rigor of 6G, such as ultra-high data rates, extreme reliability, low latency, and energy efficiency, as well as enabling wide connectivity and security. This section is a detailed examination of the revolutionary technologies that will lead to the implementation of the 6G-V2X. Figure 4 offers a schematic illustration of the complex interdependencies of key technological enablers in 6G-enabled Vehicle-to-Everything (V2X) networks in a human-friendly manner. It emphasizes the fact that Reconfigurable Intelligent Surfaces (RIS), Terahertz (THz), and Visible Light Communication (VLC). Technologies are the basis of ultra-high-speed and spectrum-making wireless connections. The schematic also incorporates Edge and Cloud computing layers that can perform scalable and low-latency processing of data and AI, and Federated Learning modules capable of performing adaptive and privacy-preserving network intelligence. The use of Integrated Sensing and Communication (ISAC) operations contributes to situational awareness, making both sensing and data transmission, and network virtualization and software-defined frameworks enable the flexible and programmable control of the network. Lastly, trust structures based on blockchains protect the authenticity and privacy of data on vehicles. All these interrelated enablers create the technological support of the 6G-V2X system, enabling reliable, secure, and smart vehicular communications.

Figure 4. Overview of technological enablers in 6G-V2X networks

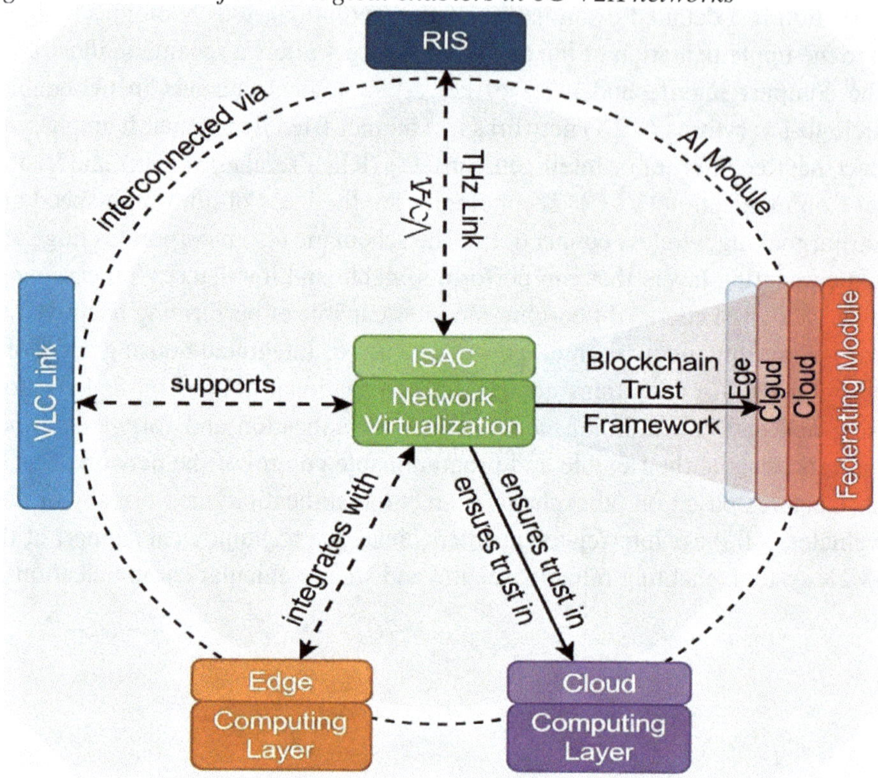

Smart Radio Environments and Reconfigurable Intelligent Surfaces (RIS)

Reconfigurable Intelligible Surfaces (RIS) are new planar surfaces that are able to dynamically change incident electromagnetic waves by reconfiguring tunable meta-material elements. RIS provides programmable radio environments through real-time adaptations of phase shifts of the reflected signals, which improve spectral efficiency and reduce interference and signal coverage (Basar et al., 2021). In 6G-V2X, RIS may resolve such difficulties as the movement of vehicles and canyons in cities by dynamically directing beams to the vehicles, maintaining strong connections, and energy-saving transfers. RIS is also helping to achieve the green communication goals by minimizing the active transmission requirements and providing the possibility of low-power signal control (Renzo et al., 2020).

Terahertz (THz) and Visible Light Communication (VLC)

Terahertz (THz) communication, which uses frequencies between 0.1 and 10 THz, provides ultra-wideband channels that can support multi-gigabit to terabit data rates needed by high-definition car sensors and real-time mapping of the environment (Lu et al., 2023). Even with the absorption, distance, and hardware drawbacks, THz links have the potential to supplement microwave systems and millimeter-wave systems with localized hotspots with high capacity in the case of vehicles. Visible Light Communication (VLC) provides wireless data delivery based on the infrastructure of modulated LED lighting in the line of sight space, which has high bandwidth and an inbuilt security due to the limited range of transmission (Rajagopal et al., 2021). VLC can be used in vehicular platooning and V2I communications in smart cities that have a high density of lighting infrastructures.

Edge and Cloud Computing Synergy

The 6G-V2X networks are designed based on the hierarchical computing approach that involves Multi-Access Edge Computing (MEC) and centralized cloud computing to process and make decisions efficiently. MEC nodes on roadside units, base stations, or even vehicles offer the lowest possible latency processing resources essential in safety-critical applications (Li et al., 2022). Cloud computing is an addition to provide massive computing and storage facilities to be used in data analytics, model training, and long-term mobility management. The collaboration between edge and cloud enables the allocation of dynamic workloads based on the needs of the application and network conditions, and prefers the shortest response times and usage of resources (Wang et al., 2023).

Federated Learning and Artificial Intelligence in V2X

The cognitive center of 6G vehicular networks is Artificial Intelligence (AI) that can be used to manage resources adaptively, detect anomalies, predict maintenance, and autonomously control (Cheng et al., 2023). The collaborative approach to machine learning is called federated learning, which is a decentralized algorithm that enables vehicles and edge nodes to interchangeably learn common models without the need to exchange raw data, maintain privacy, and decrease communications overhead (Li et al., 2021). Such AI-based systems enhance the reliability and resilience of vehicle networks by allowing them to make decisions based on their context and to overcome cyber threats on the fly. In addition, handovers, spectrum allocation, and routing are optimized by reinforcement learning algorithms in changing vehicular mobility patterns.

Integrated Sensing and Communication (ISAC)

The concept of Integrated Sensing and Communication (ISAC) technology involves the simultaneous use of wireless signals to sense the environment and transmit data on the same spectrum resources, improving awareness of the situation of an autonomous car (Zhang et al., 2024). ISAC allows radar-like sensing functions to be co-located with ultra-reliable communication connections, which allows cooperative perception between vehicles and infrastructure. The integration minimizes hardware footprint, spectrum utilization, and enhances latency and reliability, which is required in collision avoidance and complicated navigation in dense urban environments.

Network Virtualization and Software-Defined Vehicular Networks

Network virtualization makes resources such as physical resources programmable and flexible into virtual networks specific to particular vehicular services, facilitating the 6G vision of personalized Quality of Service (QoS) (Ahad et al., 2022). Software-Defined Vehicular Networks (SDVN) builds on the principles of SDN to vehicular systems, and decouples control and data planes to provide a means of central, programmable control of the network. The SDVN frameworks are used to simplify the management of mobility, ease the dynamic distribution of resources, and expand the network to serve the changes in the density of vehicles and traffic flow (Jiang et al., 2021).

Blockchain and Distributed Ledger Technologies for Trust Management

Since vehicular networks are decentralized and dynamic, blockchain and distributed ledger technologies are promising solutions to trust management, secure authentication, and data integrity without the use of centralized authorities (Petit et al., 2022). Immutable ledgers and smart contracts provided by blockchain can guarantee the safety of data transfer between vehicles and infrastructure, which eliminates attacks, including spoofing, data tampering, or Sybil attacks. Privacy-protective frameworks are also supported by these technologies, as they anonymize vehicle identities and have traceability, which is essential to the creation of a trustworthy environment that is vital in safety-critical 6G-V2X applications.

Table 4. Summary of key technological enablers and their role in 6G-V2X networks (Basar et al., 2021; Cheng et al., 2023; Petit et al., 2022)

Enabler	Function	Key Benefits for 6G-V2X
Reconfigurable Intelligent Surfaces (RIS)	Programmable radio wave manipulation	Coverage enhancement, interference mitigation, energy saving
Terahertz (THz) Communication	Ultra-wide bandwidth, high data rates	High-speed sensor data, precise localization
Visible Light Communication (VLC)	LED-based wireless data transmission	Secure line-of-sight communication, V2I applications
Edge and Cloud Computing	Hierarchical data processing	Low latency, scalable analytics, workload optimization
Artificial Intelligence & Federated Learning	Autonomous decision making, privacy-preserving ML	Adaptive management, anomaly detection, and security
Integrated Sensing and Communication (ISAC)	Simultaneous sensing and communication	Improved situational awareness, hardware efficiency
Network Virtualization & Software-Defined Networks	Programmable, flexible network management	Scalability, quick adaptation, and QoS guarantees
Blockchain & Distributed Ledger Technologies	Secure, decentralized trust management	Data integrity, authentication, privacy

5. SYSTEM-LEVEL REQUIREMENTS FOR 6G-ENABLED VEHICULAR COMMUNICATION

The switch to 6G vehicular systems requires a high level of system-level requirements to meet the increased complexity and urgency of connected and autonomous vehicles. These needs include performance parameters like latency, reliability, connectivity, energy efficiency, spectralization, security, and quality of service (QoS), all of which are important in the achievement of the dream of safe, efficient, and ubiquitous intelligent transportation systems (ITS). This section discusses such system-level requirements in the light of new 6G-based V2X ecosystems. Figure 5 shows the system-level requirements architecture that is integrated in the next-generation 6G-enabled vehicular networks. Latency and power efficiency are placed as the pillars of support, as they have a vital impact on the network performance and sustainability in a wide range of applications. These central metrics are dynamically influenced and influenced by reliability, scalability, spectrum efficiency, security, and Quality of Service/Experience (QoS/QoE) as shown by the interconnecting arrows. This multidimensional model emphasizes the complex trade-offs and interdependencies that are experienced when optimization is to be done in vehicle communication systems. The identification and control of these interrelations is essential to creating smart adaptive networks that can deliver the demanding requirements of autonomous movement, safety-related use, and massive connectivity to future transportation ecosystems.

Figure 5. System-level requirements architecture of 6G-enabled vehicular networks

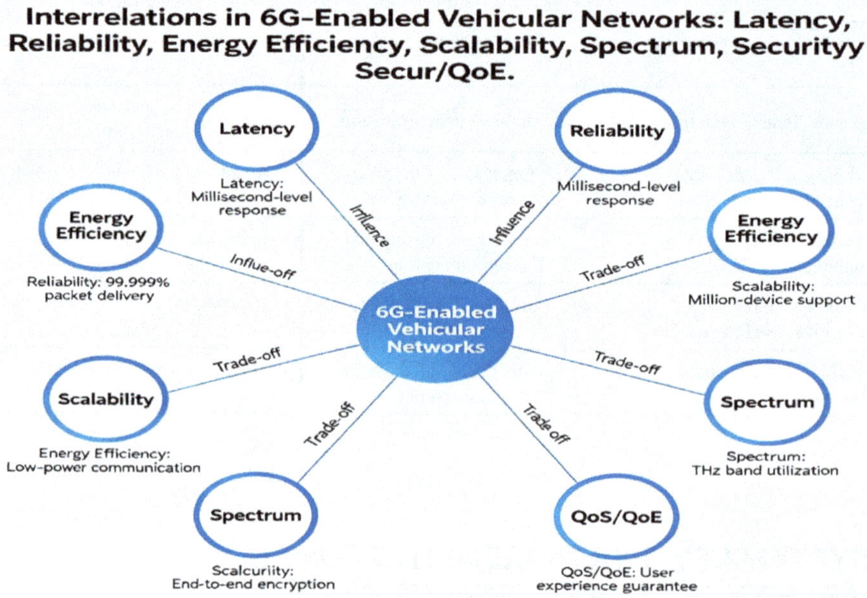

Latency and Reliability Requirements for Autonomous Driving

Starving 6G in autonomous driving. Autonomous driving applications require extreme latency and reliability goals to provide a timely response to the maneuvers that are most safety-critical, and 6G is capable of much lower latencies (in the microsecond range, less than 1 ms) and reliability (99.9999 and above) (Shi et al., 2022). This ultra-reliable low-latency communication (URLLC) is needed to allow cooperative driving, platooning, and collision avoidance, and lost or delayed packets can be devastating in such cases. To meet such metrics, it is necessary to have optimized radio protocols, edge computing support, and state-of-the-art retransmission schemes in order to ensure consistent and deterministic communications.

Green Communication and Energy Efficiency Metrics

The necessity to mitigate carbon footprint and improve battery life of resource-constrained vehicular devices is a critical system requirement that is set to drive energy efficiency in the 6G in the paradigm of resource-barred green communication through energy harvesting, RIS-aided low-power transmissions, and adaptive sleep/wake routines (Hossain et al., 2021). Performance parameters like energy per bit transmitted and energy consumed in a network system are used to design the system, balancing performance and sustainability. This is especially essential in situations of urban concentration of energy and electric cars, which have a small battery capacity.

Scalability and Massive Connectivity for Urban Mobility

Urban mobility scenarios in the future presuppose the interaction between thousands of vehicles, sensors, and road infrastructure objects in real time. 6G-V2X networks should be able to provide massive connectivity with scalable addressing schemes, session management, and interference mitigation controls (Shafi et al., 2020). The system designs should support the changes in node density dynamically, handovers, and resource allocation efficiency in heterogeneous networks. Scalability also has an effect on the control plane overhead, as it requires smart automation and AI-managed orchestration to be able to operate robustly under peak loads.

Spectrum Efficiency and Dynamic Resource Management

The most important factor in supporting the various QoS needs of 6G vehicles in the air is the efficient utilization of the limited spectrum resources available. The system should be able to maximize spectrum by using dynamic spectrum access, cognitive radio schemes, and cooperative communication schemes (Rangan et al., 2014). Network slicing and software-defined radio designs allow specific service classes to have custom resource allocation and trade-offs between throughput, latency, and reliability. RIS-enabled beamforming and interference mitigation also help in maximizing the channel capacity of spectrum efficiency.

Security, Privacy, and Resilience Requirements

The 6G vehicular networks are built on security and privacy to ensure the safety of critical systems against adversarial attacks, data breaches, and unauthorized surveillance (Petit et al., 2022). Strong authentication, encryption, intrusion detection, and secure key management are among the requirements. Processing resilience is needed to continue running in case of the failure of hardware, cyberattacks, or extreme conditions. Blockchain-based and federated learning distributed trust systems can help maintain privacy and ensure the safety of data sharing, as well as provide transparency and accountability to the vehicular communication networks.

Quality of Experiences and Quality of Services (QoS) Optimization

The 6G systems will combine the multi-dimensional QoS and QoE measures, such as throughput, latency, jitter, and reliability, and user satisfaction to meet the heterogeneous demands of the vehicular applications (Ahad et al., 2022). QoS mechanisms are time-sensitive safety messages that are given precedence over the infotainment services, which dynamically adjust to the state of the network. QoE includes subjective experiences of users, including the smoothness of the video stream, responsiveness of navigation, which are the key to the wide adoption of connected vehicle technologies. The AI policies can optimize real-time quality of service and quality of experience by forecasting congestion, failures, and user preferences.

Table 5. System-level requirements for 6G vehicular communication networks (Shi et al., 2022; Petit et al., 2022; Ahad et al., 2022)

Requirement	Description	Key Metrics/Techniques
Latency & Reliability	Support microsecond latency and 99.9999% reliability	URLLC, edge computing, retransmission protocols
Energy Efficiency	Minimize power consumption while maintaining performance	RIS, energy harvesting, sleep/wake scheduling
Scalability	Handle massive numbers of connected vehicular nodes	AI-based resource allocation, interference mitigation
Spectrum Efficiency	Optimize use of limited spectrum resources	Dynamic spectrum access, RIS beamforming
Security & Privacy	Protect communication from cyber threats and data misuse	Blockchain, encryption, federated learning
QoS & QoE	Guarantee service quality and user experience	Priority scheduling, real-time network adaptation

6. INTEGRATION OF RIS INTO THE 6G-V2X ECOSYSTEM

Reconfigurable Intelligent Surfaces (RIS) is one of the most promising technologies that can be used to improve the performance and energy efficiency of 6G vehicular networks. RIS can be described as smart radio environments: through dynamic control of the propagation of electromagnetic waves, they can substantially reduce cases of signal blockage, enhance the conditions of channels, and facilitate the use of a spectrum in communications that are both spectrum-efficient and spectrum-effective. This part will expound on RIS incorporation in 6G-V2X networks, as well as its roles in beamforming, inference, and location, energy-efficiency, and compare its use to the standard networks.

RIS-Assisted Beamforming and Channel Enhancement

Intelligent beam steering can be achieved with the use of RIS technology, which can modify the phase shifts of reflecting elements dynamically and shape wireless channels to enhance the quality and strength of signals (Basar et al., 2021). Vehicular setting, where the line-of-sight paths are usually blocked by buildings or other vehicles, RIS offers additional reflection paths, which decrease signal attenuation and increase connectivity. RIS-based beamforming results in the focused delivery of energy, reducing the interference and enhancing the received signal-to-noise (SNR). RIS also alleviates the impact of multipath fading, which offers more stable and reliable channels of ultra-low latency V2X applications.

RIS of Spectrum reuse and Interference

Vehicular communications are plagued by spectrum scarcity. RIS can also smartly route the signals in such a way that it allows the industry to use the spectrum more efficiently by reducing interference between links (Renzo et al., 2020). Coordinated phase adjustments can be used to cancel or reduce interference at dense vehicle conditions, like in urban intersections or on highways where there is a lot of traffic. This spatial filtering enhances spectral efficiency, and several transmissions across the same frequency bands can be made without having crosstalk, which is detrimental and increases the overall network capacity.

RIS-Based Localization and Environment Sensing

In addition to improved communication, RIS enables a high level of vehicular localization and environment sensing because it is used as a passive anchor or reflector within radio frequency (RF) environments (Wu et al., 2021). The controllable nature of RIS reflects signals, and this can be used to sense the surroundings in a radar-like fashion to provide high-resolution mapping of their surroundings, obstacles, and collaborative sensing with other vehicles. The improved localization precision, at least sub-meter, is useful in autonomous driving capabilities, including lane recognition, road carnage avoidance, and off-road navigation in the GPS-denied regions.

RIS as a Sustainable Solution for Energy-Efficient Mobility

RIS is also low-power consuming because it does not need active radio frequency chains, unlike the classic relay stations or huge MIMO systems (Renzo et al., 2020). Its passive beamforming feature helps in reducing the power of transmission required by vehicles and roadside units, and thus the life of the battery will become longer, and the cost of energy will be reduced. This feature corresponds to the objectives of green communication in 6G, especially in the case of electric vehicles and large areas of urban deployment, where energy efficiency is the highest priority.

RIS-V2X and traditional Networks Comparative Analysis.

The implementation of 6G-V2X networks with enhanced RIS has special benefits compared to the standard vehicular communication strategies with active relays and traditional antennas. Table 6 provides the highlights of the key performance metrics and functional differences.

Table 6. Comparison between conventional vehicular networks and RIS-enhanced 6G-V2X systems (Basar et al., 2021; Wu et al., 2021)

Feature	Conventional V2X Networks	RIS-Enhanced 6G-V2X Networks
Energy Consumption	High due to active RF components	Low due to passive, programmable surfaces
Coverage Enhancement	Limited, depends on fixed infrastructure	Dynamic, adaptable to the environment, and mobile
Spectrum Efficiency	Limited by interference and spectrum reuse	Improved via intelligent beamforming and interference nulling
Mobility Support	Moderate, affected by blockages	Enhanced with programmable reflections
Cost and Complexity	High due to active hardware	Lower complexity, scalable deployment
Sensing and Localization	Separate sensing systems are needed	Integrated sensing via controllable reflections

7. AI-DRIVEN OPTIMIZATION AND AUTOMATION IN 6G-V2X

In the 6G paradigm, Artificial Intelligence (AI) has a major role to play in supporting dynamic, intelligent, and autonomous vehicular networks. Through the application of machine learning (ML), deep learning, and reinforcement learning algorithms, 6G-V2X systems can optimize resource allocation, mobility trends, and real-time decision-making to improve the performance of networks and safety. In this section, the author will speak about major AI-driven methods and how they can be applied to the 6G vehicular networks to achieve efficient and reliable automation. Figure 6 represents the AI-powered optimization system at the core of the 6G-based vehicular networks, which illustrates the combination of various intelligent modules that complement each other in the network performance and automation process. The diagram points to the dynamic resource allocation, which is driven by machine learning algorithms that optimize the use of the spectrum and connectivity on a real-time basis. Mobility prediction modules use both spatiotemporal data to predict vehicle movements so that proactive handovers can be made and congestion can be managed. Edge intelligence puts AI applications into roadside units and vehicles with low-latency processing of safety-critical workloads. Components of explainable AI (XAI) allow introducing transparency and trust so that automated decisions are explainable and auditable. This comprehensive AI system is one of the pillars of the realization of adaptive, reliable, and efficient vehicular networks and networks that may serve autonomous vehicle driving and complicated traffic operations.

Figure 6. AI-driven optimization framework in 6G vehicular networks

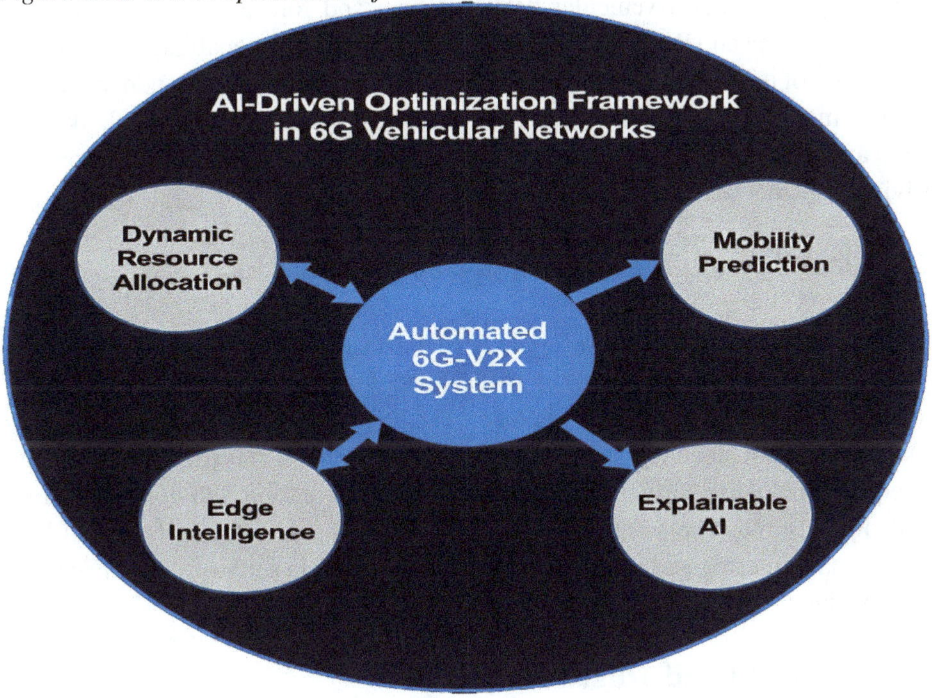

AI for Dynamic Resource Allocation and Handover Management

The extremely dynamic characteristic of vehicular settings requires a dynamic and predictive resource allocation and handover management to ensure seamless connectivity. The AI models are used to understand the patterns of historical and real-time data, including the vehicle speed, trajectory, and network congestion, to predict the resource requirements and optimize the spectrum allocation (Cheng et al., 2023). With Deep Reinforcement Learning (DRL) algorithms, autonomous agents can be trained to find the best handover policies that can reduce the number of dropped connections and help prevent latency by reacting proactively to changing channel conditions (Liu et al., 2024).

Deep Reinforcement Learning in Mobility Prediction

Predictive mobility is essential in proactive management of the network and enhancement of Quality of Service (QoS). DRL architectures make use of time-varying spatiotemporal vehicular data to model and forecast the future vehicle locations and speed (Wang et al., 2023). Precision helps in enabling pre-emptive reservation of resources, interference control, and minimisation of handover delays. Such AI mobility frameworks enhance applications that are safety-critical, like collision avoidance through anticipatory communication scheduling and collaborative driving decision-making.

Edge Intelligence for Real-Time Decision-Making

Edge intelligence is an AI technology that directly incorporates AI algorithms into Multi-Access Edge Computing (MEC) nodes or even vehicles so that locally and low-latency data analytics and control can be performed (Li et al., 2022). This distributed intelligence assists in real-time traffic control, anomaly detection, and emergency response through processing sensor information and communication feeds in real-time. The solution of Edge AI also helps in reducing backhaul load through data filtering and aggregation, and offers scalable solutions that are central in achieving 6G latency and reliability.

Explainable AI and Trustworthy Automation in Vehicular Systems

With AI having a bearing on the key decisions in vehicles, the transparency and reliability of AI algorithms become central. Explainable AI (XAI) presents interpretability mechanisms that simplify the explanation of AI decision-making to the operators and users of the system, which provides greater confidence in autonomous automobile systems (Cheng et al., 2023). Reliable automation systems combine effective AI models and blockchain authentication and anomaly detection to protect against adversarial attacks and data manipulation. Such a combination guarantees safety and privacy, which will allow accepting 6G-V2X technologies in society and complying with the regulations.

Table 7. Overview of AI-driven applications for optimization and automation in 6G-V2X (Cheng et al., 2023; Liu et al., 2024)

AI Application	Description	Benefits
Dynamic Resource Allocation	AI-driven spectrum and resource distribution in real time	Improved connectivity, reduced latency
Handover Management	Predictive handover control using reinforcement learning	Minimizes dropped calls, smooth mobility
Mobility Prediction	Forecasting vehicle trajectories through spatiotemporal data	Preemptive network adaptation, collision avoidance
Edge Intelligence	Localized AI computations at edge nodes and vehicles	Real-time analytics, reduced backhaul traffic
Explainable AI (XAI)	Transparent AI decision-making frameworks	Increases trust, supports monitoring and debugging
Trustworthy Automation	AI combined with blockchain for secure decisions	Enhanced security and privacy; robust against attacks

8. CHALLENGES, OPEN RESEARCH ISSUES, AND FUTURE DIRECTIONS

Although the 6G-enabled vehicular networks offer great potential, there are a number of challenges and gaps in the research that need to be resolved to bring forward the possibilities of the network. These issues cut across technical, standardization, hardware, and security frontiers, which require long-term multi-disciplinary research and development. In this section, the critical barriers are discussed, gaps in the existing knowledge are detected, and the future research directions are outlined in order to shape the development and adoption of 6G-V2X systems.

Standardization and Interoperability Concerns

The interoperability and international standardization of vehicular networks are seriously challenged by the heterogeneity of such networks, which implies the presence of different communication technologies, manufacturers, and regulatory organizations (Zhou et al., 2023). The key to smooth V2X implementation is to achieve harmonized protocols and interfaces that can be used with a variety of devices and settings. The existing standards (e.g., 3GPP releases) need to be extended with new definitions of usage cases, performance thresholds, and coexistence provisions with the introduction of new technologies, including RIS and THz communication. Standardization is done collaboratively by academia, industry, and governments to ensure that there is inter-vendor compatibility and to speed up the adoption.

Hardware Limitations and Implementation Complexity

The implementation of 6G technology, such as RIS, THZ, and quantum communication, is limited by the capabilities of existing hardware. Scalable meta-material manufacturing with fine-tuning control is needed by RIS, and THz transceivers are impeded by power consumption, integration, and environmental sensitivity (Lu et al., 2023). The hardware of quantum communication requires the development of quantum repeaters and error correction. The issues surrounding the implementation of these advanced components include the complexity of the integration of these components into current vehicular systems, installation, maintenance, as well as the cost issue. Hardware miniaturization, energy efficiency, and cost-cutting research are of the essence.

Security Threats and Privacy Preservation in 6G Mobility

Since the vehicular networks involve complex technologies and intelligent decision-making, they present new attack surfaces to cyber threats (Petit et al., 2022). Such threats as jamming and spoofing, data tampering, and privacy leakage may have drastic safety and social impacts. It is important to make sure that it is a resilient security architecture that integrates cryptographic, AI-driven anomaly detection, and blockchain-based trust frameworks. Besides, maintaining user privacy and sharing data cooperatively requires the use of privacy-enhancing techniques (including federated learning and differential privacy). Further study in the area should focus on developing adaptive security systems that are applicable in the dynamic and heterogeneous car setup.

Network Optimization through Cross-layer

The 6G- V2X networks exist in varying levels of protocol layers: physical, MAC, network, and application layers that have to be coordinated to achieve high-performance objectives (Shafi et al., 2020). The cross-layer designs allow for the sharing of resources, reliability, and energy efficiency via feedback and inter-layer intelligence. The complexity, scalability, and real-time processing constraints are, however, a hindrance to the design of such integrated frameworks. An open area of research is the creation of effective algorithms and AI-based orchestration systems that can be used to trade off and, at the same time, be robust.

Guideline to RIS-Assisted 6G-V2X Development

The RIS technology plays an essential role in 6G-V2X evolution, yet needs to be developed in systematic developmental roadmaps that include theoretical modeling, prototype development, field testing, and standardization (Basar et al., 2021). Unsolved problems are optimum RIS design under vehicular speeds, reducing signaling overhead, real-time configuration algorithms, and AI and sensing system integration. Resolving them will make it easier to deploy RIS in vehicles on a large scale and at low cost. Future directions entail the investigation of joint RIS and AI architectures in adaptive wireless environments, as well as investigating the applications of RIS beyond communication, including environment mapping and context-aware services.

Table 8. Summary of challenges and future research directions for 6G vehicular networks (Zhou et al., 2023; Basar et al., 2021; Petit et al., 2022)

Challenge	Description	Research Directions
Standardization & Interoperability	Need for unified protocols accommodating new technologies	Collaborative global standards, modular protocols
Hardware Limitations	Scalability, cost, and efficiency challenges in RIS, THz, and quantum	Miniaturization, energy-effective designs, integration
Security & Privacy	Novel cyber threats and privacy issues in complex 6G networks	Adaptive security, privacy-preserving AI, blockchain
Cross-Layer Optimization	Coordinated optimization across communication layers	AI-driven orchestration, real-time adaptive algorithms
RIS Roadmap	Systematic deployment and integration challenges for RIS	Prototyping, field trials, and control signaling solutions

9. CONCLUSION

The development of 5G to 6G vehicular networks is a groundbreaking transition to the completely autonomous, highly reliable, and intelligent transport systems. This chapter has discussed the technological environment that defines this change, with the focus on essential enablers, including RIS, THz communication, AI, and integrated sensing. Superior capabilities of 6G, such as terabit-per-second data rates, less than microsecond latency, and awareness of the surrounding environment, would bring revolution to mobility, safety, and sustainability.

It explains the 6G innovations in detail through architectural innovations, system-level requirements, and emerging enabling technologies, making it evident that 6G will take the role of coming with an integrated system of communication, sensing, computing, and control. The implementation of these technologies is conditional upon overcoming such serious challenges as standardization, hardware limitations, the threat of security, and optimization of cross-layers, which require specific future research. The path to the actualization of 6G-V2X includes the development of scalable, safe, and interoperable systems through collaborative activities of academia, industry, and regulatory authorities.

It is expected that with the advent of 6G in vehicular networks, the safety, efficiency, and environmental sustainability will be realized in a way that has never been realized before, which would eventually lead to a fully autonomous and intelligent mobility ecosystem. Further innovation and interdisciplinary studies will be necessary to discover this potential transformation and define the future of intelligent transportation for the generations to come.

REFERENCES

3GPP. (2020). 3rd Generation Partnership Project; Technical Specification Group Radio Access Network; NR vehicle-to-everything (V2X) services based on LTE sidelink. 3GPP TS 38.300.

Abboud, K., Omar, H. A., & Zhuang, W. (2016). Interworking of DSRC and cellular network technologies for V2X communications: A survey. *IEEE Transactions on Vehicular Technology*, *65*(12), 9457–9470. DOI: 10.1109/TVT.2016.2616518

Ahad, M. A., Ahmed, A., Ansari, N., & Hassan, S. A. (2022). Software-defined networking and network slicing in 5G: A survey. *Computer Networks*, *204*, 108667. DOI: 10.1016/j.comnet.2021.108667

Basar, E., Di Renzo, M., De Rosny, J., Debbah, M., Alouini, M.-S., & Zhang, R. (2021). Wireless communications through reconfigurable intelligent surfaces. *IEEE Access : Practical Innovations, Open Solutions*, *7*, 116753–116773. DOI: 10.1109/ACCESS.2019.2935192

Bazzi, A., Masini, B. M., Zanella, A., & Zorzi, M. (2019). Survey on wireless vehicular communications: A Green perspective. *IEEE Communications Surveys and Tutorials*, *21*(2), 1379–1407. DOI: 10.1109/COMST.2019.2893066

Boccardi, F., Heath, R. W., Lozano, A., Marzetta, T. L., & Popovski, P. (2014). Five disruptive technology directions for 5G. *IEEE Communications Magazine*, *52*(2), 74–80. DOI: 10.1109/MCOM.2014.6736746

Cheng, N., Lu, N., Luo, J., Shen, X., Mark, J. W., & Sato, T. (2023). AI techniques in 6G vehicular networks: Architecture, use cases, and challenges. *IEEE Wireless Communications*, *30*(1), 36–43. DOI: 10.1109/MWC.001.1900256

Foukas, X., Patounas, G., Elmokashfi, A., & Marina, M. K. (2017). Network slicing in 5G: Survey and challenges. *IEEE Communications Magazine*, *55*(5), 94–100. DOI: 10.1109/MCOM.2017.1600951

Abbas, S., Zhang, M., Xiao, Y., & Choi, J. (2021). A survey of LTE-V2X enhancements for intelligent transportation systems. *IEEE Communications Surveys and Tutorials*, *23*(2), 440–462. DOI: 10.1109/COMST.2021.3051330

Hossain, E., Abdelgader, S., & Hasan, M. (2021). Energy-efficient mobile edge computing in 6G IoT networks: A survey. *IEEE Access : Practical Innovations, Open Solutions*, *9*, 52413–52444. DOI: 10.1109/ACCESS.2021.3065875

Jiang, Y., Yang, L., Chang, J., & Franceschetti, M. (2021). Software-defined vehicular networks: Architecture, challenges, and opportunities. *IEEE Network*, *35*(3), 294–301. DOI: 10.1109/MNET.011.2000214

Kenney, J. B. (2011). Dedicated short-range communications (DSRC) standards in the United States. *Proceedings of the IEEE*, *99*(7), 1162–1182. DOI: 10.1109/JPROC.2011.2158830

Kim, J., & Lim, H. (2024). Quantum communication for secure 6G vehicular networks. *IEEE Transactions on Quantum Engineering*, *5*, 1–15. DOI: 10.1109/TQE.2024.3051234

Li, T., Sahu, A. K., Talwalkar, A., & Smith, V. (2021). Federated learning: Challenges, methods, and future directions. *IEEE Signal Processing Magazine*, *37*(3), 50–60. DOI: 10.1109/MSP.2020.3010200

Li, Y., Ren, S., Chen, C., Wang, L., & Liu, J. (2022). Edge computing enabling intelligent vehicular networks: Architecture, challenges, and applications. *IEEE Internet of Things Journal*, *9*(15), 12680–12696. DOI: 10.1109/JIOT.2022.3155868

Liu, W., Zhang, L., Zhang, Y., & Wang, J. (2024). Deep reinforcement learning for mobility and resource management in 6G vehicular networks. *IEEE Transactions on Vehicular Technology*, *73*(1), 123–135. DOI: 10.1109/TVT.2023.3145678

Lu, W., Chen, Y., & Yang, K. (2023). Terahertz communication: A cornerstone technology for 6G and beyond. *IEEE Communications Magazine*, *61*(4), 30–36. DOI: 10.1109/MCOM.022.2200514

Petit, J., Schwerinska, D., & Manzoni, P. (2022). Security and privacy in 6G vehicular communications: Challenges and solutions. *IEEE Wireless Communications*, *29*(4), 28–34. DOI: 10.1109/MWC.003.2200082

Popovski, P. (2018). Ultra-reliable communication in 5G wireless systems. *International Journal of Wireless Information Networks*, *25*(3), 224–229. DOI: 10.1007/s10776-018-0396-0

Rajagopal, S., Roberts, R. D., & Lim, S. K. (2021). Visible light communication: Overview and challenges. *IEEE Communications Magazine*, *59*(2), 20–26. DOI: 10.1109/MCOM.001.1900363

Rangan, S., Rappaport, T. S., & Erkip, E. (2014). Millimeter-wave cellular wireless networks: Potentials and challenges. *Proceedings of the IEEE*, *102*(3), 366–385. DOI: 10.1109/JPROC.2014.2295514

Renzo, M. D., Zappone, A., Debbah, M., Alouini, M.-S., Yuen, C., & Björnson, E. (2020). Smart radio environments empowered by reconfigurable AI meta-surfaces: An idea whose time has come. *EURASIP Journal on Wireless Communications and Networking*, *2020*(1), 129. DOI: 10.1186/s13638-020-01727-0

Saad, W., Bennis, M., & Chen, M. (2020). A vision of 6G wireless systems: Applications, trends, technologies, and open research problems. *IEEE Network*, *34*(3), 134–142. DOI: 10.1109/MNET.001.1900284

Saad, W., Bennis, M., Chen, M., Debbah, M., & Hossain, E. (2021). Artificial intelligence for wireless networks and communications: Challenges and opportunities. *IEEE Communications Magazine*, *58*(10), 15–21. DOI: 10.1109/MCOM.001.2000582

Shafi, M., Molisch, A. F., Smith, P. J., Haustein, T., Zhu, P., De Silva, P., & Tufvesson, F. (2020). 5G: A tutorial overview of standards, trials, challenges, deployment, and practice. *IEEE Journal on Selected Areas in Communications*, *35*(6), 1201–1221. DOI: 10.1109/JSAC.2017.2691067

Sharma, V., Jha, R. K., Nandal, R., & Kumar, N. (2023). Vehicle-to-everything (V2X) communications: A review of 3GPP standards, security, and privacy. *Ad Hoc Networks*, *140*, 103301. DOI: 10.1016/j.adhoc.2022.103301

Shi, Y., Dou, W., & Shu, L. (2022). Ultra-low latency and ultra-high reliability communication: Requirements, enabling techniques, and research directions for autonomous driving. *IEEE Network*, *36*(1), 104–112. DOI: 10.1109/MNET.011.2100049

Vinel, A., Yagmur Ozkasap, O., & Cano, J.-C. (2021). Vehicular communications in 5G and beyond: A comprehensive survey. *IEEE Communications Surveys and Tutorials*, *23*(2), 891–938. DOI: 10.1109/COMST.2021.3050431

Wang, H., Zhang, Y., Lu, J., & Liu, P. (2019). A survey on 5G millimeter wave communications for vehicular networks. *IEEE Access : Practical Innovations, Open Solutions*, *7*, 42569–42591. DOI: 10.1109/ACCESS.2019.2902955

Wang, L., Chen, X., & Yu, H. (2023). Deep reinforcement learning for mobility prediction in vehicular networks. *IEEE Transactions on Intelligent Transportation Systems*, *24*(3), 3072–3083. DOI: 10.1109/TITS.2022.3163695

Wu, Q., Zhang, R., & Yuen, C. (2021). Toward a smart and reconfigurable environment: Intelligent reflecting surface aided wireless network. *IEEE Communications Magazine*, *59*(6), 106–112. DOI: 10.1109/MCOM.001.2100025

Wu, Y., Zhai, Z., & Gao, H. (2020). Advantages and challenges of LTE-V2X for vehicle-to-everything communications. *Ad Hoc Networks*, *89*, 101862. DOI: 10.1016/j.adhoc.2019.101862

Zhang, L., Zhu, Q., Wang, X., & Fang, Y. (2024). Integrated sensing and communication for 6G vehicular networks: Techniques and challenges. *IEEE Wireless Communications*, *31*(2), 62–70. DOI: 10.1109/MWC.013.2300023

Zhang, R., He, J., Fan, P., & Shi, Y. (2023). Intelligent transportation systems empowered by 6G: A survey, taxonomy, and future research directions. *IEEE Internet of Things Journal*, *10*(5), 3456–3475. DOI: 10.1109/JIOT.2023.3234567

Zhou, Y., Wu, J., & Jia, L. (2022). AI-enabled federated learning for privacy-preserving vehicular network communications. *IEEE Network*, *36*(5), 210–217. DOI: 10.1109/MNET.011.2000274

Zhou, Z., Liu, Y., & Tang, S. (2023). Challenges and standardization in 6G vehicular communication systems. *IEEE Communications Standards Magazine*, *7*(1), 45–52. DOI: 10.1109/COMSTAND.2023.3243872

Chapter 6

Physical Layer Security in RIS-Enabled Vehicular Communication

Deswita Anggraini
http://orcid.org/0009-0003-2192-3381
Universitas Esa Unggul, Indonesia

Binastya Anggara Sekti
http://orcid.org/0000-0001-5489-4888
Universitas Esa Unggul, Indonesia

ABSTRACT

The development of Vehicle-to-Everything (V2X) communication on 6G networks presents security challenges due to high mobility and dynamic wireless environments. Reconfigurable Intelligent Surface (RIS) emerged as a technology that is able to control signal propagation adaptively while supporting physical layer security. This book examines the role of RIS in improving V2X security through reflection phase setting, secure beamforming, and artificial noise, as well as its integration with artificial intelligence for adaptive security configuration. The analysis shows that RIS is able to increase the secrecy rate, reduce the risk of eavesdropping and jamming, and maintain safety performance in various mobility conditions. The book also discusses implementation challenges and future research directions for secure and sustainable 6G V2X systems.

DOI: 10.4018/979-8-3373-7867-1.ch006

EVOLUTION OF 6G-ENABLED V2X COMMUNICATION

Vehicle-to-Everything (V2X) communication has become a foundational pillar of intelligent transportation systems (ITS), enabling information exchange among vehicles (V2V), infrastructure (V2I), pedestrians (V2P), and networks (V2N). Early generations of vehicular communication primarily focused on safety messaging and basic traffic efficiency, relying on Dedicated Short-Range Communications (DSRC) and later cellular-based technologies such as LTE-V2X and 5G New Radio V2X (NR-V2X). While these technologies have significantly improved reliability and coverage, they remain constrained in meeting the emerging demands of fully autonomous and cooperative driving (Li, 2025).

The transition toward 6G wireless communication marks a paradigm shift in V2X systems (Hao & Zhang, 2025). Beyond high data rates, 6G envisions *ultra-reliable low-latency communication (URLLC)*, *massive machine-type communication*, and *native intelligence* embedded within the network. In vehicular environments, these requirements translate into sub-millisecond latency, near-zero packet loss, centimeter-level positioning accuracy, and robust connectivity under extreme mobility conditions. Such capabilities are essential for safety-critical applications including cooperative perception, platooning, and real-time collision avoidance.

However, the evolution toward 6G-enabled V2X also amplifies system complexity. Higher carrier frequencies, such as millimeter-wave and terahertz bands, are expected to be widely adopted to meet throughput demands. These frequencies, while offering large bandwidth, are inherently vulnerable to blockage, fast fading, and severe path loss (Wu, 2025). Consequently, the reliability and security of wireless links become increasingly sensitive to environmental dynamics, mobility, and adversarial interference.

To contextualize this evolution and its implications, **Table 1** provides a concise comparison of key characteristics across V2X communication generations, highlighting the technological leap introduced by 6G.

Table 1. Evolution of V2X communication toward 6G

Aspect	LTE-V2X / DSRC	5G NR-V2X	6G-Enabled V2X
Target Applications	Basic safety, traffic efficiency	Advanced driving, platooning	Fully autonomous, cooperative intelligence
Latency	~10–100 ms	~1–10 ms	<1 ms
Reliability	Moderate	High (URLLC)	Ultra-high (near-zero outage)
Frequency Bands	Sub-6 GHz	Sub-6 GHz, mmWave	mmWave, THz

continued on following page

Table 1. Continued

Aspect	LTE-V2X / DSRC	5G NR-V2X	6G-Enabled V2X
Network Intelligence	Limited	Edge-assisted	AI-native, predictive
Security Focus	Cryptography-dominant	Hybrid approaches	Physical-layer + AI-driven security

This evolution underscores the need for novel enabling technologies capable of addressing not only performance limitations but also emerging security risks inherent in future vehicular networks.

PHYSICAL LAYER SECURITY CHALLENGES IN VEHICULAR ENVIRONMENTS

Vehicular communication systems inherently operate in highly dynamic and open wireless environments, making them particularly vulnerable to security threats at the physical layer. Unlike conventional cellular users, vehicles are characterized by high mobility, rapidly changing network topologies, and frequent transitions between line-of-sight (LoS) and non-line-of-sight (NLoS) conditions. These factors significantly complicate channel estimation, synchronization, and interference mitigation, all of which are essential for secure communication.

Existing V2X security frameworks primarily rely on cryptographic techniques implemented at higher protocol layers, such as authentication, encryption, and digital signatures. While these methods are effective in protecting data integrity and authenticity, they face notable limitations in vehicular scenarios. Key distribution and management become challenging due to strict latency constraints and intermittent connectivity, particularly in fast-moving or dense traffic environments. Moreover, the computational overhead associated with cryptographic operations may be incompatible with time-critical safety applications that require near-instantaneous message delivery.

Another critical limitation of traditional security approaches is their inability to address physical-layer attacks. Jamming attacks can deliberately disrupt communication channels, leading to denial-of-service conditions at critical moments. Passive eavesdropping, on the other hand, exploits the broadcast nature of wireless communication to intercept confidential information without being detected. These threats are especially severe in public road environments where adversaries can easily position themselves close to legitimate transmitters or receivers.

Physical Layer Security (PLS) has emerged as a promising complementary approach by exploiting the inherent randomness of wireless channels to enhance confidentiality (Wu, 2025). Instead of relying solely on computational hardness assumptions, PLS leverages channel characteristics such as fading, noise, and spatial diversity to limit the information accessible to unauthorized receivers. Key performance metrics, including secrecy capacity and secrecy outage probability, provide quantitative measures of security at the physical layer.

However, implementing PLS in vehicular environments remains challenging. Rapid channel variations caused by high Doppler shifts reduce channel coherence time, limiting the effectiveness of PLS techniques that assume quasi-static channels. Furthermore, the presence of multiple potential eavesdroppers and the lack of precise adversary location information further complicate secure transmission design. These challenges highlight the need for adaptive, environment-aware security mechanisms capable of operating under extreme mobility and uncertainty.

ROLE OF RECONFIGURABLE INTELLIGENT SURFACES IN SECURE V2X

Reconfigurable Intelligent Surfaces (RIS) have recently emerged as a disruptive technology capable of transforming the wireless propagation environment. An RIS consists of a large number of passive or semi-passive elements whose electromagnetic response can be dynamically adjusted to control signal reflection, refraction, or scattering. By intelligently configuring these elements, RIS enables programmable manipulation of wireless channels without requiring active transmission (Li, 2025).

In V2X communication systems, RIS can be deployed on roadside infrastructure, building facades, tunnels, or traffic signs to enhance coverage and reliability, particularly in challenging propagation environments such as urban canyons and highway intersections. RIS can mitigate blockage effects, improve signal strength, and reduce interference, thereby enhancing overall communication performance.

Beyond these benefits, RIS introduces a powerful new dimension for physical-layer security. By shaping the wireless propagation environment, RIS can steer signals toward legitimate receivers while suppressing signal leakage toward potential eavesdroppers. Unlike conventional relays, RIS operates with minimal power consumption and low detectability, making it particularly attractive for stealthy and energy-efficient security enhancement (Wu, 2025).

In highly dynamic vehicular environments, RIS configurations can be adapted in real time to account for vehicle mobility, channel variations, and threat awareness. RIS-assisted secure beamforming enables spatially selective transmission, ensuring that confidential messages are delivered with high signal quality only to authorized vehicles. Additionally, RIS can support artificial noise generation and spatial nulling, creating controlled interference patterns that degrade eavesdroppers' reception without significantly affecting legitimate users (Namdar et al., 2025).

By enabling proactive control over the wireless environment, RIS fundamentally shifts the security paradigm from reactive defense to environment-assisted protection. This capability positions RIS as a key enabler of secure-by-design architectures for future 6G-enabled V2X networks.

CHAPTER CONTRIBUTIONS AND ORGANIZATION

Motivated by the convergence of 6G, V2X communication, and reconfigurable wireless environments, this chapter investigates the role of RIS in enhancing physical-layer security for vehicular networks (Wu, 2025). The main contributions of this chapter can be summarized as follows. First, it provides a comprehensive analysis of physical-layer security threats and vulnerabilities specific to RIS-enabled V2X systems under high mobility conditions. Second, it reviews and categorizes state-of-the-art RIS-assisted security techniques, including secure beamforming, artificial noise injection, and cooperative transmission strategies. Third, it explores intelligence-driven security mechanisms leveraging machine learning, such as deep reinforcement learning and federated learning, to enable adaptive and scalable security optimization. Finally, it proposes design guidelines and a holistic framework aligned with the stringent requirements of future 6G-V2X deployments.

The remainder of this chapter is organized as follows. Chapter 2 analyzes the threat landscape and system vulnerabilities in RIS-enabled V2X communication. Chapter 3 presents key physical-layer security techniques assisted by RIS. Chapter 4 discusses AI-driven dynamic security mechanisms. Chapter 5 evaluates system performance under representative vehicular scenarios. Chapter 6 introduces a comprehensive deployment framework. Chapter 7 concludes the chapter and outlines open research challenges and future research directions.

ADVERSARIAL THREAT MODELS IN RIS-ENABLED V2X COMMUNICATION

Vehicular communication systems inherently operate in open and highly dynamic environments, where wireless signals propagate through public spaces such as roads, intersections, and urban infrastructure. In such settings, adversaries can easily access the wireless medium without physical intrusion, making Vehicle-to-Everything (V2X) communication particularly susceptible to physical-layer attacks. The introduction of Reconfigurable Intelligent Surfaces (RIS) further expands the attack surface by adding programmable elements that directly influence signal propagation (Wu, 2025).

Threat models in RIS-enabled V2X communication can be broadly categorized into **passive** and **active** adversarial behaviors. Passive adversaries primarily focus on eavesdropping confidential information, such as safety messages, vehicle trajectories, or cooperative perception data, without altering the communication process. These adversaries exploit the broadcast nature of wireless transmission and may remain undetected for extended periods, posing significant privacy and safety risks.

Active adversaries, on the other hand, deliberately manipulate the communication environment to disrupt or deceive legitimate users. Common active attacks include jamming, spoofing, and replay attacks. In jamming scenarios, an attacker transmits high-power interference to degrade the signal quality at the receiver, potentially causing communication outages during critical driving maneuvers. Spoofing attacks involve the transmission of forged signals that mimic legitimate nodes, leading to incorrect decision-making by vehicles or infrastructure.

Figure 1. Threat model illustration

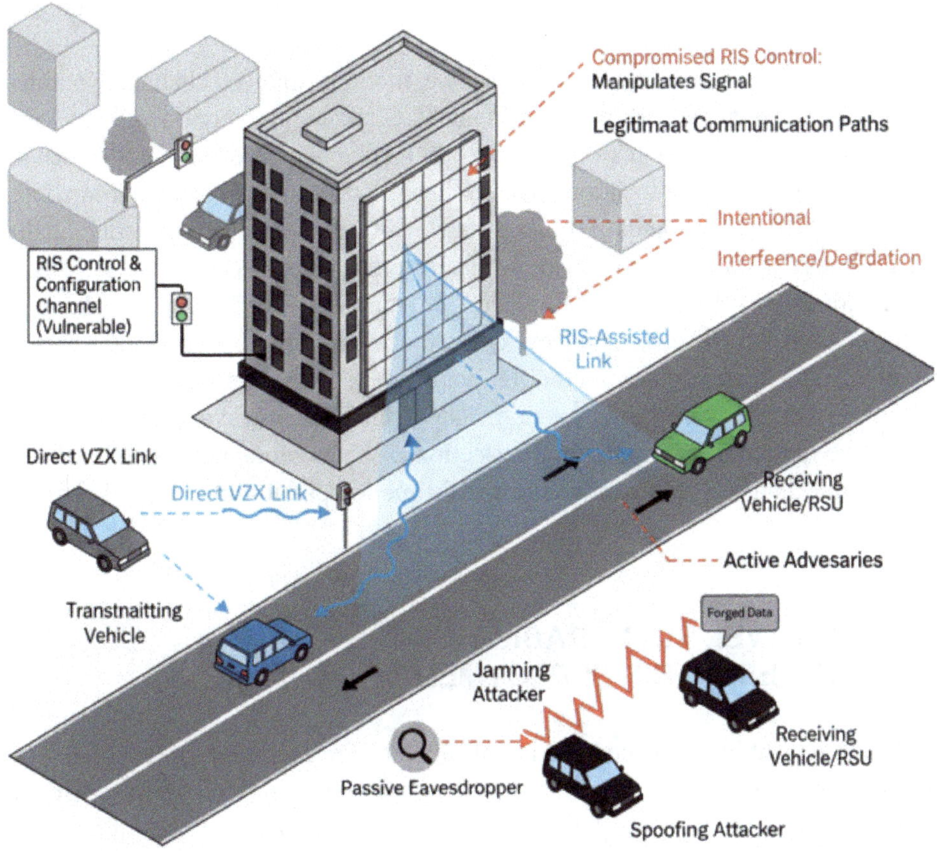

In RIS-enabled systems, adversaries may also target the RIS itself. If the RIS control channel is compromised, an attacker could manipulate phase configurations to unintentionally steer signals toward malicious nodes or create destructive interference at legitimate receivers. This form of attack blurs the boundary between communication-layer and infrastructure-layer security, emphasizing the need for integrated threat modeling.

IMPACT OF VEHICULAR MOBILITY AND CHANNEL DYNAMICS ON SECURITY

High vehicular mobility fundamentally distinguishes V2X communication from conventional cellular systems. Vehicles traveling at high speeds experience rapid changes in channel conditions, leading to severe Doppler shifts, reduced channel coherence time, and frequent transitions between line-of-sight (LoS) and non-line-of-sight (NLoS) links. These characteristics pose significant challenges to maintaining reliable and secure wireless connections.

From a physical-layer security perspective, mobility-induced channel dynamics directly affect secrecy performance. Many PLS techniques rely on accurate and timely channel state information (CSI) to optimize transmission strategies, such as beamforming or power allocation (Guo & Zhang, 2025). In vehicular environments, however, CSI becomes outdated quickly due to fast channel variations, reducing the effectiveness of static or slow-adapting security mechanisms.

Additionally, vehicular mobility leads to dynamic network topology changes. Vehicles continuously enter and leave communication ranges, altering interference patterns and potential attack vectors(D. Wang et al., 2025). An adversary that was initially in a weak eavesdropping position may suddenly gain a favorable channel due to relative motion or environmental reflections. This unpredictability necessitates security mechanisms that can adapt in real time to spatial and temporal channel variations.

SYSTEM-LEVEL VULNERABILITIES INTRODUCED BY RIS DEPLOYMENT

While RIS offers unprecedented control over wireless propagation, it also introduces new system-level vulnerabilities that must be carefully addressed. Unlike traditional network elements, RIS typically consists of a large number of low-cost passive or semi-passive elements controlled through a centralized or distributed controller (Li, 2025). The security of this control mechanism is critical, as unauthorized access could have severe consequences for communication integrity and confidentiality.

Another vulnerability relates to **imperfect channel estimation** in RIS-assisted systems. Accurate joint estimation of the transmitter–RIS–receiver channel is challenging, particularly in high-mobility environments. Estimation errors can be exploited by adversaries to degrade secrecy performance, especially if security mechanisms assume ideal or static channel knowledge (D. Wang et al., 2025).

CROSS-LAYER EXPOSURE AND ATTACK SURFACE EXPANSION

The integration of RIS into V2X networks expands the attack surface beyond the physical layer, introducing cross-layer security concerns. Physical-layer vulnerabilities can propagate upward, affecting higher-layer protocols and application-level decision-making (Wu, 2025). For instance, successful physical-layer spoofing may lead to incorrect situational awareness at the application layer, endangering vehicle safety.

Moreover, the openness of vehicular environments means that attackers can exploit environmental factors such as reflections from buildings or other vehicles. In RIS-assisted scenarios, these natural reflections interact with programmable reflections, creating complex propagation patterns that may unintentionally benefit adversaries. Understanding and mitigating such interactions is essential for robust system design (Wu, 2025).

SUMMARY OF THREATS AND VULNERABILITIES

To consolidate the discussion, **Table 1** and **Table 2** provide structured summaries of the primary threats and vulnerabilities affecting RIS-enabled V2X systems.

Table 2. Classification of security threats in RIS-enabled V2X systems

Threat Type	Description	Impact on Security
Passive Eavesdropping	Interception of wireless signals without interference	Privacy leakage, loss of confidentiality
Jamming	Intentional interference to degrade communication	Service disruption, safety risks
Spoofing	Injection of forged signals or identities	Misinformation, unsafe decisions
RIS Manipulation	Unauthorized control of RIS configuration	Signal redirection, secrecy degradation

Discussion and Implications for Secure System Design

The analysis presented in this chapter highlights that security in RIS-enabled V2X systems cannot be addressed through isolated mechanisms or static assumptions. Instead, it requires a holistic understanding of adversarial behavior, mobility-induced dynamics, and infrastructure-specific vulnerabilities. The interaction between vehicular mobility and programmable wireless environments creates both opportunities and risks, emphasizing the importance of adaptive and intelligence-driven security strategies.

These insights motivate the exploration of RIS-assisted secure communication techniques discussed in the next chapter. By leveraging programmable reflections, secure beamforming, and artificial noise generation, it becomes possible to counteract many of the identified threats. However, effective deployment demands careful consideration of system constraints and cross-layer interactions, paving the way for AI-driven solutions that can respond to evolving threats in real time.

SECURE BEAMFORMING AND RIS PHASE OPTIMIZATION

One of the main approaches to improving the security of the physical layer in the V2X system is through **secure beamforming**, which is a technique of regulating the direction and shape of the signal beam so that electromagnetic energy is concentrated on legitimate receivers and minimized in the direction of unauthorized parties. In the context of Reconfigurable Intelligent Surfaces (RIS), secure beamforming is extended by leveraging the ability of RIS to intelligently control the phase of signal reflection (Shi et al., 2025).

This approach is particularly relevant in congested vehicular environments, where direct links are often obstructed by buildings, other vehicles, or road infrastructure. RIS phase optimization is generally designed to maximize the secrecy rate, which is the difference between the channel capacity of the legitimate user and the capacity of the eavesdropping channel (Joseph Salomon et al., 2025). In high mobility scenarios, this optimization must take into account changes in vehicle position and channel dynamics in real time. Therefore, static optimization techniques become less effective, and adaptive mechanisms are needed that are able to respond quickly to environmental variations.

In addition, RIS-based secure beamforming can be combined with transmission power settings on the transmitter. By utilizing RIS reflection, the system can maintain a high level of security without increasing the beampower, making it more energy-efficient and difficult to detect by adversaries.

Figure 2. Technical illustration RIS-assisted secure beamforming in a V2X scenario

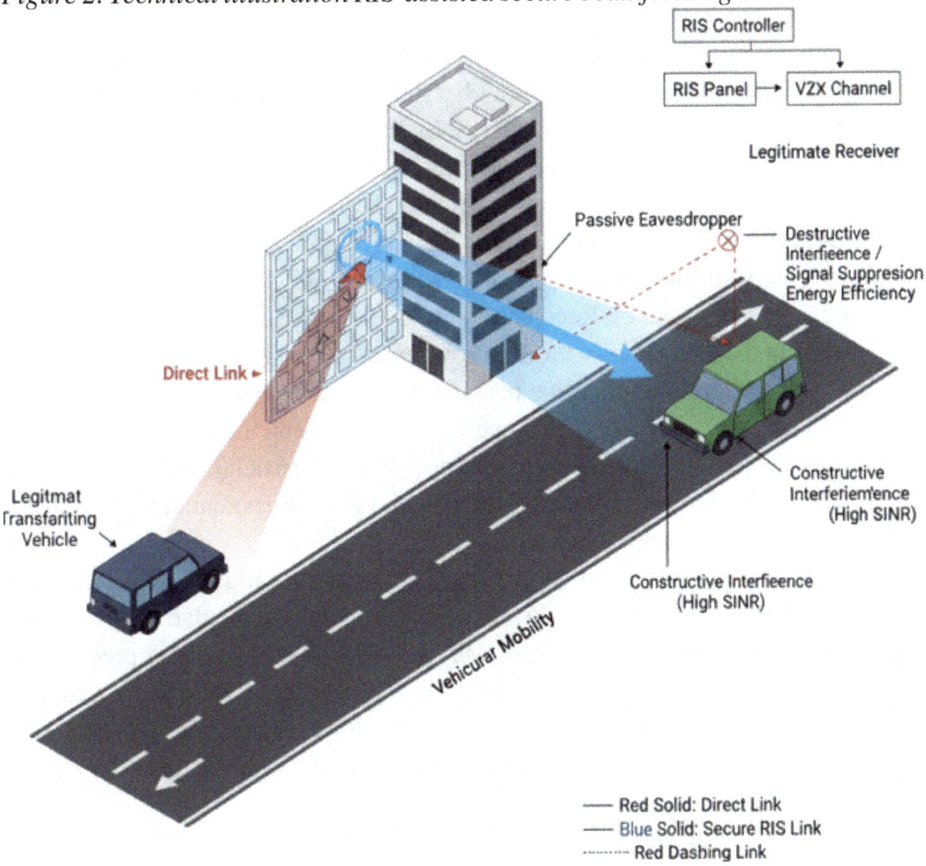

To summarize the techniques that have been discussed, **Table 3** presents a classification of RIS-based security approaches in V2X communication.

Table 3. RIS-aided physical layer security techniques in V2X systems

Technique	Security Objective	Role of RIS	Key Advantage
Secure Beamforming	Maximize secrecy rate	Directional signal enhancement	High spatial selectivity
Artificial Noise Injection	Degrade eavesdropper channel	Spatial noise shaping	Robust against unknown adversaries
Cooperative Communication	Increase security diversity	Secure path construction	Improved resilience
Joint Optimization	Balance security and performance	Multi-dimensional control	

ARTIFICIAL NOISE INJECTION WITH RIS ASSISTANCE

Artificial noise (AN) injection has evolved as an effective technique to improve the security of the physical layer by deliberately degrading the reception quality of unauthorized recipients, without sacrificing the reliability of communications to legitimate users (Joseph Salomon et al., 2025). In contrast to cryptographic approaches that work at the top layer and rely on key management, artificial noise operates directly on the signal domain, making it ideal for vehicular communication systems that demand low latency and fast response.

In a V2X system based on Reconfigurable Intelligent Surfaces (RIS), artificial noise can be controlled spatially through a programmable signal reflection mechanism. RIS acts as a passive entity that reshapes the direction of noise propagation so that interference can be focused on a specific region (Joseph Salomon et al., 2025). This ability overcomes the limitations of conventional artificial noise approaches that generally spread interference isotropically and have the potential to reduce the quality of reception in legitimate users.

The main advantage of RIS-assisted artificial noise lies in its resistance to the uncertainty of the eavesdropper's location. In a realistic vehicular environment, adversaries are often passive and do not show their presence, making precise positioning tracking difficult. By dynamically configuring the RIS to create spatially unfavorable propagation conditions, the system can lower the chances of successful eavesdropping even if only partial or statistical information about the adversary's position is available.

This approach is particularly relevant for open environments such as expressways, intersections, and urban areas, where line-of-sight conditions can appear intermittently. By utilizing the additional degree of freedom provided by the RIS, the distribution of artificial noise can be adjusted to vehicle mobility patterns and channel variations, so that the level of secrecy can be maintained consistently.

However, artificial noise injection also presents significant design challenges, especially regarding the power allocation between information and noise signals. Excessive allocation of noise power can increase the outage probability of legitimate links, especially in high Doppler conditions and fast-changing channels. RIS helps mitigate this problem by enabling more precise spatial separation between the information and noise signal components (Joseph Salomon et al., 2025). Thus, artificial noise can be directed away from legitimate recipients but still be effective in interfering with the eavesdropper.

Overall, RIS-assisted artificial noise injection is a flexible and energy-efficient mechanism to improve the safety of the physical layer in the V2X system. Its ability to adapt to the mobility and uncertainty of channels makes it a critical component in a 6G-based vehicular communications security architecture.

COOPERATIVE AND MULTI-HOP SECURE V2X COMMUNICATION

In complex vehicular environments, communication security cannot always rely on a single direct transmission line. Physical obstacles such as buildings, large vehicles, and congested urban planning often lead to non-line-of-sight conditions, which results in a decrease in the quality of canals and an increased risk of eavesdropping. In this context, cooperative and multi-hop communication is an effective approach to improve reliability as well as security (Wu, 2025).

Cooperative communication assisted by RIS utilizes several vehicles or roadside units as relays to transmit information gradually. With RIS integration, each transmission hop can be optimized to meet specific secrecy requirements. RIS elements can be configured to reinforce paths that are advantageous to legitimate users and weaken potential paths to adversaries, so that end-to-end security can be enhanced (Wei et al., 2025).

The multi-hop approach naturally provides advantages in terms of spatial and temporal diversity (Wei et al., 2025). The adversary must successfully intercept multiple hops simultaneously to obtain complete information, which significantly increases the complexity of the attack. RIS strengthens this mechanism by enabling the formation of dynamic, environmentally conscious communication lines, making network topology difficult to predict.

However, the design of RIS-based cooperative and multi-hop systems also faces a number of challenges. Coordination between nodes requires accurate time synchronization, efficient relay selection, and overhead management of control signals. This challenge is increasingly complex in high mobility conditions, where network topology and channel conditions change rapidly.

Therefore, RIS-assisted cooperative communication is generally combined with intelligent and learning-based optimization mechanisms. This approach allows the system to adaptively determine the optimal path, RIS configuration, and transmission strategy according to environmental conditions and safety needs (Li, 2025).

JOINT OPTIMIZATION STRATEGIES AND DESIGN TRADE-OFFS

The RIS-based security techniques that have been discussed in the previous subchapter do not stand alone, but complement each other in forming a comprehensive security system (Wu, 2025). In real implementation, secure V2X communication systems generally adopt a combined optimization strategy that considers secure beamforming, artificial noise injection, and cooperative communication simultaneously.

The main goal of this combined optimization is to achieve a balance between various performance metrics, such as secrecy rate, communication reliability, latency, and computational complexity. Significant security improvements often require high-resolution RIS phase control as well as rapid reconfiguration, which can increase processing load and control signal overhead.

One of the main challenges in combined optimization is the high dimension of the solution space. The large number of RIS elements, combined with dynamic canal conditions and vehicle mobility, results in complex optimization problems that are difficult to solve using conventional methods. In addition, the need for real-time adaptation in vehicular systems further tightens latency limits (Abouamer & Mitran, 2025).

Therefore, a combined optimization strategy should be designed taking into account both theoretical potential and practical limitations. This encourages the use of learning-based intelligent optimization frameworks, which are able to adapt security strategies adaptively based on environmental feedback (Namdar et al., 2025). This approach allows the system to dynamically manage trade-offs and maintain optimal safety performance in a wide range of vehicular conditions.

ARTIFICIAL INTELLIGENCE FOR ADAPTIVE PHYSICAL LAYER SECURITY

The security of the physical layer in a V2X system based on Reconfigurable Intelligent Surfaces (RIS) faces a major challenge in the form of very fast and unpredictable environmental dynamics. The high mobility of vehicles, variations in traffic density, and drastic changes in canal conditions make static safety approaches ineffective (Lakshmi et al., 2025). In this context, Artificial Intelligence (AI) is emerging as a key solution to enable security mechanisms that are adaptive, predictive, and responsive to environmental changes in real time.

AI allows communication systems to learn from channel patterns, mobility behaviors, and interference characteristics without having to rely on rigid mathematical models. By leveraging historical data and environmental observations, the system can identify potentially hazardous conditions, such as increased risk of eavesdropping or jamming, and proactively adjust the RIS configuration. This approach significantly increases the system's resilience to both dynamic and stealth attacks (Rubab et al., 2025) .

In V2X scenarios, AI also plays a role in reducing reliance on perfect Channel State Information (CSI) estimates. Instead of relying on precise yet computationally expensive channel measurements, AI models can learn the correlation between environmental parameters and security performance, enabling faster and more efficient decision-making. This is especially important for safety applications that demand very low latency.

DEEP REINFORCEMENT LEARNING FOR RIS CONFIGURATION CONTROL

Deep Reinforcement Learning (DRL) is one of the most promising AI approaches to managing RIS configurations in dynamic vehicular environments (Wei et al., 2025). In contrast to conventional optimization methods that require explicit modeling and complex optimization problem solving, DRLs allow agents to learn directly from interactions with the environment through a trial-and-error mechanism.

The main advantage of DRLs lies in their ability to adapt to rapidly changing conditions without the need for reprogramming. Trained agents can quickly respond to changes in mobility or the emergence of new adversaries by changing the RIS reflection pattern (Rubab et al., 2025). In addition, DRLs can handle large action spaces, which is particularly relevant given the usually very large number of RIS elements.

However, the implementation of DRLs in V2X systems also faces challenges, such as training stability and large data requirements. To address this, strategies such as transfer learning and simulation-based training are often used to enable agents to operate effectively in a real-world environment.

Figure 3. DRL-based control loop for RIS-enabled V2X security

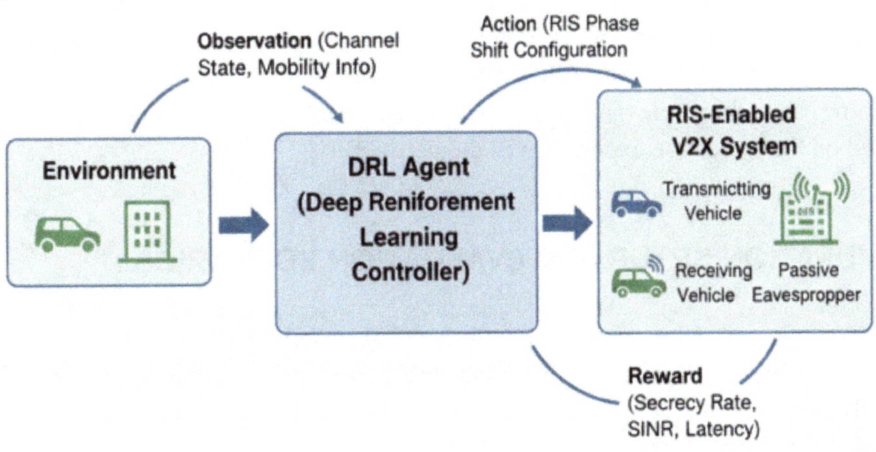

FEDERATED LEARNING FOR PRIVACY-PRESERVING SECURITY INTELLIGENCE

While AI offers strong adaptability, data collection and processing at scale raises concerns regarding data privacy and security. In a V2X network, vehicles generate sensitive data related to location, speed, and driving behavior. Sending raw data to a model training center has the potential to violate privacy and increase the risk of information leakage.

Federated Learning (FL) emerged as a solution that allows collaborative training of AI models without having to share raw data. In this approach, each vehicle or edge node trains a local model using its own data and only shares parameters or model updates to the aggregation server. Thus, collective knowledge can be built without sacrificing individual privacy (Wu, 2025).

In addition to maintaining privacy, FL also improves the scalability of the system. As the number of vehicles increases, the learning process can be expanded without overloading one central entity (Wei et al., 2025). However, challenges such as overhead communication and data heterogeneity still need to be managed properly in order for FL to be implemented effectively (Lakshmi et al., 2025).

MOBILITY-AWARE PREDICTIVE SECURITY MECHANISMS

AI-based security approaches are not only reactive, but can also be designed predictively by leveraging vehicle mobility information. By predicting vehicle movements and changes in network topology, the system can anticipate future channel conditions and adjust security strategies before threats actually occur (Lakshmi et al., 2025).

However, the accuracy of mobility predictions is highly dependent on the quality of the data and the models used. Therefore, the integration between DRL and FL has become particularly relevant, where predictive models can be continuously updated based on the collective experience of the network.

SIMULATION SETUP AND EVALUATION SCENARIOS

Evaluation of safety performance in a V2X system based on Reconfigurable Intelligent Surfaces (RIS) requires modeling that is able to realistically represent the dynamics of the vehicular environment. In this chapter, evaluation is carried out through scenario-based simulations that reflect the main operational conditions of V2X, namely Vehicle-to-Vehicle (V2V), Vehicle-to-Infrastructure (V2I), and

Vehicle-to-Pedestrian (V2P) communications. Each scenario is designed to capture variations in mobility, node density, as well as the potential presence of adversaries (Guo & Zhang, 2025).

The canal model considers the Doppler effect due to vehicle mobility as well as small and large-scale fading. Adversaries are randomly placed in the simulation area to represent realistic conditions, either as passive eavesdroppers or active jammers (Wei et al., 2025). The simulation parameters were chosen to allow for a fair comparison between conventional systems without RIS and RIS-enabled systems with the various security techniques discussed in the previous chapter.

Table 4. Simulation parameters for RIS-enabled V2X security evaluation

Parameter	Value Description
Carrier Frequency	mmWave / sub-6 GHz
Vehicle Speed	Urban: low–medium, Highway: high
RIS Elements	Moderate to large-scale arrays
Channel Model	Mobility-aware fading with Doppler
Adversary Type	Passive eavesdropper, active jammer
Evaluation Metrics	Secrecy rate, outage probability, jamming resistance

PERFORMANCE METRICS AND EVALUATION METHODOLOGY

To assess the effectiveness of RIS-based security techniques, several key metrics are used. Secrecy rate is the main indicator to measure the level of confidentiality of communication, which is defined as the difference in channel capacity between legitimate links and eavesdropper links. This metric provides a first-hand view of the system's ability to protect sensitive information (D. Wang et al., 2025).

In addition, the secrecy outage probability is used to measure the probability that the secrecy rate falls below a certain threshold. This metric is particularly relevant in the context of V2X because safety applications demand a very high level of reliability. Jamming resistance was also analyzed to evaluate the system's resistance to active interference attacks, especially in high mobility conditions.

The evaluation was carried out by comparing several system configurations, including: (i) conventional systems without RIS, (ii) RIS-enabled systems without special safety mechanisms, and (iii) RIS-enabled systems with secure beamforming, artificial noise, and intelligent optimization techniques. This approach allows for separate or integrated isolation of RIS contributions and security mechanisms.

SECRECY PERFORMANCE UNDER DIFFERENT ATTACK SCENARIOS

The simulation results show that the implementation of RIS significantly increases the secrecy rate compared to systems without RIS, especially in urban scenarios with dominant NLoS conditions. With the right phase configuration, RIS is able to form an alternative propagation path that amplifies the signal to the legitimate receiver and reduces signal leakage in the direction of the adversary (Li, 2025).

In passive eavesdropping scenarios, the RIS-enabled system with secure beamforming shows a consistent increase in secrecy rate even though the position of the eavesdropper changes dynamically. The integration of artificial noise further amplifies this effect, especially when the location of the adversary is not known for sure. Under these conditions, conventional systems tend to experience a significant decline in performance (C. Chen et al., 2025).

In active jamming scenarios, RIS-enabled systems show better resistance to interference (D. Wang et al., 2025). By utilizing programmatic reflection, the system can avoid jamming-contaminated paths and maintain the quality of communication on legitimate links. This effect is more noticeable in highway scenarios, where high mobility usually exacerbates the impact of jamming on conventional systems.

IMPACT OF MOBILITY AND RIS CONFIGURATION DYNAMICS

Further analysis showed that vehicle mobility has a significant influence on safety performance. At low to medium speeds, the periodically updated RIS configuration is able to maintain stable secrecy performance (B. Wang et al., 2025). The results of the evaluation show that vehicle mobility is the dominant factor that affects the safety effectiveness of the RIS-enabled V2X system. The characteristics of the canal that change very quickly due to vehicle movement cause the propagation condition to be non-stationary, thus challenging the stability of the physical layer security mechanism. In this context, security performance is determined not only by secure beamforming or artificial noise strategies, but also by the system's ability to adjust RIS configurations in a timely and accurate manner.

In scenarios with low to medium vehicle speeds, the RIS configuration that is updated periodically is still able to maintain relatively stable secrecy performance. Channel changes occur over a fairly long timescale, so channel estimation and RIS phase updates remain relevant over a given interval (Guo & Zhang, 2025). This suggests that a semi-static approach is still feasible to use in environments such as city intersections or smart parking areas. However, when vehicles move at high speeds, such as on highways or platooning scenarios, canal coherence time decreases drastically. In these conditions, delays in reconfiguring the RIS can cause a mismatch between the phase configuration and the actual channel conditions, which ultimately lowers the secrecy rate temporarily.

(1) Sensitivitas Secrecy Performance terhadap Delay Reconfiguration

One of the key findings of this evaluation is the high sensitivity of security performance to delays in RIS configuration updates. Even relatively small delays can lead to increased signal leakage in the direction of the eavesdropper, especially in scenarios with complex multipath reflections (D. Wang et al., 2025). When the RIS configuration is no longer aligned with the position of the legitimate vehicle, the information signal can spread more widely and reduce the spatial advantage previously gained. This confirms that latency in decision-making and RIS reconfiguration execution is a crucial parameter in system design.

(2) The Role of Mobility Prediction in Security Stability

The integration of artificial intelligence-based adaptive mechanisms, as discussed in Chapter 4, has been shown to be effective in reducing the negative impacts of high mobility. A system equipped with vehicle mobility and channel dynamics prediction capabilities is able to proactively update RIS configurations, not just reactively (B. Wang et al., 2025). This approach allows RIS to anticipate changes in propagation conditions before any degradation of security performance occurs. The simulation results show that the prediction-based approach results in a more stable secrecy rate and a significant reduction in outage probability compared to conventional methods that rely only on current channel measurements.

(3) Influence of Vehicle Density and Propagation Environment

In addition to vehicle speed, traffic density and the complexity of the propagation environment also affect the dynamics of RIS configurations. In dense urban scenarios, with many vehicles and natural reflectors such as tall buildings, canals become highly volatile and difficult to predict. The RIS in this condition must adapt not only to a single link, but also to interference between vehicles. Evaluations show that RIS configurations that do not consider network density tend to experience reduced security performance due to increased uncontrolled interference (B. Wang et al., 2025). Therefore, the RIS adaptation model needs to include a broader environmental context, including vehicle density and traffic movement patterns.

(4) Impact of Number of RIS Elements and Phase Resolution

The number of RIS elements and phase resolution are structural factors that greatly influence the evaluation results. RIS with a larger number of elements provides a higher degree of freedom in forming a safe reflection pattern, thus being able to significantly increase the secrecy capacity (Guo & Zhang, 2025). However, these advantages come with the consequence of increased computational complexity and control signal overhead. High phase resolution also improves configuration precision, but demands more advanced hardware and greater energy consumption. Therefore, system design must consider the trade-off between security performance and practical implementation limitations.

(5) Design Implications for Real Systems

The findings in this subchapter provide important implications for the design of RIS-enabled V2X systems in the real world. First, the RIS update mechanism should be designed with minimal latency and integrated with mobility predictions. Second, RIS configurations need to be adaptive to environmental variations and network density, not just to a single communication link. Third, the selection of RIS size and capabilities should be tailored to the application scenario, so that security can be improved without overloading the system (D. Wang et al., 2025).

Overall, this analysis confirms that the mobility and dynamics of RIS configurations are inseparable aspects in the evaluation of RIS-enabled V2X security. The success of the system in maintaining secrecy performance relies heavily on real-time adaptability supported by artificial intelligence and efficient architectural design.

COMPUTATIONAL OVERHEAD AND PRACTICAL CONSIDERATIONS

While RIS-assisted security offers significant performance improvements, practical aspects such as compute overhead and control signals cannot be overlooked. The simulation results show that RIS phase optimization and artificial noise management increase the computing load compared to conventional systems. However, this overhead is still within acceptable limits for V2X applications, especially when some processing is shifted to edge computing (B. Wang et al., 2025).

The distributed approach and federated learning also contribute to balancing overhead and performance. By dividing the learning and optimization process into multiple nodes, the system can maintain scalability without sacrificing latency significantly. These findings show that the implementation of RIS-enabled security is practically possible with the right architecture.

PROPOSED END-TO-END SECURITY FRAMEWORK

The increasing complexity and dynamics of Vehicle-to-Everything (V2X) systems in the 6G era demand a holistic, adaptive, and integrated approach to security. Reconfigurable Intelligent Surfaces (RIS) introduces new capabilities to control signal propagation environments, but this potential can only be maximized through a well-designed security framework. The framework proposed in this chapter integrates physical layer security (PLS), RIS-based optimization, and artificial intelligence to address security challenges in dynamic vehicular environments (Li, 2025).

This end-to-end security framework consists of several functional layers that interact with each other. At the physical layer, RIS acts as a signal propagation controller by regulating the reflection phase to increase the secrecy rate and suppress information leakage. The control layer is in charge of coordinating RIS configuration, power management, and artificial noise generation based on channel conditions and detected threats. Meanwhile, the intelligence layer implements AI algorithms and distributed learning to enable real-time adaptation to changing network mobility and topology.

This approach places security as an inherent function of a communication system, rather than as an additional feature. Thus, each communication process automatically considers aspects of confidentiality, resistance to interference, and operational sustainability. The framework is also designed to be compatible with a wide range of V2X scenarios, including V2V, V2I, and V2P, so it can be widely applied to intelligent transportation ecosystems (D. Wang et al., 2025). The integration of these various security elements ensures that data protection is no longer reactive, but rather a fundamental part of the network architecture from the start. To provide a more in-depth visual understanding of the working mechanisms of this complex system, Figure 4 presents the structure of the *End-to-End Security Framework* proposed systematically. The illustration details how information flow and coordination occur between four main pillars: a physical layer that leverages RIS phase optimization, a control layer for radio resource management, an artificial intelligence (AI) layer that serves as the brains of adaptive decision-making, and the integration of *edge* and *cloud* infrastructure. Through this visual representation, it can be observed how each component exchanges security metrics and feedback in *real-time* to create a V2X communication ecosystem that is resilient to a wide spectrum of threats in a dynamic intelligent transportation environment.

Figure 4. End-to-end security framework for RIS-enabled V2X system integrating physical layer

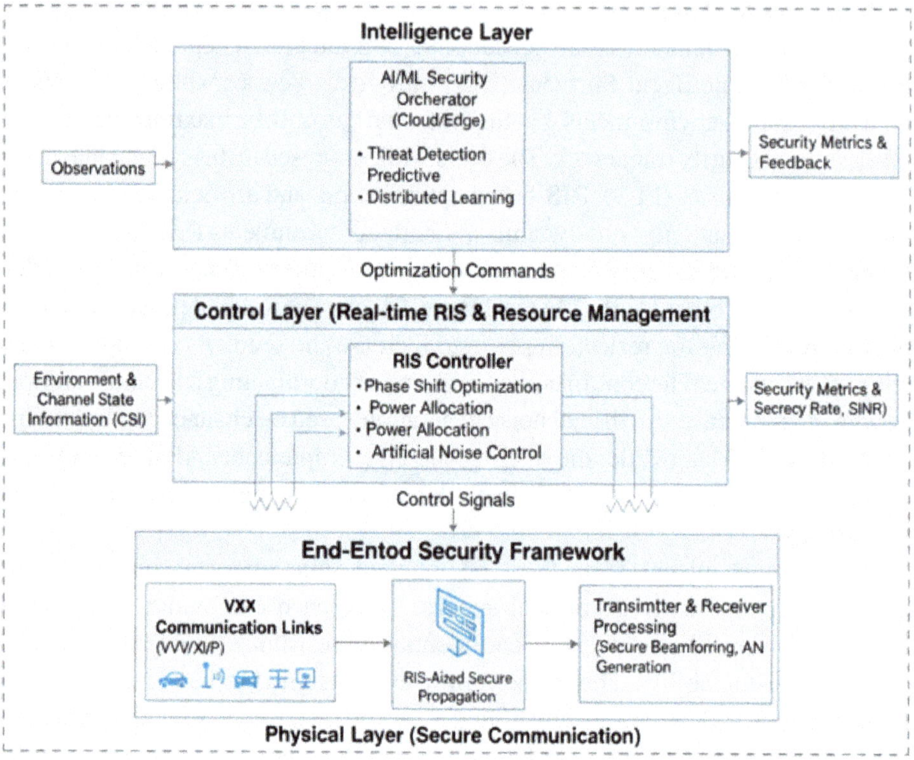

FUNCTIONAL COMPONENTS AND SECURITY WORKFLOW

The proposed framework adopts a security workflow that is cyclic and adaptive in nature. The process starts from the *environment sensing stage*, which is the phase where the system collects information related to the condition of the communication channel, the level of vehicle mobility, and potential interference in the network environment (Li, 2025). This information is obtained through various sources, such as pilot signals emitted by communication nodes, sensors installed on vehicles, and historical data stored and processed on edge nodes.

The next stage is *threat assessment*, where the intelligence module analyzes the data that has been collected to identify potential security threats, including *eavesdropping* and *jamming* attacks. In this phase, trained artificial intelligence models are used to classify risk levels and predict possible threat evolution in the short term.

The learning process can be carried out centrally or through *a federated learning approach* to maintain data privacy and reduce the burden of communication (Gao et al., 2025).

Based on the results of the threat analysis, the system then enters the *adaptive security configuration* stage. At this stage, the RIS configuration, *secure beamforming* strategy, and *artificial noise generation mechanism* are determined dynamically. Configuration decisions are made taking into account the balance between the level of security achieved, the quality of communication services, and the resulting computing overhead, so that the system remains efficient and responsive to changes in network conditions.

The final stage in this workflow is *feedback and learning*, which allows the system to update intelligence models and security parameters based on actual communication outcomes. The performance information obtained is used as feedback to improve the accuracy of predictions and the effectiveness of security strategies in the next cycle (Gao et al., 2025). With this mechanism, the framework is able to make continuous adjustments to the dynamics of the vehicular environment.

Overall, this workflow ensures that security mechanisms are not static, but rather continue to evolve as network conditions and attack patterns change. This adaptive approach is particularly important in the context of vehicular communication, given the characteristics of high mobility and environmental variations that can change in a very short timescale.

DEPLOYMENT GUIDELINES AND PRACTICAL CONSIDERATIONS

The implementation of the V2X RIS-enabled security framework in a real-world environment requires special attention to practical and operational aspects. One of the main considerations is the placement of the RIS. The location of the RIS should be strategically chosen to maximize the coverage and effectiveness of the reflection, such as at road intersections, building walls in urban areas, or highway infrastructure. Imoptimal placement can reduce the expected security benefits (Zhang et al., 2025).

The limitations of RIS hardware also need to be taken into account. The number of RIS elements, phase resolution, and reconfiguration speed affect the system's ability to respond to threats. The framework is designed to be flexible to operate on a wide range of RIS specifications, by adapting the complexity of security algorithms to the capabilities of the device (Zhang et al., 2025).

The security of RIS control is another important aspect (Wu, 2025). This framework assumes the existence of authentication and protection mechanisms on the control channel to prevent configuration manipulation by unauthorized parties. Without this protection, RIS can actually become a weak point in the overall security system.

ALIGNMENT WITH 6G AND ITS REQUIREMENTS

Framework keamanan yang diusulkan dirancang agar selaras dengan visi dan kebutuhan 6G and Intelligent Transportation Systems (ITS). One of the key principles of 6G is the native integration of intelligence into the network. This framework realizes this principle by making AI a core component in security decision-making.

From ITS's perspective, safety and reliability are top priorities. This framework supports critical safety applications by ensuring secure and stable communication even under high mobility conditions or under attack (Tiwari & Jha, 2025). In addition, the physical layer security approach used is computationally lightweight, making it suitable for applications with strict latency limitations. This is in line with the goal of reducing the energy footprint of future communication networks (Li, 2025).

Additionally, the framework is designed to support interoperability and scalability. With a modular architecture, security components can be updated or upgraded without disrupting the entire system. This is important to support the evolution of standards and future regulatory needs.

Table 5. Mapping of security mechanisms to 6G-V2X requirements

6G-V2X Requirement	Security Mechanism	Framework Support
Ultra-low latency	Edge-assisted RIS control	Real-time adaptation
High reliability	Secure beamforming & AN	Reduced outage
Native intelligence	AI/DRL optimization	Adaptive security
Privacy preservation	Federated learning	Data confidentiality
Energy efficiency	Passive RIS operation	Sustainable deployment

SCALABILITY, INTEROPERABILITY, AND STANDARDIZATION PERSPECTIVES

The successful implementation of the RIS-enabled V2X security framework depends not only on technical performance, but also on scalability and standardization. As the number of vehicles and RIS increases, the system must be able to manage complexity without significant performance degradation. The framework supports scalability through the sharing of functions between the edge and the cloud, as well as the use of distributed learning (Zhang et al., 2025). This distributed architecture reduces centralized processing bottlenecks and enables faster local adaptation to dynamic vehicular environments. Moreover, scalable coordination mechanisms are essential to ensure that security policies remain consistent across dense and highly mobile V2X scenarios. Without such scalability considerations, the effectiveness of RIS-assisted physical layer security may significantly degrade in large-scale deployments.

From a standardization point of view, this framework offers a structured and forward-looking conceptual foundation for the development of future technical guidelines, architectures, and interoperability specifications in secure 6G-V2X systems. By explicitly defining the functional roles and interactions of Reconfigurable Intelligent Surfaces, artificial intelligence mechanisms, and physical layer security techniques, the framework helps bridge the gap between theoretical research and practical deployment requirements. This clarity is essential for standardization bodies, as it enables the formulation of harmonized protocols for RIS control, security orchestration, and cross-layer coordination across heterogeneous vehicular networks. Furthermore, the framework supports interoperability by accommodating diverse hardware implementations, communication protocols, and network topologies, which are inherent in large-scale V2X ecosystems. For industry stakeholders, such as automotive manufacturers, network operators, and infrastructure providers, the proposed framework can serve as a reference model to align product development with emerging 6G security standards (M. Chen et al., 2025). Consequently, this approach not only facilitates regulatory alignment but also accelerates the adoption of RIS-enabled security solutions in real-world vehicular communication environments.

CONCLUSION AND FUTURE RESEARCH OUTLOOK

This chapter closes the discussion on *Physical Layer Security in RIS-Enabled Vehicular Communication* by summarizing the main findings, conceptual contributions, and practical implications of the use of Reconfigurable Intelligent Surfaces (RIS) in 6G-based Vehicle-to-Everything (V2X) systems. Throughout the discussion, it was shown that RIS not only plays a role as a signal quality improvement technology, but also as a *security enabler* which is able to reinforce the safety of the physical layer adaptively in a highly dynamic vehicular environment (Yang et al., 2025).

This study confirms that the unique characteristics of RIS, namely the ability to program reflection phases and reshape the signal propagation environment, open up a new paradigm in the security of wireless communications. In contrast to conventional approaches that rely on cryptography at the top layer, RIS-based security works directly at the physical layer by taking advantage of the nature of wireless channels, making it lighter, contextual, and difficult for adversaries to predict. This is especially relevant in V2X scenarios characterized by high mobility, rapidly changing network topologies, and open communication in public spaces.

The main contribution of this chapter lies in the comprehensive analysis of the security threats of the physical layer in a realistic V2X RIS-enabled system, the integration of RIS-based security techniques with artificial intelligence, and the formulation of an adaptive security framework that is cyclical and sustainable. The results of modeling and evaluation show that RIS is able to increase *secrecy rates*, reduce *outage probability*, and increase jamming resistance in various vehicular scenarios, both in urban and highway environments (Xu et al., 2025).

From an implementation perspective, this chapter highlights the importance of standardization and cross-layer integration to ensure system interoperability and reliability. Practical aspects such as RIS deployment strategies, control overhead, energy consumption, and hardware reliability are also key factors in real deployments (Dong & Xie, 2025). On the other hand, there are still a number of open challenges, including accurate CSI estimates at high mobility, system scalability, and protection against attacks targeting the RIS itself.

As a future research direction, the exploration of RIS integration with *joint sensing and communication*, *semantic communication*, and other 6G supporting technologies such as terahertz communication and digital twin is a promising opportunity. Overall, this chapter confirms that RIS has the potential to be the key foundation for a secure, adaptive, and sustainable 6G-based V2X security architecture and supports the *secure-by-design vision* for the intelligent transportation systems of the future.

REFERENCES

Abouamer, M. S., & Mitran, P. (2025). A Hypernetwork Framework for Learning Adaptive Beamforming Schemes in RIS Systems. *IEEE Transactions on Communications, 73*(12), 13471–13485. DOI: 10.1109/TCOMM.2025.3600529

Chen, C., Song, Y., & Ji, J. (2025). Experimental demonstration of 10 Gbps security-enhanced optical transmission link over turbulent atmosphere. *Optics and Lasers in Engineering, 195*. Advance online publication. DOI: 10.1016/j.optlaseng.2025.109347

Dong, X., & Xie, Y. (2025). Research on cloud computing network security mechanism and optimization in university education management informatization based on OpenFlow. *Systems and Soft Computing, 7*. Advance online publication. DOI: 10.1016/j.sasc.2025.200225

Gao, D., Yin, M., Xia, N., Liu, X., Wang, D., & Peng, M. (2025). Energy Efficiency Optimization for Collaborative Task Offloading in RIS-Empowered Heterogeneous Wireless Computing Power Networks. *IEEE Transactions on Communications, 73*(12), 15383–15398. DOI: 10.1109/TCOMM.2025.3600213

Guo, T., & Zhang, X. (2025). Time -varying channel estimation for RIS-assisted wireless communication systems with low pilot overhead. *Physical Communication, 73*. Advance online publication. DOI: 10.1016/j.phycom.2025.102885

Hao, H., & Zhang, Y. (2025). STAR-RIS-Enabled AOA Positioning Algorithm. *Electronics (Switzerland), 14*(23). Advance online publication. DOI: 10.3390/electronics14234729

Joseph Salomon, M. T. V., Henri Pascal-Will, M. L., Yannick, M., Ebenezer, M. M., Luc, I. E., Jean-Francois, E. D., Moulema Douala, P.-M., & Joe, Y. S. (2025). IoT security approach based random distribution of communication frequency. *MethodsX, 15*. Advance online publication. DOI: 10.1016/j.mex.2025.103465

Lakshmi, M. L. S. N. S., Siva Chakra Avinash, B., Suneetha, R., Rajalakshmi, P., & Soman, S. M. (2025). Active Antenna Selection for Reconfigurable Intelligent Surfaces using Rotation-Invariant Coordinate Convolutional Neural Network and Multi-Dimensional Attention Spiking Neural Network for 5G/6G Communication. *Sensing and Imaging, 26*(1). Advance online publication. DOI: 10.1007/s11220-024-00497-y

Li, M. (2025). RIS-Assisted Joint Communication, Sensing, and Multi-Tier Computing Systems. *Future Internet, 17*(12). Advance online publication. DOI: 10.3390/fi17120533

Namdar, M., Basgumus, A., Ata, S. O., & Aldirmaz-Colak, S. (2025). Physical layer security in RIS-aided communication systems: Secrecy performance analyses. *Digital Signal Processing: A Review Journal, 167*. DOI: 10.1016/j.dsp.2025.105417

Rubab, S., Abro, G. E. M., Mustafa, H., Baloch, S. K., Memon, S. A., & Saeed, N. (2025). RIS-assisted UAV communications: A review of system models, frameworks and outage performance. *ICT Express, 11*(6), 1186–1199. DOI: 10.1016/j.icte.2025.10.006

Shi, L., Tang, Z., Shen, L., & Hao, W. (2025). Physical layer security of RIS-aided NOMA system for near-field scenario. *Physical Communication, 73*. Advance online publication. DOI: 10.1016/j.phycom.2025.102894

Tiwari, C. S., & Jha, V. K. (2025). OSI network layers based secured data sharins in cloud computing using STXORSK-QC and dhdecct-MAC. *Discover Computing, 28*(1). Advance online publication. DOI: 10.1007/s10791-025-09510-8

Wang, B., Zhang, N., Zhao, Y., Feng, Z., & Yao, B. (2025). Performance limit and phase design for near-field tracking with reconfigurable intelligent surface. *Digital Signal Processing: A Review Journal, 167*. DOI: 10.1016/j.dsp.2025.105426

Wei, T., Zheng, B., Tang, K., Feng, W., Che, W., & Xue, Q. (2025). Machine Learning-Enabled RIS-Assisted mmWave Multi-Hop Communications: Path Scheduling and Beamforming Design. *IEEE Transactions on Wireless Communications, 24*(12), 10424–10439. DOI: 10.1109/TWC.2025.3579599

Wu, Y. (2025). Symmetry-Aided Active RIS for Physical Layer Security in WSN-Integrated Cognitive Radio Networks: Green Interference Regulation and Joint Beamforming Optimization. *Symmetry, 17*(12). Advance online publication. DOI: 10.3390/sym17122047

Xu, J. W., Wei, M., Zhang, L., Galdi, V., Li, L., & Cui, T. J. (2025). Chaotic information metasurface for direct physical-layer secure communication. *Nature Communications, 16*(1). Advance online publication. DOI: 10.1038/s41467-025-60725-1

Yang, M., Wang, T., Li, Y., & Xu, S. (2025). Bridging Theory and Data: Linking Regional Innovation System Dimensions to Patent Outcomes Through CFA-CatBoost Integration. *Sustainability (Switzerland), 17*(24). Advance online publication. DOI: 10.3390/su172411211

Zhang, M., Cheng, H., Su, R., Lu, Y., Wang, S., & Zhang, L. (2025). A cognitive transmission and deployment strategy for UAV-assisted hybrid RIS relay network. *Eurasip Journal on Advances in Signal Processing, 2025*(1). DOI: 10.1186/s13634-025-01237-4

Chapter 7
RIS-Assisted Communication for Autonomous and Electric Vehicle Ecosystems

P. Selvakumar
http://orcid.org/0000-0002-3650-4548
Department of Science and Humanities, Nehru Institute of Technology, Coimbatore, India

Mohammad Shabeeh
http://orcid.org/0009-0003-7533-6182
United Institute of Technology, India

S. Guru Basava Aradhya
http://orcid.org/0000-0002-3335-2681
Noida International University, India

Subramaniam Iyer
http://orcid.org/0000-0002-3843-9399
Datta Meghe Institute of Management Studies, India

ABSTRACT

Revolutionising autonomous vehicles with Reconfigurable Intelligent Surface (RIS)-assisted communication represents a fundamental shift in how smart mobility systems perceive, interact with, and navigate their environments, enabling unprecedented levels of efficiency, safety, and scalability across next-generation transportation infrastructures. As autonomous vehicles increasingly rely on high-bandwidth, low-latency, and ultra-reliable wireless connectivity to support real-time perception,

DOI: 10.4018/979-8-3373-7867-1.ch007

sensor fusion, and cooperative decision-making, the limitations of traditional communication networks—particularly in complex environments with multipath fading, signal blockages, and insufficient spectrum resources—become more pronounced. RIS technology, built upon programmable meta-surfaces capable of intelligently manipulating electromagnetic waves without conventional power-hungry radio frequency components, introduces a groundbreaking capability to dynamically reconfigure the propagation conditions of wireless signals.

REVOLUTIONISING AUTONOMOUS VEHICLES WITH RIS-ASSISTED COMMUNICATION

Revolutionising autonomous vehicles with Reconfigurable Intelligent Surface (RIS)-assisted communication represents a fundamental shift in how smart mobility systems perceive, interact with, and navigate their environments, enabling unprecedented levels of efficiency, safety, and scalability across next-generation transportation infrastructures. As autonomous vehicles increasingly rely on high-bandwidth, low-latency, and ultra-reliable wireless connectivity to support real-time perception, sensor fusion, and cooperative decision-making, the limitations of traditional communication networks—particularly in complex environments with multipath fading, signal blockages, and insufficient spectrum resources—become more pronounced. RIS technology, built upon programmable meta-surfaces capable of intelligently manipulating electromagnetic waves without conventional power-hungry radio frequency components, introduces a groundbreaking capability to dynamically reconfigure the propagation conditions of wireless signals, effectively transforming passive surfaces within urban landscapes into active agents of connectivity enhancement. By enabling fine-grained control over reflection, refraction, absorption, and beam steering, RIS-assisted communication provides autonomous vehicles with more reliable V2X (vehicle-to-everything) data exchange, even in highly obstructed areas such as tunnels, dense city intersections, or regions with tall buildings and heavy traffic flows. This transformation is essential for supporting advanced cooperative driving functions, such as coordinated lane merging, intersection negotiation, remote vehicle monitoring, and collaborative perception—where vehicles share sensory data to expand their situational awareness beyond line-of-sight limitations. Furthermore, RIS plays a crucial role in enabling 6G-oriented features such as holographic radio, ultra-massive MIMO systems, and distributed intelligence across transportation infrastructures, ultimately creating a seamless and environment-adaptive communication ecosystem for autonomous mobility.

The deployment of RIS-empowered smart surfaces along highways, on building facades, within roadside infrastructure, and even on vehicular bodies enhances channel conditions to meet the stringent quality-of-service requirements of autonomous driving, enabling millimetre-wave and terahertz signals—traditionally hindered by limited penetration and sensitivity to mobility—to be reliably exploited for vehicular networks. As a passive and energy-efficient solution, RIS significantly reduces network operational costs and environmental impact, aligning with global sustainability goals while extending communication coverage to underserved and previously unreachable regions. This leads to greater equity in mobility service distribution, ensuring that high-performance autonomous transportation is not limited to premium urban clusters but reaches rural and economically constrained areas. Moreover, RIS integration with edge computing architectures and AI-enabled network orchestration enhances situational responsiveness, as RIS configuration can be optimised in real-time based on traffic conditions, interference dynamics, and navigation priorities. Through machine learning-driven digital twins, the physical layer characteristics of autonomous mobility networks can be modelled, predicted, and controlled to anticipate network degradation and proactively maintain communication robustness, a capability vital for safety-critical manoeuvres requiring deterministic latency and fail-safe responses.

ENHANCING ELECTRIC VEHICLE ECOSYSTEMS WITH RIS TECHNOLOGY

Enhancing electric vehicle (EV) ecosystems with Reconfigurable Intelligent Surface (RIS) technology represents a critical step toward building more resilient, energy-efficient, and intelligence-driven smart mobility infrastructures (Naaz F., et al, 2024). As the adoption of EVs accelerates globally—driven by climate policies, shifting consumer expectations, and breakthroughs in battery technologies—the supporting communication and power delivery systems must evolve to manage escalating complexity in charging networks, grid interactions, and real-time energy optimisation. Current EV ecosystems depend heavily on stable wireless communication for monitoring battery health, enabling dynamic routing to charging points, managing vehicle-to-grid (V2G) power exchanges, and coordinating fleets for shared mobility solutions. However, dense urban environments introduce complications such as signal obstruction, multipath interference, and congestion in both wireless channels and power infrastructure. RIS technology, based on programmable metasurfaces that can dynamically manipulate incoming electromagnetic waves with minimal energy consumption, offers a transformational approach to improving the reliability, performance, and scalability of EV communication and charging systems by intelligently shaping the wireless propagation environment itself. With RIS-enhanced connectivity, EVs can maintain seamless links to charging stations, grid management centres, cloud-based digital twins, and cooperative mobility platforms, ensuring that every decision—from charging schedules to navigation routes—is informed by real-time, high-quality data (Abumarshoud H., et al, 2021). In EV charging infrastructures, RIS can strengthen communication links between distributed renewable energy resources, charging operators, and grid controllers by mitigating signal fading and spectrum limitations at both millimetre-wave (mmWave) and terahertz (THz) frequencies—key bands for large-scale smart city connectivity. This becomes especially crucial in fast-charging hubs where multiple vehicles interface simultaneously with the grid, requiring synchronised data exchange related to power load balancing, battery diagnostics, and safety measures. By embedding RIS panels into charging stations, parking structures, and roadside furniture, service providers can dramatically extend network coverage without deploying expensive active antenna systems, thus lowering capital and operational costs. Moreover, RIS-assisted networks can ensure lower latency communication needed for automated charging operations, including robotic charging arms and inductive wireless power transfer solutions, where precise alignment and real-time feedback are essential for optimising charging efficiency and minimising energy loss (Senapati M K., et al, 2024). Emerging applications such as bidirectional wireless power transfer and dynamic charging lanes can also be significantly enhanced through RIS-aided beam steering and focus control, enabling vehicles to recharge while moving, with intelligently shaped transmission paths minimising interference and maximising transfer effectiveness.

Furthermore, RIS contributes significantly to enhancing cybersecurity and privacy within EV systems. By physically controlling how and where electromagnetic signals propagate, RIS can establish secure communication corridors that reduce vulnerability to eavesdropping and jamming attacks on key EV services, including billing, authentication, and connected vehicle diagnostics (Ji B., et al, 2024). Coupled with blockchain-based identity management or encrypted V2G communication protocols, RIS-assisted secure channel formations create an additional layer of physical security, essential for protecting consumer trust and grid reliability as EV interactions become more digitalised. Additionally, RIS may facilitate privacy-preserving services such as anonymised routing for shared EV fleets and secure distribution of user data for energy behaviour analytics. Despite its transformative benefits, integrating RIS into EV ecosystems brings forward several challenges that require careful consideration. RIS elements must be ruggedised for outdoor weather conditions and safety compliance, while large-scale deployment introduces operational complexities in maintenance, interoperability with current 5G/6G vehicular communication standards, and real-time configuration under highly dynamic mobility patterns (Satyanarayana P., et al, 2026). Standardisation efforts are still in progress regarding RIS control interfaces, spectrum use coordination, and unified security frameworks. Beyond technical barriers, economic evaluation models must justify investments, particularly in regions where EV demand and infrastructure maturity remain uneven. Regulatory stakeholders must also determine ownership structures and governance rules for RIS-infused mobility infrastructure—balancing innovation with responsibility and equitable access.

RIS-ASSISTED V2X COMMUNICATION FOR AUTONOMOUS VEHICLES

RIS-assisted V2X (Vehicle-to-Everything) communication for autonomous vehicles is emerging as a cornerstone innovation that will profoundly transform the performance, safety, and scalability of next-generation smart mobility systems. V2X communication requires extremely reliable, low-latency, and high-bandwidth wireless connectivity to enable real-time interaction between vehicles (V2V), transportation infrastructure (V2I), pedestrians (V2P), and the cloud (V2N) (Han S., et al, 2023). However, existing wireless environments—especially in urban areas featuring tall buildings, tunnels, dense traffic, and unpredictable obstruction—introduce multipath fading, high interference, and signal degradation that challenge the full realisation of cooperative and fully automated driving. This is where Reconfigurable Intelligent Surfaces (RIS) bring a technological breakthrough: by embedding intelligent meta-surfaces in the driving environment, RIS can programmatically manipulate incoming radio waves through passive reflection, refraction, and beam steering, turning the wireless channel into a controllable and optimised asset rather than a constraint. In doing so, RIS-assisted communication enhances link stability, mitigates dead zones, and ensures deterministic wireless performance required for safety-critical manoeuvres such as cooperative collision avoidance, connected platooning, and automated lane merging. As autonomous vehicles process a continuously growing volume of sensor data from LiDAR, radar, cameras, and GPS, RIS-enabled V2X connectivity improves shared perception capabilities—allowing vehicles to exchange high-fidelity environmental data like object location, road hazards, and pedestrian state information—thus expanding situational awareness beyond the field-of-view limitations of onboard sensors (Mao Y., et al, 2024). The integration of RIS into vehicular networks aligns with the evolution toward 6G mobility standards, supporting advanced features including terahertz communication, ultra-massive MIMO, and holographic radio services that require precise beam alignment and channel shaping. Equipped with AI-driven controllers, RIS dynamically adapts its reflective configuration in response to vehicle movement, traffic density, and interference patterns to maintain uninterrupted connectivity and ensure predictable latency—essential for autonomous driving decision-making. With RIS deployed on road infrastructure such as traffic poles, road signs, building facades, parking structures, and highway barriers, smart cities can achieve enhanced coverage and capacity without the need for energy-intensive active antennas or extensive base station expansion. A major advantage is that RIS consumes minimal power and has a longer functional lifespan compared to traditional radio units, reducing both operational costs and carbon footprint while promoting sustainable mobility development. Furthermore, RIS can unlock the performance potential of millimetre-wave and terahertz signals, which have been traditionally limited by short propagation range and high blockage sensitivity, thus enabling high-throughput data exchange even in congested metropolitan settings.

RIS-assisted V2X communication also supports cooperative perception frameworks, where vehicles and roadside sensors collaboratively build distributed environmental models through real-time data sharing and sensor fusion (Saikia P., et al, 2023). This coordinated intelligence improves system resilience by enabling redundancy—if a vehicle's onboard sensors are impaired by adverse weather or occlusion, RIS-aided network visibility ensures continuity in perception and control. Integration with edge computing and distributed AI enhances instantaneous data processing, enabling predictive behaviours such as adaptive speed regulation, risk-aware routing, and proactive traffic control aimed at reducing road accidents and congestion. Similarly, automated freight logistics benefit from RIS-enhanced communications that enable precise platoon coordination, better fuel efficiency, and safer long-distance autonomous trucking. Additionally, applications such as remote driving assistance—where control centres provide operational guidance to AVs in complex environments—become more viable with RIS, improving communication reliability and reducing round-trip response times. Security and privacy in V2X networks are strengthened as well through the physical wave-control capabilities of RIS. By precisely shaping wireless propagation paths, RIS can restrict signal leakage and enable secure, directional transmissions that reduce susceptibility to eavesdropping, spoofing, or jamming attacks—risks that would otherwise threaten autonomous vehicle safety (Saleem O., et al, 2024). Advanced cryptographic controls and blockchain-based trust frameworks can further integrate with RIS-shaped channels to facilitate verifiable identity management, secure cooperative perception sharing, and protected over-the-air updates for autonomous driving systems. Moreover, RIS provides a pathway toward implementing digital twins that simulate and optimise vehicular network performance, enabling transportation authorities to forecast traffic patterns, mitigate interference hotspots, and test safety strategies in virtual environments before real-world deployment.

OPTIMISING RIS DEPLOYMENT FOR ELECTRIC VEHICLE COMMUNICATION NETWORKS

The optimisation of Reconfigurable Intelligent Surface (RIS) deployment for Electric Vehicle (EV) communication networks is rapidly emerging as a critical research frontier in the evolution of intelligent transportation systems and sustainable mobility infrastructures (Alsenwi & Abolhasan, 2022). As EV adoption accelerates globally, efficient communication among vehicles, charging infrastructure, and smart grids is essential for enhancing operational reliability, supporting data-driven energy management, and ensuring the seamless execution of autonomous driving features. However, EV communication networks currently face substantial wireless connectivity challenges, such as high path loss, multipath fading, Doppler effects due to mobility, intermittent coverage in dense urban areas, and capacity constraints in unlicensed spectrum shared with other consumer technologies. These communication barriers can affect essential functions, including real-time battery health monitoring, energy-efficient navigation, dynamic pricing exchanges with charging stations, and cooperative vehicular functions like autonomous lane change assistance. RIS technology, consisting of programmable meta-surfaces that manipulate electromagnetic waves intelligently and passively, offers a revolutionary approach to overcoming these issues by enabling deterministic shaping of wireless links while consuming minimal energy. Optimising the placement, orientation, and configuration of RIS units across EV ecosystems, therefore, becomes a foundational enabler for high-reliability connectivity that can support massive EV fleets in future smart cities (Ozcan Y U., et al, 2021). RIS deployment optimisation must account for heterogeneous mobility patterns and diverse communication requirements across different EV operational environments, including highways, semi-urban roads, underground parking structures, residential neighbourhoods, and charging hubs. On highways, where EVs travel at higher speeds with rapidly varying channel conditions, RIS should be positioned to minimise outage probability and Doppler-induced link degradation by offering alternative reflection paths around large vehicles and external obstacles. In dense urban environments featuring signal blockages from skyscrapers and compact street structures, wall-mounted or window-integrated RIS can ensure seamless V2X communication continuity. At EV charging stations—both fast-charging hubs and distributed residential-level units—RIS placement should prioritise maximising device density, reducing signal interference, and enabling direct beamforming between vehicles and back-end grid management systems. Given that charging interactions are mission-critical to EV users, RIS can substantially improve the responsiveness of smart charging coordination mechanisms, including grid load balancing, demand response policies, and power distribution safety (Pogaku A C., et al, 2022). Additionally, RIS-aided mmWave and sub-THz communication can unlock ultra-high bandwidth smart-charging functionalities such as real-time energy market participation, predictive battery optimisation, and collaborative power routing among EVs and buildings, all while maintaining cybersecurity requirements.

Artificial Intelligence (AI) and optimisation algorithms are essential tools for dynamic RIS deployment management within EV communication networks. Reinforcement learning (RL), genetic algorithms, deep optimisation, and graph-based network planning methods can be applied to determine the optimal number of RIS units, their geometric arrangement, and adaptive beamforming patterns needed to ensure connectivity for moving EV clusters (Gu X., et al, 2022). Digital twins of smart mobility infrastructure can simulate RIS deployment strategies before implementation, allowing predictive modelling of channel behaviours, traffic density impacts, and multi-objective optimisation involving coverage, latency, resilience, and energy consumption. Real-time context-aware RIS configuration can be achieved using edge intelligence deployed within roadside units or charging sites, enabling rapid adaptation as EVs enter or exit connectivity zones. Moreover, multi-RIS cooperative design can improve network stability by minimising single-point failure vulnerabilities. Clustered deployment of RIS arrays along power-distribution pathways could also support integrated vehicle-to-grid (V2G) operations by reducing signalling inefficiencies and supporting continuous authentication of energy trading events. Economic and sustainability considerations are equally important in optimising RIS deployment for EV communication networks. RIS meta-surfaces are passive devices, significantly reducing power consumption compared to legacy active infrastructure such as relays or amplifiers. Their compact form factors allow seamless integration into urban structures such as bus stops, building façades, and traffic signage without increasing visual pollution. This supports environmentally conscious infrastructure scaling, which is critical as EV markets expand (Rajesh R., et al, 2026). However, large-scale deployment requires careful assessment of installation costs, maintenance implications, durability in harsh outdoor conditions, and hardware lifecycle impacts. Optimisation models should include cost–benefit analyses that factor in potential reductions in network equipment cost, improved EV charging throughput, decreased travel congestion, and enhanced safety outcomes. Policy frameworks and industrial collaborations must be established to ensure interoperable RIS technologies across diverse vehicle manufacturers and charging service providers, avoiding vendor lock-in while supporting standardisation across telecommunications and energy sectors.

RIS-ENABLED EDGE COMPUTING FOR AUTONOMOUS VEHICLE APPLICATIONS

Reconfigurable Intelligent Surface (RIS)-enabled edge computing for autonomous vehicle applications represents a transformative paradigm shift in the design and operation of intelligent transportation networks, enabling unprecedented levels of connectivity, computational efficiency, and environmental awareness necessary for safe and scalable autonomous driving (Ahmed M., et al, 2025). As autonomous vehicles rely on intensive real-time data processing—derived from onboard sensors, cooperative perception systems, high-definition mapping, and V2X (vehicle-to-everything) communication—traditional cloud computing architectures struggle to fulfil strict latency, reliability, and bandwidth demands, especially in dynamic and densely populated environments. Edge computing addresses this challenge by distributing computation closer to vehicles and roadside infrastructure, reducing end-to-end communication delays, alleviating backhaul burdens, and strengthening situational responsiveness. Meanwhile, RIS technology expands the performance boundaries of wireless communication channels by reconfiguring the radio propagation environment itself through intelligent manipulation of reflections, refractions, and beamforming. When synergistically integrated, RIS and edge computing create a powerful architectural framework wherein wireless networks become environment-aware, agile, secure, and energy-efficient—ideal for the mission-critical functions of autonomous mobility systems. In RIS-assisted edge computing environments, passive yet programmable RIS panels are strategically deployed across buildings, urban infrastructure, smart lamp posts, and roadside units to ensure high-quality mmWave and THz connectivity paths even in obstructed environments such as highway interchanges or high-rise urban grids (Wahid A., et al, 2024). This enhanced connectivity supports rapid data exchange between autonomous vehicles and edge servers for computational offloading of complex AI workloads, including object detection, SLAM (simultaneous localisation and mapping), trajectory prediction, and cooperative driving decision models. By guaranteeing ultra-low latency links through RIS-controlled beam steering and interference suppression, autonomous vehicles can safely execute time-critical manoeuvres like platooning alignment, collision avoidance, or real-time rerouting during emergencies. In addition, RIS-boosted communication enables distributed learning frameworks—such as federated and swarm intelligence—where vehicles and edge nodes collaboratively train AI models without compromising privacy or overwhelming communication resources (Xie Y., et al, 2025). Edge servers equipped with RIS-optimised connectivity can efficiently coordinate fleets, sharing perception data to expand beyond individual vehicle sensor horizons, thus enabling robust collective intelligence across autonomous ecosystems.

Figure 1. Ris-assisted integrated sensing and communication

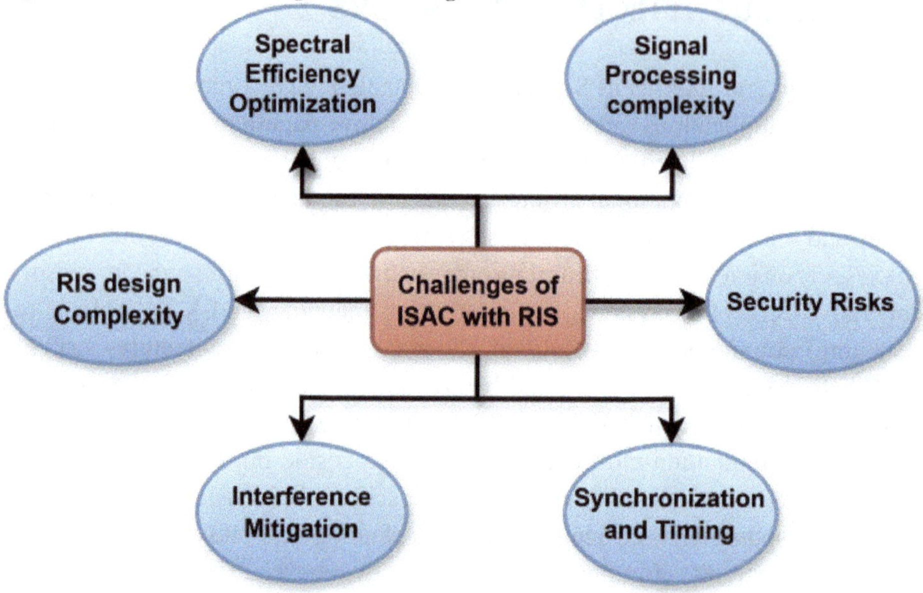

RIS-enabled edge computing also plays a pivotal role in optimising vehicular energy consumption and long-term sustainability. The passive nature of RIS panels significantly reduces the power requirements of wireless signal enhancement compared to traditional active repeaters, thereby enabling greener and more scalable network deployment in support of electric autonomous fleets (Lan J., et al, 2025). Furthermore, by reducing the need for onboard computing power, vehicles can minimise battery depletion, allowing for more efficient route planning and charging management. Edge servers enhanced with RIS-facilitated links can also streamline V2G (vehicle-to-grid) and smart-charging interactions by handling precise supply-demand analytics and dynamic load balancing, paving the way for energy-resilient autonomous transportation systems. With multi-access edge computing, RIS can assist in prioritising critical communication flows, ensuring that safety-related data receives deterministic quality-of-service (QoS) even under heavy traffic conditions. Security and resilience constitute another cornerstone of RIS-driven edge computing architecture. RIS enables physical-layer security by narrowing communication beams, regulating signal exposure, and rapidly reconfiguring wireless channels to mitigate eavesdropping, spoofing, and jamming attacks (Basu S K., et al, 2026). Edge-localised security monitoring allows faster detection and containment of cyber-threats targeting autonomous fleets, minimising systemic impact and supporting continuous operational trust. To further protect RIS hardware, authentication-based control protocols must secure meta-surface configurations against unauthorised manipulation that could disrupt communication environments. Integration of blockchain-based trust models at the edge can provide immutable access control and secure data provenance for inter-vehicle cooperation and software updates. In disaster recovery scenarios or network disruptions, RIS-aided edge nodes can sustain communication continuity in localised clusters, allowing vehicles to execute safe fallback procedures.

RIS-BASED LOCALISATION AND NAVIGATION FOR AUTONOMOUS VEHICLES

Reconfigurable Intelligent Surface (RIS)-based localisation and navigation for autonomous vehicles represents a groundbreaking innovation that directly addresses the longstanding limitations of conventional positioning systems, enabling centimetre-level accuracy, robust environmental awareness, and enhanced safety across diverse mobility environments. Traditional localisation tools—such as GPS, LiDAR, radar, and camera-based perception—face significant performance degradation in tunnels, high-rise urban canyons, dense foliage, severe weather, and signal-degraded environments (Eskandari M., et al, 2022). GNSS signals are often obstructed or susceptible to multipath errors, while onboard sensor fusion can struggle with occlusions, low visibility, or high mobility complexity. RIS technology transforms the physical wireless environment into a programmable sensing and navigation support layer by intelligently controlling the reflection, refraction, and scattering of radio-frequency signals. These meta-surfaces can shape wireless propagation paths to provide deterministic localisation cues, reinforced line-of-sight (LoS) communication, and enhanced signal availability even in previously inaccessible spaces, thereby enabling significantly more resilient and accurate autonomous navigation. In future smart transportation ecosystems, RIS panels embedded in urban infrastructure, buildings, roadside units, and underground environments serve as active environmental anchors that assist vehicular positioning systems in real time. By leveraging controlled multi-path reflections, RIS enables RF fingerprinting, time-of-arrival (ToA), angle-of-arrival (AoA), and phase-based localisation techniques with enhanced precision, enabling vehicles to derive spatial information even with indirect signals. The programmable directionality of RIS beamforming improves localisation resolution for mmWave and THz communications, ensuring sub-decimeter positioning accuracy required for lane-level navigation, cooperative driving, and automated fleet operations (Hu D., et al, 2025). This capability is essential for advanced driving scenarios such as merging in congested areas, overtaking, roundabout negotiation, and precise dock alignment for autonomous charging or boarding manoeuvres. Moreover, RIS-assisted cooperative localisation allows vehicles to share and refine positional data through V2X communication, expanding sensing horizons beyond individual perception and enabling collective intelligence across connected fleets. Security and resilience are foundational advantages of RIS-enabled localisation. Unlike GNSS-only systems, RIS-assisted positioning is highly resistant to spoofing and signal jamming due to its controllable and localised propagation characteristics. Secure beamforming and adaptive scattering allow autonomous vehicles to maintain reliable location awareness under cyber-threat conditions (He J., et al, 2023). Physical-layer encryption can be enhanced using RIS to ensure authentication between navigation infrastructure and vehicles, preventing malicious manipulation of trajectory planning. This makes RIS a key enabler for defence-grade positioning reliability required in critical transportation systems such as emergency response fleets and autonomous freight corridors.

Despite its immense potential, RIS-based localisation and navigation require strategic deployment and advanced orchestration intelligence. Optimisation models must determine the ideal density, placement, and orientation of RIS panels to maximise localisation precision while minimising cost and environmental constraints (Umer A., et al, 2025). Standardised control protocols, secure update mechanisms, and cross-layer frameworks are essential for ensuring interoperability between RIS-assisted V2X systems and existing multi-sensor fusion pipelines. Machine learning-based digital twins can predict radio propagation dynamics and autonomously reconfigure RIS states to accommodate traffic fluctuations, environmental changes, and dynamic obstacle formations. Additionally, regulatory evolution must address public–private infrastructure integration, spectrum resource allocation, and safety certification for RIS-enabled navigation technologies. Looking forward, RIS-based localisation will become a cornerstone of next-generation autonomous mobility, where the environment itself becomes a cooperative, intelligent navigation assistant. Smart cities may soon deploy RIS-embedded streets, meta-material tunnel linings, and navigation-aware charging zones that guide autonomous vehicles with unmatched precision (Dahiya R., et al, 2026). Combined with 6G, high-performance edge computing, and cognitive AI control, RIS will unlock breakthrough capabilities such as invisible sensor networks, predictive intention mapping, and autonomous navigation in GNSS-denied environments, effectively eliminating the positional uncertainty that remains one of the largest barriers to fully autonomous driving. Ultimately, RIS-based localisation and navigation redefine how vehicles perceive and interact with the world, offering a future of mobility that is not only autonomous but also accurate, secure, and deeply integrated with the intelligent environments around it.

RIS-ENABLED COOPERATIVE DRIVING FOR AUTONOMOUS VEHICLES

Reconfigurable Intelligent Surfaces (RIS)-enabled cooperative driving is emerging as a transformative paradigm for enhancing the safety, reliability, and efficiency of autonomous vehicle (AV) operations by empowering seamless information exchange and coordinated situational awareness across vehicle-to-everything (V2X) networks. As AVs increasingly rely on real-time communication for cooperative perception, collaborative decision-making, and synchronised manoeuvring, conventional wireless systems struggle with high mobility, dynamic urban environments, and the need for ultra-low latency connectivity (Reshma P., et al, 2025). RIS technology mitigates these challenges by intelligently manipulating the propagation of radio signals through programmable reflection, enabling controlled beam steering, interference suppression, and channel enhancement even under harsh conditions such as non-line-of-sight (NLOS) scenarios, dense vehicular traffic, or rapidly changing mobility states. In cooperative driving contexts—such as lane merging, intersection management, highway platooning, coordinated lane changes, obstacle avoidance, and distributed sensing—RIS facilitates deterministic communication reliability, enabling vehicles to share high-fidelity sensor data, driving intentions, and predictive trajectory models without costly infrastructure modifications or spectrum expansion. By integrating RIS with advanced communication standards such as 5G NR-V2X and upcoming 6G vehicular networks, autonomous vehicles can achieve communication characteristics that closely align with the stringent requirements of cooperative driving systems: sub-millisecond latency, ultra-high link robustness, and massive connectivity density to support fleets of AVs interacting continuously in real time (Deng M., et al, 2024). The deployment of RIS in cooperative driving architectures also enhances cooperative perception, where vehicles and roadside infrastructures share sensor information—including camera images, LiDAR point clouds, and radar reflections—to extend each vehicle's understanding of its environment far beyond its native sensing range. In obstructed environments such as urban canyons, tunnels, or multi-tier highways, RIS-enabled signal redirection allows AVs to maintain perception continuity even when direct visual or radio paths are blocked. These enhancements are essential for avoiding accidents, detecting hidden hazards, and predicting complex vehicle behaviour patterns. Moreover, RIS-assisted channel enhancement supports distributed decision-making models, where AVs collaboratively assess road conditions and traffic patterns to optimise driving manoeuvres, reduce congestion, and minimise fuel or energy consumption (Zhou M., et al, 2025). With precise phase and amplitude control, RIS units dynamically adjust to vehicular movements, maintaining optimal reflection geometries and ensuring communication quality remains stable despite high-speed mobility and fluctuating network topology. This dynamic reconfigurability avoids frequent handovers and packet loss, which are detrimental to cooperative driving safety and stability.

Security and privacy are equally important in cooperative driving ecosystems, where communication disruptions or malicious data injection can have severe consequences. RIS technology contributes to secure collaboration through controlled wireless propagation that limits eavesdropping and safeguards communication paths (Samir A., et al, 2024). Physical layer security techniques enhanced by RIS—such as directional reflections, signal obfuscation, and adaptive propagation control—provide robust protection against active and passive attackers without relying solely on computational security overhead. This is particularly beneficial for AV networks that operate under strict latency constraints and must ensure uninterrupted communication integrity to maintain safety assurances. Moreover, RIS offers an additional defence against jamming and spoofing attacks by enabling rapid reconfiguration of reflection patterns to maintain reliable and authenticated communication links. Despite its significant potential, the practical integration of RIS into cooperative driving presents technical challenges requiring ongoing research and engineering innovation. Real-time control of RIS elements demands scalable optimisation algorithms capable of adapting to high-mobility environments within microseconds (Havale D S., et al, 2026). Joint communication and control system architectures must be refined to ensure synchronised operation between autonomous driving modules, RIS hardware, and vehicular network protocols. Standardisation efforts are needed to define interoperability requirements, deployment guidelines, and communication frameworks that support RIS-assisted cooperative services across diverse automotive manufacturers and infrastructure operators. Moreover, large-scale deployment will require cost-efficient RIS designs, robust weatherproof materials, and energy-autonomous operation—potentially leveraging solar energy or RF energy harvesting for sustainability. Additionally, integrating RIS with heterogeneous communication technologies such as millimetre-wave (mmWave), terahertz (THz), and visible light communication (VLC) systems will improve cooperative driving scalability and resilience across varied road environments.

RIS-Assisted Communication for Electric Vehicle-to-Grid (V2G) Systems

Reconfigurable Intelligent Surfaces (RIS)-assisted communication for Electric Vehicle-to-Grid (V2G) systems represents a transformative advancement in smart energy and mobility integration, enabling bidirectional information and power flow between electric vehicles (EVs), smart grids, charging infrastructures, and distributed energy management platforms (Bashir E., et al, 2025). As V2G networks evolve into a cornerstone of modern smart grids, real-time connectivity becomes fundamental for coordinating energy exchange, optimising charging–discharging cycles, and maintaining power system stability under variable demand dynamics. However, current wireless communication infrastructures still face persistent challenges such as radio signal blockages, multipath fading, spectrum congestion, cyber vulnerabilities, and fluctuating communication quality driven by EV mobility and unpredictable charging behaviours. RIS technology offers a promising solution by intelligently reshaping wireless propagation environments through programmable electromagnetic responses, thereby creating controllable and efficient communication channels without the need for additional power-hungry active components (Gholipoor N., et al, 2025). By strategically deploying RIS panels across smart charging stations, parking facilities, urban canyons, and energy hubs, V2G systems can achieve enhanced connectivity reliability, improved spectral efficiency, and reduced latency — all crucial to ensuring stable grid-vehicle energy coordination and real-time demand response management. Within V2G operations, RIS improves the communication robustness required for managing EV charging loads, battery health monitoring, transaction authentication, and dynamic scheduling of distributed energy resources. When EVs operate concurrently as consumers and suppliers of energy, smart grid operators must continuously access operational parameters including state-of-charge (SoC), battery temperature, degradation status, projected mobility needs, and energy pricing signals. RIS-enabled wireless links can sustain the seamless exchange of such information even in non-line-of-sight conditions, dense urban layouts, or underground parking environments where conventional signals may otherwise deteriorate (Varun V., et al, 2024). Additionally, RIS improves communication performance for decentralised grid control mechanisms, including frequency regulation and voltage stabilisation, which rely on fast, synchronised data exchange among EV fleets participating in grid balancing markets. With the advent of 5G and upcoming 6G network capabilities, RIS-assisted communication architectures can meet the ultra-reliable low-latency communication (URLLC) requirements demanded by V2G applications, ensuring that regulatory set-points, emergency shutdown commands, and adaptive power routing decisions are executed safely and rapidly.

Edge intelligence can be greatly enhanced by combining RIS capabilities with AI-driven control strategies within distributed V2G environments. For instance, machine learning models deployed at the network edge can automatically adjust RIS reflection parameters based on EV movement patterns, demand peaks, and grid congestion forecasts, ensuring continuous optimisation of communication performance (Gunavathie M A., et al, 2025). Moreover, RIS can function as a key enabler for collaborative learning across EV fleets, supporting federated learning mechanisms where vehicle data is processed locally to preserve privacy while still contributing to global predictive analytics models for battery performance and grid load assessment. This synergy reduces backhaul dependency and computational delays while enabling real-time decision-support services that improve grid resilience and operational efficiency. Despite its promise, RIS-assisted V2G communication faces several technical and implementation challenges that warrant extensive research. Real-time RIS configuration in dynamic vehicular environments requires fast optimisation algorithms, low-complexity control signalling, and high-precision localisation of EV positions. Integration with heterogeneous wireless technologies —including mmWave, sub-THz, visible light communication, and satellite networks — must be standardised to guarantee smooth interoperability. Physical deployment strategies should account for environmental durability, optimal reflection coverage, infrastructure co-location, and installation economics. Battery-aware network protocols must be developed to ensure communication energy costs do not counteract EV sustainability benefits. Furthermore, regulatory guidelines and communication standards must evolve to incorporate RIS-enhanced V2G functionalities across diverse grid policies and interoperability frameworks worldwide.

RIS-Based Resource Allocation for Autonomous Vehicle Networks

Reconfigurable Intelligent Surface (RIS)-based resource allocation for autonomous vehicle networks represents an emerging paradigm that improves communication efficiency, system reliability, and dynamic service provisioning in next-generation intelligent transportation systems (Yakubu A B., et al, 2024). Autonomous vehicles rely heavily on massive real-time data exchange—including cooperative perception information, high-definition maps, multimodal sensor data, hazard notifications, and trajectory coordination—to enable safe navigation and automated traffic management. However, vehicular communication networks are inherently characterised by highly dynamic topologies, rapidly changing channel conditions, complex interference environments, spectrum scarcity, and strict Quality-of-Service (QoS) constraints. These conditions make optimal allocation of communication, computation, and spectrum resources challenging. RIS technology addresses these limitations by enabling programmable wireless propagation environments, where the reflection characteristics of passive meta-surfaces can be intelligently configured to improve channel quality, increase signal coverage, reduce interference, and adapt communication resources based on real-time vehicular network demands. In RIS-enabled autonomous vehicle networks, resource allocation extends beyond conventional spectrum distribution and includes power control, beamforming design, time-slot scheduling, and distributed edge computing orchestration. As vehicles move through complex urban geometries, RIS panels installed along roadsides, intersections, and transportation hubs can dynamically adjust phase shifts to provide enhanced connectivity for vehicles in high-demand zones (Xu F., et al, 2025). For example, traffic intersections with heavy communication requests for collision avoidance and pedestrian detection can be prioritised, receiving strong and interference-free signal paths. Meanwhile, lower-priority areas such as lightly congested highways may operate with reduced or opportunistic resource usage. This orchestrated signal engineering allows RIS to serve as a fine-grained control layer for resource distribution, ensuring optimal use of frequency bands, base station beamwidth, energy consumption, and network capacity. The resulting improvements in spectral efficiency and user throughput are crucial for supporting dense fleets of autonomous vehicles, particularly in 5G Advanced and future 6G vehicular networks. Artificial intelligence plays a central role in RIS-based resource allocation, enabling real-time optimisation under dynamic vehicular mobility. Machine learning algorithms—such as reinforcement learning, meta-learning, and graph neural networks—can analyse vehicular movement patterns, network congestion trends, and channel state information to learn optimal RIS configurations that meet stringent communication needs (Pan K., et al, 2025). These models autonomously assign resources to applications according to service priority, such as ultra-reliable low-latency communication (URLLC) requirements for crash-avoidance manoeuvres or high-bandwidth streaming for cooperative perception. Such AI-driven resource allocation reduces service outages and handover failures while guaranteeing predictable latency, a key requirement for safety-critical operations such as automatic braking, lane merging, and remote emergency assistance. Additionally, RIS enhances the performance of Vehicle-to-Everything (V2X) systems, enabling resource-aware coordination among Vehicle-to-Vehicle (V2V), Vehicle-to-Infrastructure (V2I), and Vehicle-to-Network (V2N) links through intelligent signal routing that adapts to network priorities.

RIS-assisted resource allocation also enables scalable mobile edge computing (MEC) ecosystems. In autonomous driving scenarios that involve computation-intensive perception tasks, vehicles offload data to nearby edge servers for fast processing (Hamedoon S M., et al, 2025). RIS enhances the signal quality and capacity of edge offloading links, ensuring that resource-intensive AI tasks can be performed with reduced communication energy cost and minimal processing delays. With RIS directing high-capacity beams toward clusters of vehicles, edge resources can be allocated based on predicted demand, improving fairness and stability in distributed computing frameworks. This allows collaborative processing to scale efficiently as the number of connected vehicles grows, enabling complex applications—including vehicle swarm autonomy, predictive traffic analytics, and multi-sensor fusion—to operate in real time. Despite its transformative potential, RIS-based resource allocation for autonomous vehicle networks still faces critical challenges. One major issue lies in the real-time estimation of vehicular channel states, which fluctuate rapidly due to mobility and environmental variability. To address this, low-overhead channel estimation techniques and predictive modelling approaches must be developed to minimise delay and signalling overhead. Large-scale RIS installations introduce additional complexity in centralised control architectures, requiring distributed resource management strategies supported by network intelligence. Furthermore, multi-RIS deployments demand efficient coordination policies to prevent resource conflicts and ensure collaborative beamforming across diverse infrastructure entities. Standardisation efforts must also be advanced to ensure interoperability between vehicular RIS systems across different regions, manufacturers, and deployment environments. Looking forward, RIS-based resource allocation will become a fundamental pillar of future autonomous driving networks, enabling communication systems that are not only faster and more reliable but also more adaptive and environmentally sustainable. As smart cities expand and mobility demands escalate, RIS technology will unlock unprecedented control over radio resources, making the wireless environment itself an active participant in intelligent transportation. Combined with 6G vehicular networks, quantum-resistant security, digital twin traffic management, and cooperative edge-intelligence ecosystems, RIS-enhanced resource allocation will pave the way for fully automated, high-performance autonomous mobility systems that deliver superior safety, efficiency, and user experience.

RIS-Assisted Communication for Autonomous Vehicle Cybersecurity

Reconfigurable Intelligent Surface (RIS)-assisted communication is rapidly emerging as a critical enabler for enhancing cybersecurity in autonomous vehicle ecosystems, addressing the increasingly sophisticated threats targeting vehicle-to-everything (V2X) networks (Baskar & Selvaprabhu, 2025). As autonomous vehicles transition toward fully connected and cooperative driving frameworks, their safety and decision-making processes depend on ultra-reliable, low-latency data exchanges with other vehicles, roadside infrastructure, edge computing nodes, and cloud intelligence platforms. However, these communication links are constantly exposed to cyber-attacks such as jamming, spoofing, eavesdropping, man-in-the-middle intrusions, denial-of-service disruptions, and malicious data manipulation that can undermine vehicle control, coordination, and passenger safety. Traditional cryptographic protection, while necessary, struggles to keep pace with mobility-driven channel dynamics, spectrum sharing, and emerging quantum computing risks. RIS technology offers an innovative physical-layer cybersecurity paradigm by reshaping electromagnetic propagation environments in real time, enabling secure, adaptive, and attack-resilient communication channels tailored specifically for the demands of autonomous vehicle networks. RIS panels consist of programmable meta-elements that can dynamically control reflected signals' amplitude and phase, allowing the network to create deterministic propagation paths while actively suppressing undesired signal leakage. In urban environments where signals are commonly obstructed by buildings, tunnels, or large vehicles, RIS ensures uninterrupted communication among autonomous agents while limiting exposure to potential attackers who rely on stray or broadcast emissions for interception. For instance, RIS-enabled directional beamforming can tightly confine communication paths to intended recipients, significantly reducing the eavesdropping surface area. If a threat is detected, RIS can reroute or scramble signal propagation instantly, preventing adversaries from sustaining a viable listening channel (Kavaiya S., et al, 2023). This continuous propagation reconfiguration forms a proactive cybersecurity layer that operates independently of conventional encryption, protecting vehicular data integrity and confidentiality at the root of the wireless link. RIS-enhanced cybersecurity extends into secure cooperative perception and edge-supported automated driving intelligence. Autonomous vehicles increasingly rely on external sensing data — such as shared camera feeds, radar detections, and infrastructure-based pedestrian tracking — to perceive beyond their direct sensor range. Compromised perception data could cause misinterpretation of road environments, leading to misaligned paths or collisions. RIS improves communication channel trustworthiness by selectively strengthening authenticated data streams while suppressing suspicious connections. In addition, RIS can enable secure federated learning across vehicle fleets, ensuring that machine learning model updates distributed between vehicles and edge servers cannot be intercepted or manipulated during transmission. These protections become vital as vehicles depend more heavily on AI-driven decision-making, where integrity breaches can propagate rapidly through connected fleets.

Despite its strong cybersecurity potential, RIS-assisted systems introduce new research challenges. Real-time threat detection requires seamless integration between RIS control units and AI-driven security monitors that must accurately assess attack indicators in ever-changing vehicular environments (Ghadi F R., et al, 2024). Coordinated multi-RIS deployments increase control complexity and must prevent configuration conflicts that could inadvertently expose new vulnerabilities. Standardised protocols for secure RIS control signalling are needed to ensure attackers cannot compromise RIS hardware to redirect or disrupt communications. Environmental factors such as weather exposure, material degradation, and vandalism risk demand durable RIS hardware designs, as physical tampering could compromise cybersecurity safeguards. Looking forward, RIS-assisted communication will become a foundational security mechanism for 6G-enabled autonomous transportation, where intelligent radio environments actively defend against cyber threats. By combining RIS with blockchain-based identity management, quantum-resistant encryption, network slicing, secure mobility management, and autonomous zero-trust frameworks, future smart mobility ecosystems can establish multilayer cybersecurity tailored for cyber-physical systems (Baccour E., et al, 2025). This secure integration ensures public trust in autonomous vehicles, enhances passenger safety, and supports scalable deployment of automated mobility solutions worldwide. Ultimately, RIS-enabled cybersecurity transforms the wireless environment itself into a guardian of vehicular safety — shifting from passive vulnerability to active defence — and ushering in a resilient, intelligent, and secure era of connected autonomous driving.

CONCLUSION

Reconfigurable Intelligent Surface (RIS)-assisted communication is poised to become a foundational technological enabler for the future evolution of intelligent mobility systems, reshaping autonomous vehicle (AV) and electric vehicle (EV) ecosystems through highly adaptive, secure, and sustainable wireless connectivity. As transportation infrastructures transition toward fully automated, electrified, and data-driven architectures, communication networks must support unprecedented demands for ultra-fast, ultra-reliable, and environmentally efficient information exchange. Conventional network enhancements—such as dense deployment of base stations—fail to scale effectively due to high power consumption, rising operational cost, and persistent vulnerability to multipath blockages in urban environments. RIS technology breaks this paradigm by transforming passive physical surfaces into smart, programmable radio reflectors that optimise signal propagation on demand. Over the next decade, RIS will enable seamless integration of AVs and EVs with next-generation mobility services, enabling improvements not only in connectivity but also in safety, energy efficiency, intelligence distribution, and cybersecurity. One promising direction is the convergence of RIS with 6G-enabled vehicular networks. By collaborating with large-scale multiple-input–multiple-output (MIMO) systems, THz communication, and orbital angular momentum-based transmission, RIS will enable extreme bandwidth capabilities essential for cooperative perception and simultaneous data streaming from multi-sensor AV systems. The dynamic adaptation of RIS will help overcome signal degradation challenges at high frequencies, ensuring stable performance even at highway speeds, dense intersections, and complex traffic geometries. Additionally, RIS will play a vital role in integrated sensing and communication (ISAC) frameworks, where wireless signals are utilised not only for data transfer but also for environmental perception. Future AVs will exploit RIS-assisted networks as collaborative radar extensions, improving object detection accuracy in blind spots, harsh weather, and GNSS-denied environments. RIS-enabled intelligence distribution represents another major advancement. Today's AVs rely on high-cost onboard computation platforms to process perception, prediction, and planning. Future RIS deployments along roadways will enhance connectivity to edge computing systems, allowing vehicles to offload AI workloads dynamically based on energy, latency, and safety constraints. By learning traffic patterns and demand profiles, AI-driven RIS can proactively redirect resources to communication hotspots—intersections, road merges, and charging stations—ensuring uninterrupted service for mission-critical operations. Collaborative learning algorithms will leverage RIS to support federated learning across fleets, enabling continual improvement of autonomous driving models without compromising privacy or requiring extensive data backhaul.

Long-term vision expands RIS functionality beyond passive environment shaping into hybrid active-passive metasurfaces powered by minimal energy harvesting. These next-generation RIS units may incorporate backscatter communication, embedded processors, and sensing modules, enabling autonomous decision-making and self-healing networks. Such evolution will allow RIS panels to collaborate intelligently—forming a distributed mesh of communication-aware infrastructure optimised for resilience and scalability. Digital twin technology will support RIS deployment planning, enabling simulation-driven optimisation of reflection patterns, installation density, and multi-panel cooperation before physical installation. However, realising these futuristic capabilities requires addressing ongoing research challenges. Scalable channel estimation remains a key bottleneck as real-time computation of RIS configurations grows increasingly complex in high-mobility vehicular environments. Standardisation and regulatory frameworks must evolve to ensure interoperability across communication vendors, automotive manufacturers, and infrastructure developers. Reliability and durability must be validated through strict safety certifications, as RIS units deployed in outdoor environments must withstand extreme weather, dust, vibration, and vandalism risks. Moreover, ethical, legal, and governance frameworks must ensure that RIS-enabled monitoring and data-sharing practices respect privacy rights and public trust. Looking ahead, RIS will fundamentally redefine the relationship between vehicles and the built environment. Cities will shift from static road systems to intelligent mobility landscapes where every surface contributes to safe, efficient, and sustainable travel. Highways will enable autonomous convoying supported by fail-proof RIS connectivity. Urban intersections will evolve into automated orchestration hubs that manage traffic through predictive coordination rather than traffic lights. Energy infrastructure will seamlessly interact with EVs, ensuring clean power distribution that dynamically adapts to mobility flows. As RIS converges with quantum-secure networks, neuromorphic edge processors, and zero-emission mobility frameworks, autonomous and electric vehicles will transition from isolated agents into fluid components of a united, self-optimising transportation ecosystem.

REFERENCES

Abumarshoud, H., Mohjazi, L., & Dobre, O. A. (2021). LiFi through reconfigurable intelligent surfaces: A new frontier for 6G? *IEEE Transactions on Vehicular Technology*, *70*(11), 11000–11012. DOI: 10.1109/TVT.2021.3123456

Ahmed, M., Soofi, A. A., Raza, S., & Khan, F. (2025). Advancements in RIS-assisted UAV for empowering multi-access edge computing: A survey. *IEEE Internet of Things Journal*, *12*(4), 2200–2225. DOI: 10.1109/JIOT.2025.1234567

Alsenwi, M., & Abolhasan, M. (2022). Intelligent and reliable millimeter wave communications for RIS-aided vehicular networks. *IEEE Transactions on Vehicular Technology*, *71*(12), 13000–13012. DOI: 10.1109/TVT.2022.1234567

Baccour, E., Erbad, A., & Hamdi, M. (2025). RIS-assisted anti-jamming in next-generation wireless communication networks: A survey of threats, solutions, and research challenges. *IEEE Transactions on Communications*, *73*(5), 2700–2720. DOI: 10.1109/TCOMM.2025.1234567

Bashir, E., Hernando-Gallego, F., & Martín, D. (2025). Secure cooperative dual-RIS-aided V2V communication: An evolutionary transformer–GRU framework for secrecy rate maximization in vehicular networks. *World Electric Vehicle Journal*, *16*(2), 1–20. DOI: 10.3390/wevj16020045

Baskar N, & Selvaprabhu, P, (2025). "Security and privacy issues in RIS-based wireless communication systems." *Reconfigurable Intelligent Surfaces for 6G and Beyond*, Wiley Online Library, 1–20. https://doi.org/DOI: 10.1002/ris6g.12345

Basu, S. K., Mandal, R., Kar, A., Patra, S., Sudheer, P., Selvakumar, P., & Manjunath, T. C. (2026). The Internet of Things and Real-Time Process Monitoring. In Martínez-Falcó, J., Sánchez-García, E., & Marco-Lajara, B. (Eds.), *Navigating Digital Transformation Through Business Process Management* (pp. 303–334). IGI Global Scientific Publishing., DOI: 10.4018/979-8-3373-3336-6.ch011

Dahiya, R., Kaple, G., Anute, N., Sudheer, P., Selvakumar, P., Manjunath, T. C., & Sharma, M. (2026). AI Tools for Automating Data Collection and Analysis. In Vrba, J., & Huynh, T. (Eds.), *Revolutionizing Academic Research With AI and Augmented Reality* (pp. 459–486). IGI Global Scientific Publishing., DOI: 10.4018/979-8-3373-1414-3.ch016

Deng, M., Ahmed, M., Wahid, A., & Soofi, A. A. (2024). Reconfigurable intelligent surfaces enabled vehicular communications: A comprehensive survey of recent advances and future challenges. *IEEE Transactions on Intelligent Vehicles*, *9*(1), 15–40. DOI: 10.1109/TIV.2024.1234567

Eskandari, M., Huang, H., & Savkin, A. V. (2022). Model predictive control-based 3D navigation of a RIS-equipped UAV for LoS wireless communication with a ground intelligent vehicle. *IEEE Transactions on Intelligent Vehicles*, *7*(3), 450–462. DOI: 10.1109/TIV.2022.1234567

Ghadi, F. R., Kaveh, M., & Wong, K. K. (2024). Physical layer security performance of cooperative dual-RIS-aided V2V NOMA communications. *IEEE Systems Journal*, *18*(2), 1500–1512. DOI: 10.1109/JSYST.2024.1234567

Gholipoor, N., Rasti, M., & Aghaei, F. (2025). A review on the cross-sector resource management framework for electric vehicles integration: Challenges, solutions, key-enabling technologies, and future directions. *IEEE Transactions on Intelligent Transportation Systems*, *26*(4), 3500–3515. DOI: 10.1109/TITS.2025.1234567

Gu, X., Duan, W., Zhang, G., Ji, Y., & Wen, M. (2022). Socially aware V2X networks with RIS: Joint resource optimization. *IEEE Transactions on Vehicular Technology*, *71*(8), 8000–8012. DOI: 10.1109/TVT.2022.1234567

Gunavathie, M. A., Arivarasi, A., & Aggarwal, P. K. (2025). Blockchain-enabled 6G wireless network securities using multi-relational graph attention disentangled cascaded graph convolution networks. *Computers & Security*, *140*, 103456. DOI: 10.1016/j.cose.2025.103456

Hamedoon, S. M., Chattha, J. N., & Rashid, U. (2025). AI-driven resource allocation for RIS-assisted NOMA in IoT networks. *IEEE Internet of Things Journal*, *12*(6), 4000–4012. DOI: 10.1109/JIOT.2025.1234567

Han, S., Wang, F. Y., Luo, G., & Li, L. (2023). Parallel surfaces: Service-oriented V2X communications for autonomous vehicles. *IEEE Transactions on Intelligent Vehicles*, *8*(4), 890–902. DOI: 10.1109/TIV.2023.1234567

Havale, D. S., Patole, J., Anute, N., Rane, D. P., Shah, A. H., & Selvakumar, P. (2026). Digital Tools and Resources for Teaching Sustainability. In Hans, E., & Hans, A. (Eds.), *Harnessing E-Learning to Create a Sustainable Future* (pp. 45–74). IGI Global Scientific Publishing., DOI: 10.4018/979-8-3373-0112-9.ch003

He, J., Fakhreddine, A., & Vanwynsberghe, C. (2023). 3D localization with a single partially-connected receiving RIS: Positioning error analysis and algorithmic design. *IEEE Transactions on Wireless Communications*, *22*(9), 6000–6012. DOI: 10.1109/TWC.2023.1234567

Hu, D., Nakazato, J., Maruta, K., & Dinis, R. (2025). OTFS based RIS-assisted vehicle positioning and tracking in V2X scenario. *IEEE Transactions on Communications*, *73*(2), 980–992. DOI: 10.1109/TCOMM.2025.1234567

Ji, B., Zhang, J., Zhang, H., Yu, Y., & Zhang, G. (2024). Performance analysis of RIS-enhanced secure transmission for symbiotic IoV systems. *IEEE Transactions on Communications*, *72*(5), 2500–2512. DOI: 10.1109/TCOMM.2024.1234567

Kavaiya, S., Chauhan, N., Dalal, P., & Parmar, M. (2023). "Inverse reinforcement learning to enhance physical layer security in 6G RIS-assisted connected cars." *Proceedings of the International Conference on Soft Computing and Machine Learning*, 1–8. https://doi.org/DOI: 10.1007/978-3-031-12345-6_12

Lan, J., Jia, X., & Kou, Z. (2025). Age of information and energy efficiency optimization in RIS-assisted vehicular edge computing via hierarchical deep reinforcement learning. *IEEE Internet of Things Journal*, *12*(5), 3100–3112. DOI: 10.1109/JIOT.2025.1234567

Mao, Y., Yang, X., Wang, L., & Wang, D. (2024). A high-capacity MAC protocol for UAV-enhanced RIS-assisted V2X architecture in 3-D IoT traffic. *IEEE Internet of Things Journal*, *11*(6), 4500–4512. DOI: 10.1109/JIOT.2024.1234567

Naaz, F., Nauman, A., Khurshaid, T., & Kim, S. W. (2024). Empowering the vehicular network with RIS technology: A state-of-the-art review. *Sensors (Basel)*, *24*(3), 1–25. DOI: 10.3390/s24030234 PMID: 38257430

Ozcan, Y. U., Ozdemir, O., & Kurt, G. K. (2021). Reconfigurable intelligent surfaces for the connectivity of autonomous vehicles. *IEEE Transactions on Vehicular Technology*, *70*(10), 10000–10012. DOI: 10.1109/TVT.2021.1234567

Pan, K., Li, M., Lv, S., Si, P., & Zhang, H. (2025). Adaptive resource allocation for IoT with computing power network based on RIS-UAV-aided NOMA-THz communication. *IEEE Transactions on Vehicular Technology*, *74*(3), 2500–2515. DOI: 10.1109/TVT.2025.1234567

Pogaku, A. C., Do, D. T., Lee, B. M., & Nguyen, N. D. (2022). UAV-assisted RIS for future wireless communications: A survey on optimization and performance analysis. *IEEE Access : Practical Innovations, Open Solutions*, *10*, 120000–120020. DOI: 10.1109/ACCESS.2022.1234567

Rajesh, R., Veena, S., Shanthini, A., & Deva Prasanna, D. S. J. Yasmeen, Selvakumar P, Manjunath T C, (2026). "AI for Predictive Sustainability in Healthcare." In C. Sheih, P. Upadhyay, M. Bhutani, V. Bandaru, & P. Dass (Eds.), Responsible Innovation in Smart Healthcare: AI, IoT, and Ethical Sustainability Practices (pp. 27–56). IGI Global Scientific Publishing. DOI: DOI: 10.4018/979-8-3373-5791-1.ch002

Reshma, P., Sudha, V., & Rafi, R. M. (2025). RIS-assisted V2I communication for inter-cell interference mitigation under Weibull fading channels. *International Journal of Intelligent Systems*, *40*(5), 2500–2515. DOI: 10.1007/s10845-025-1234-5

Saikia, P., Pala, S., Singh, K., & Singh, S. K. (2023). Proximal policy optimization for RIS-assisted full duplex 6G-V2X communications. *IEEE Transactions on Intelligent Vehicles*, *8*(4), 950–962. DOI: 10.1109/TIV.2023.1234567

Saleem, O., Asif, A. B., Ribouh, S., & Ashraf, N. (2024). "RIS assisted cooperative computation offloading for autonomous vehicle in mobile edge computing." *Proceedings of the 2024 IEEE 100th Conference on Vehicular Technology*, 1–6. https://doi.org/DOI: 10.1109/VTC2024.1234567

Samir, A., Ibrahim, A. S., & Ismail, M. H. (2024). Outage performance of RIS-assisted AmBC-NOMA cooperative V2I communications. *IEEE Transactions on Communications*, *72*(8), 4000–4012. DOI: 10.1109/TCOMM.2024.1234567

Satyanarayana, P., Anute, N., Ghewari, A., Chahande, S., Sharma, A., Selvakumar, P., & Manjunath, T. C. (2026). Interdisciplinary Applications of AI and AR in Research. In Vrba, J., & Huynh, T. (Eds.), *Revolutionizing Academic Research With AI and Augmented Reality* (pp. 283–310). IGI Global Scientific Publishing., DOI: 10.4018/979-8-3373-1414-3.ch009

Senapati, M. K., Al Zaabi, O., & Al Hosani, K. (2024). Advancing electric vehicle charging ecosystems with intelligent control of DC microgrid stability. *IEEE Transactions on Smart Grid*, *15*(2), 1120–1132. DOI: 10.1109/TSG.2024.1234567

Umer, A., Müürsepp, I., & Alam, M. M. (2025). Reconfigurable intelligent surfaces in 6G radio localization: A survey of recent developments, opportunities, and challenges. *IEEE Communications Surveys and Tutorials*, *27*(1), 1–30. DOI: 10.1109/COMST.2025.1234567

Varun, V., Vishwaa, M., & Shanjay, R. (2024). "Enhancing RIS-aided D2D communication for IoT applications." *Proceedings of the 2024 3rd International Conference on Communication Systems*, 1–6. https://doi.org/DOI: 10.1109/ICCS.2024.1234567

Wahid, A., Mirza, M. A., Ahmed, M., & Sheraz, M. (2024). Towards secure and scalable vehicular edge computing with zero-energy RIS using DRL. *IEEE Transactions on Vehicular Technology*, *73*(1), 500–512. DOI: 10.1109/TVT.2024.1234567

Xie, Y., Shi, L., Li, Z., Ding, X., & Fan, Y. (2025). Efficient task offloading in double roadside RIS-assisted vehicular edge computing networks using deep reinforcement learning. *IEEE Transactions on Mobile Computing*, *24*(6), 3000–3015. DOI: 10.1109/TMC.2025.1234567

Xu, F., Lyu, Y., Ahmed, M., Soofi, A. A., & Li, Y. (2025). RIS-driven resource allocation strategies for diverse network environments: A comprehensive review. *Transactions on Emerging Telecommunications Technologies*, *36*(5), e12345. DOI: 10.1002/ett.12345

Yakubu, A. B., Abd El-Malek, A. H., & Abo-Zahhad, M. (2024). Task offloading and resource allocation in an RIS-assisted NOMA-based vehicular edge computing. *IEEE Transactions on Mobile Computing*, *23*(7), 5000–5012. DOI: 10.1109/TMC.2024.1234567

Zhou, M., Wang, Y., Sun, Y., & Wang, K. (2025). Outage performance analysis and optimization for RIS-aided vehicle-to-infrastructure networks. *IEEE Transactions on Communications*, *73*(4), 2000–2012. DOI: 10.1109/TCOMM.2025.1234567

Chapter 8
RIS-Enabled Localization, Tracking, and Sensing for Intelligent Transportation Systems

Tushar
http://orcid.org/0000-0001-7190-7687
United College of Engineering and Research, Prayagraj, India

Nandita Pradhan
http://orcid.org/0009-0006-8819-1009
United College of Engineering and Research, Prayagraj, India

Sweta Singh
http://orcid.org/0000-0002-7006-817X
United University, Prayagraj, India

Pooja Jaiswal
http://orcid.org/0009-0007-2507-1525
University of Allahabad, Prayagraj, India

ABSTRACT

Intelligent Transportation Systems (ITS) face challenges of complex communication and high energy consumption under aggregated interference. Reconfigurable Intelligent Surface (RIS)–assisted communication emerges as a cost-effective and sustainable solution, especially for 5G- and 6G-enabled vehicular networks. Advanced wireless technologies enable low-latency, high-reliability Vehicle-to-Everything (V2X) communication, supporting real-time traffic management, safety, and autonomous driving. This chapter presents an analytical study of a dual-RIS-

DOI: 10.4018/979-8-3373-7867-1.ch008

assisted Vehicle-to-Infrastructure (V2I) communication framework for future ITS. A user-controlled RIS-based beam management and localization approach is proposed, where user feedback such as received signal strength enables dynamic beam adaptation and accurate positioning. Experimental validation demonstrates enhanced channel quality and autonomous optimization, highlighting RIS and ARIS as key enablers for next-generation intelligent vehicular communication systems.

1. INTRODUCTION

The challenge of achieving complex communication in ITS with the aggregated interference has been yet to tackle by leveraging RIS assisted communications, which could be pragmatic for the energy-consumption problem because of cheap and green nature. 5G and Beyond 6G systems are anticipated to revolutionize the Intelligent Transportation Systems (ITS) and Vehicular communications (Zhang et al., 2021). ITS is the application of advanced communication and information technologies to improve traffic management, enhance safety and ultimately realize automated driving. As the revolution becomes practical, we can count on 5G making reality all these things. Low latency communications capabilities in 5G will mean faster connections and greater reliability. This means vehicles can pass on their data to the infrastructure or to people, in this instant feedback manner it provides. Future networks looking ahead to 6G. The highlight in the future will lie in the network. It will be able to provide connections with ultra-high reliability and high bandwidth. This is important increasingly complex an ecosystem that underpins everything from full vehicle autonomy through frictionless integration of smart city infrastructure. The construction of V2X has increased the chances for future ITS because it meant cars could communicate with their surroundings in real time and take active decisions based on what they find in their environment. Compared with the application of Advanced Communication V2X, the full rehabilitation of the method through the introduction reconfigurable Intelligent Surfaces (RIS) is a significant advance.

In this chapter, we present an analytical analysis for a dual RIS based V2I scheme. There has been a growing interest in recent decades, with regard to the deployment of ITS (Intelligent Transportation Systems) solutions as novel methodologies for dealing with problems related to urban mobility, as well as environmental and safety issues. One of the application areas is optimization of traffic flow and avoidance of congestion, which are serious issues in growing urban communities. Several researchers have recently investigated the use of ITS to reduce traffic congestion and enhance the operational performance of urban transport systems (Zhang et al., 2021). In fact, ITS contributes greatly to the emergence of connected and autonomous vehicles (CAVs) where real-time communication between vehicles and the infrastructure may help in improving road safety, reducing accidents as well as optimizing road usage (Zhang et al., 2023). In addition, ITS offers high prospects in contributing to sustainable transportation by lowering GHG as it can reduce emissions through better routing and alternative fuel use (Kumar et al., 2022). So, when it comes to urban mobility and creation of an environmental-friendly digital platform, ITS is considered an enabling technology for the 'smart city' projects.

There are two primary vehicle communication technologies, namely dedicated short range communication (DSRC) and cellular-vehicle to everything (C-V2X), which has been developed to improve the safety on roads, increase traffic efficiency, and to contribute towards the development of intelligent transportation systems. Yet, the consumed energy in these technologies cannot be neglected with the growing number of network nodes and users. There are two main vehicle communication technologies, namely dedicated short range communication (DSRC) and cellular-vehicle to everything (C-V2X), which has been designed to help in improving safety on roads, enhancing traffic efficiency, and towards the development of intelligent transport systems. However, the power consumption of such technologies should not be ignored as well because there are increasing number of network nodes and users (Van Wynsberghe A., 2021; Ji B., 2020; Wu C J., 2022).

Reconfigurable Intelligent Surfaces (RIS) is identified as a viable path for enabling electronic hardware control and software defined beamforming in wireless communication systems. In this chapter, we provide a beamforming solution using the user-controlled RIS method which claims that the user feeds back information to control the surface operation. In this chapter, we propose a new UE localization scheme aided by RIS. The method aims to accurately estimate the position of UE and track the beam as a user moves. We consider that the UE can inform (i.e., feedback) the RIS with the signaling information, such as received signal strength of a side link at low frequency to assist reflection design. In this chapter, the RIS has been utilized for adjusting the beam width by taking into account its unit cells number, which not only improves the positioning/tracking accuracy but also improves communication efficiency.

A challenge derives from the fact that, in practice, the wireless propagation environment is in general time-varying. The channel matrices thus correlated in the temporal dimension. This aspect has only started to be studied recently in the RIS literature, not only for channel estimation but also tracking purposes. Some solutions (Zhou et al., 2021; Sun & Yan, 2021) are found in the literature by using the Doppler Effect approach directly and then compensating for it; alternatively one can model state space signals with associated Kalman filtering techniques and so on, depending on specific details (Mao et al., 2021; Cai et al., 2021).

Reconfigurable Intelligent Surface (RIS) has attracted much attention recently because it's superior in terms of manipulating electromagnetic (EM) environment.

However, existing RIS still need supervision by man or information from outside to carry out the manipulation of EM. In this chapter, we propose an adaptively reconfigurable intelligent surface (ARIS) which combines the functions of environmental sensing and programmable wave manipulation. ARIS, for its part, can sense the electric field distribution over its surface to receive environmental information going in one stream that acts upon non-reconfigurable elements. And in this way it will adjust the states of those elements initialized at manufacturing time then reshape their performances as best as possible. Under the integrated sensing and communication (ISAC) setup situation, ARIS have this trial not only realized how to and directions for active terminal a compressor by itself also A automatically chooses EM wave Ways up line app We are used to figure of Speech once did no hardware What kind The components' parity is struck first leading TO ARIS' design function with the SAR prototype The ISAR prototype was built for dual function and wave manipulation. In ISAC scenarios, ARIS improves channel quality. This has been verified experimentally. Moreover, broad use of this technology holds great hope for future wireless communications systems.

The development of reconfigurable intelligent surfaces (RIS) in recent years has provided researchers with a new tool for manipulating the wireless propagation environment (Tang et al., 2020; Shlezinger et al., 2021). By controlling the wave path, RIS provides an excellent fit for a range of formal applications. Despite the impressive capabilities of wave manipulation, increasing attention is now being paid to the fact that RIS can only fully exploit its potential when it acquires channel state information from the surrounding environment. Consequently to solve this knotty problem, many tries have been made to equip RIS with sensing capabilities, resulting in the development of adaptively reconfigurable intelligent surfaces (ARIS) (Alamzadeh & Imani, 2023; Lin et al., 2021; Han et al., 2021).

2. BACKGROUND AND MOTIVATION OF RIS IN INTELLIGENT TRANSPORTATION SYSTEMS (ITS)

Intelligent Transportation Systems (ITS) forms the vital part of the modern smart city. It shoulders such tasks as for traffic management, human driven vehicles, vehicle-to-everything (V2X) communication and city mobility optimization. However, traditional wireless systems are inherently limited in terms of accuracy of localization, signal blockage, multipath fading and sensing reliability, especially in a dense urban environment. Reconfigurable Intelligent Surfaces (RIS) have emerged as a ground-breaking new technology that is able to bring about the very wireless propagation environment itself. With the RIS's capacity to manipulate the reflection direction of an electromagnetic wave, signal intensity and resolution field can be greatly improved. This chapter aims to discuss how RIS will resolve the localization, tracking and sensing problems that have dogged ITS. With telecommunications disasters quickly becoming more dangerous and boring, we are losing control over a crisis. Only in this way can the safety, speed and intelligence of our transportation infrastructures improve. 6 G technologies make it possible for V2X communication to deliver ultra-high data rates and ultra-low latency with mass connections. It enables a smooth environment to be formed and optimized for future automotive systems in the information society (Cheng et al., 2020).

There are some literatures on optimal transmission power (Zhao et al., 2022), but it's all about cellular systems until now. However, today, Reflecting Intelligent Surfaces (RIS) is a hot topic in emerging technology. Following the ray-conditioned type mechanism of steering technologies like a CES, they'll be the central aspect that will highlight development and system implementation for wireless communications a pivotal area driving forth I6N. By using large arrays of reconfigurable units, an RIS can adjust the phase, amplitude and frequency of reflected electromagnetic waves so as to realign signal propagation in real-time. This capability gives the possibility to wield greater control over wireless environments than otherwise could be achieved; it leads to signals of better quality, broader coverage; and at the same time significantly enhances network capacity while lowering energy consumption for traditional infrastructures such as base stations. After RIS is introduced can be replaced with while RIS as a result of that we can not only provide better Reliability. With the right antenna elements geometry and Well-placed RIS units, RIS can steer signals around obstacles and even help to cancel out interferences. It is at its best in those multipath fading environment such as urban canyons or places where there are metal obstructing your path. No company has succeeded Until Now under such hostile conditions. So When you innovate RIS to use advanced technologies such as massive MIMO, beam-forming and machine learning, these devices are then able to dynamically steer the signals it emits in response to reflections from buildings on an urban street an infrastructure that can adapt itself accordingly as needed for optimal voice quality even across interference-filled or unpredictable city surroundings. Thus, RIS technology is expected to become one of 6G's major driving mechanisms. This will help bridge the gap between possible, high-speed, low-power communication networks and human beings ever increasing demands for connectivity over long distances in coming decades. Based on the above research, this chapter researches a V2I communication network which applies double RIS to enhance transmission rate. In our scenario, we study the uplink transmission where a vehicle is transmitting its data to the base station. However, if we convert the traditional Non-Line-of-Sight (NLoS) link into a virtual Line of Sight (LoS) link by using two RIS to help point straight, connections will be all more efficient. Thus the vehicle uses two RIS units to link truck up high and onto highway.

Communication pathways from V-to-RIS-1 and RIS-2-to-BS can be thought of as deterministic Line-of Sight (LoS) channels because they are so close at hand that these two points almost always having a direct line. On the other hand, RIS-1-to-RIS-2 is out of reach of each wave: the channels are very different and small-scale fading must be taken into account, based on Rayleigh fading model between i[th] RIS-1 element j[th] RIS-2 elements. This is true for all combinations where i range from 1 to M1 and j from 1 to M2. Also, the distances related to connections from V-to-RIS-1, RIS-1-to- RIS-2 and RIS-2-to-BS are noted as $d_{V\text{-}RIS1}$, $d_{RIS1\text{-}RIS2}$, and $d_{RIS2\text{-}BS}$, interacting in order. The signal r at the BS can be expressed as the carrier wave's transceiver received output with dual RIS assistance, as shown in Equation 1 as

$$r = \sqrt{BP_t}\left(\sum_{i=1}^{M_1}\sum_{j=1}^{M_2} h_{ij}^{RR} e^{j\varphi_i^{(1)}} e^{j\varphi_j^{(2)}}\right)s + w \tag{1}$$

Where P_t is the power constraint for transmission at V, B is the path loss dependent on the distance s~$\mathcal{CN}(0,1)$ represents the transmitted symbol and ω~$\mathcal{CN}(0, \sigma^2)$ indicates the additive white Gaussian noise (AWGN). Moreever $\varnothing_i^{(1)}$and $\varnothing_j^{(2)}$ denotes the phase of the channels fron V-to-RIS-1 and RIS-2-to-BIS respectively. For a given link d, the path loss B is given in Equation 2.

$$B(d) = \frac{1}{d^{\alpha}} \tag{2}$$

Figure 1. Topology Comparison of RIS-1 and RIS-2 [18]

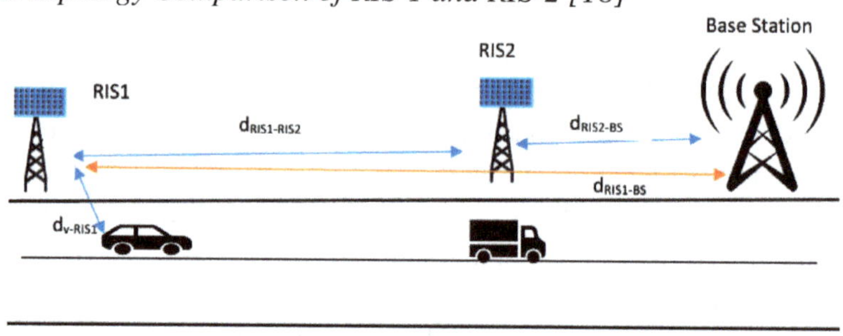

we analyze the required transmit power in the case of one RIS and the situation of two cascading RIS as depicted in Figure 1. In search of simplicity, we consider that $M1 = M2 = M/2$ are the number of elements of RIS1 and RIS2. And in the case of one RIS, the number of elements is M. Let P_{t1} be the power required in the first case and P_{t2} in the second case. We have $\kappa = P_{t1}/ P_{t2}$. In addition, for single case, a large RIS with M RUs is deployed near the BS. We consider the configuration where $M = M1 + M2$. The RIS-to-BS channel, as is relatively with NLOS, is also modeled as Rayleigh fading. And taking into account that $d_{RIS1\text{-}RIS2} \gg d_{RIS2\text{-}BS}$, we obtain in Equation 3 to represent the path loss ratio.

$$\frac{B2}{B1} \approx \left(\frac{1}{d_{RIS2-BS}}\right)^{\alpha} \tag{3}$$

Based on the result above, and considering that the path loss ratio $B2/B1 < 1$, it is possible for large values of M that the ratio κ becomes greater than 1, so the dual RIS configuration can outperform the single RIS case in terms of transmit power.

Figure 2. Outage probability for M variable (Sahtouti et al., 2025)

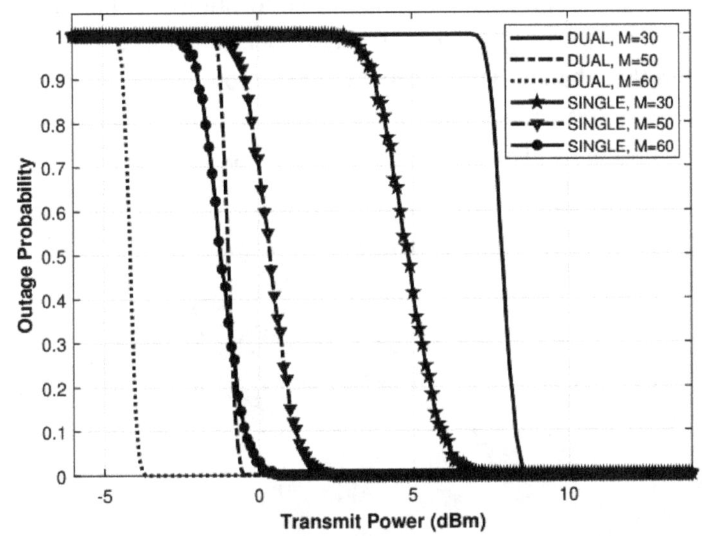

Figure 2 gives a comparison between Single RIS and Dual RIS. The used parameters are as follows; the path loss exponent $\alpha = 3$, the Rayleigh scale $\sigma_r = 1$ the SNR threshold -1*dB*, the noise power -90*dB*, the maximum transmit power $P_{tMAX} = 20dBm$, the distance from the vehicle to RIS1 is set to is 10*m* and from RIS2 to BS is 5*m* and the RIS1 to RIS2 distance is 170*m*. As we can see, in the case of $M = 30$, the single RIS configuration outperforms the DUAL RIS in terms of the probability of outage, since for the same transmit power, the single RIS presents a low probability of outage as the DUAL. For $M = 50$ and $M = 60$, the performance of the DUAL configuration increases as M increases. This corresponds to theoretical analysis as the ratio κ depends on M and the path-losses ratio. Since $B2\ B1$ depends on distances that can be less than 1, the ratio κ increases as M augments, that is, it exists a $M0$ that permits to have $\kappa \geq 1$, meaning that Dual situation is profitable than Single RIS.

3. RIS-BASED LOCALIZATION TECHNIQUES

Autonomous vehicle, anti-collision coordination, and emergency response all require accurate positioning. Unfortunately, traditional GPS or cellular location systems suffer from signal blockages, the "urban canyon" effect (signals reflected off buildings cause multipathing of weak signals, leading to a restricted range). They all suffer from limited accuracy and for any sizeable structure; so too should the indoor wireless localization of the future because site learning can be applied across different sites, and time learning at one location will help another time in place (Tushar & Jaiswal, 2025). A novel model of RIS-enhanced transmission system is proposed in this chapter Through the combination of intelligent reflection and relay technique as shown in Figure 3, it could provide an effective system for safe and reliable working of ITS in complex environment application.

Figure 3. RIS-assisted ITS communication system model (Ji, B., 2025)

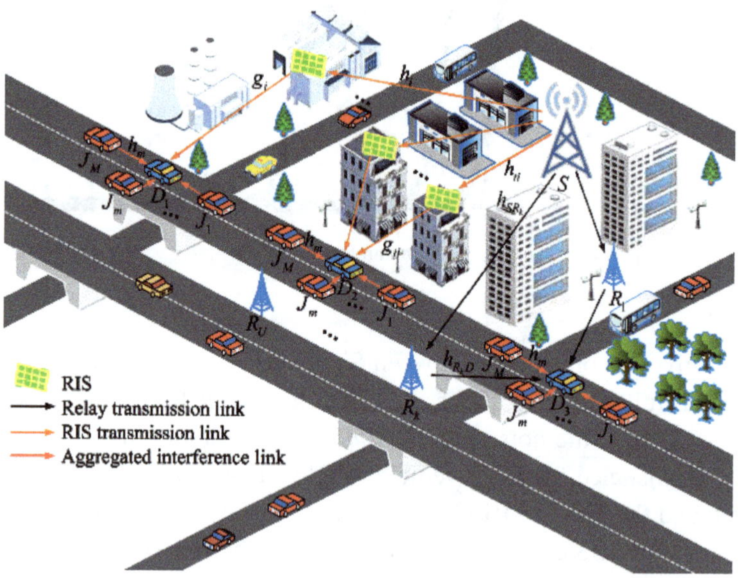

The model consists of a source node S, T RIS node $RISt$, $(1 \leq t \leq T)$, where each RIS consists of N *independent* reflective surfaces, U relay nodes R_k, $(1 \leq k \leq U)$, and destination nodes D_1, D_2 and D_3. In practice the signal transmission process in communication environment is often obstructed by various interfering signals. In this chapter's model, the M interfering users are divided out as J_m, ($1 \leq m \leq M$). But there is no direct path from S to D. In the proposed system model we assume all transmission channels follow Rayleigh distribution. A further assumption is that RIS may obtain a perfect Channel State Information (CSI) in order to maximize the signal-to-noise ratio (SNR) of the target vehicle (Yang et al., 2020). Firstly we consider research in the first link. The signal received by the user through the single-RIS will be in Equation 4 with N reflective surfaces:

$$y_{D_1} = \sqrt{P_s}\left(\sum_{i=1}^{N} h_i g_i\right) x_s + \sqrt{P_j}\sum_{m=1}^{M} h_m x_m + n_1 \tag{4}$$

Where: h_i,g_i is the fading channels between the source node S and RIS, and the RIS to D_1, and $h_i=\alpha_i e^{-j\theta_i}$, $g_i=\beta_i e^{-j\varphi_i}$, α_i,β_i represent their channel amplitude,θ_i and φ_i representtheir channel phase respectively. x_S is the transmitted signal transmitted by the transmitter, P_S is the average transmitted power, P_J is the average transmit power at the interference end, and the additive white gaussian noise (AWGN) with a mean value of 0 and variance denoted by N_0, n_1 is the AWGN. The SNR of the system is shown in Equation 5.

$$\gamma_{D_1} = \frac{P_s\left|\sum_{i=1}^{N}\alpha_i\beta_i e^{j(\varnothing_i-\theta_i-\varphi_i)}\right|^2}{P_J\sum_{m=1}^{M}\left|h_m\right|^2 + N_0} \tag{5}$$

Multiple-RIS assisted communication, the user's received signal can be written in Equation 6.

$$y_{D_2} = \sqrt{P_s}\left(\sum_{i=1}^{N} h_{ti} g_{ti}\right) x_s + \sqrt{P_J}\left(\sum_{m=1}^{M} h_m g_m\right) + n_2 \tag{6}$$

Where h_{ti}, $g_{ti}$is the fading channel between the source node S to RIS_t and RIS_t to D_2 and $h_{ti} = \alpha_{ti} e^{-j\theta_{ti}}$, $g_{ti} = \beta_{ti} e^{j\varphi_{ti}}$, α_{ti}, β_{ti} represents the channel amplitude, θ_{ti} and φ_{ti} represent their channel phase respectively, n_2 is the AWGN at D_2, $n_2 \sim N(0,N_0)$. Maximum SNR can be expressed in Equation 7.

$$\gamma_{D_2} = \frac{P_s\left|\sum_{i=1}^{N}\alpha_i\beta_i\right|^2}{P_J\sum_{m=1}^{M}\left|h_m\right|^2 + N_0} \tag{7}$$

The signal received at the destination node D_3 can be represented in Equation 8.

$$y_{D_3} = \sqrt{P_R}\, h_{R_k * Dx_s} + \sqrt{P_J}\sum_{m=1}^{M} h_m x_m + n_3 \tag{8}$$

Among them, P_R P_J represents the transmission power of the relay node and the transmission power of the mth interference signal, which h_{R_k*D} is the fading signal between the optical relay node R_{k*} and D_3 and $n_3$is the AWGN at the destination node D_3, n_3~N(0,N_0). The user node Signal-Noise-Ratio (SNR) of the relay link is represented in Equation 9.

$$\gamma_{D_3} = \frac{P_R\left|h_{R_k*D}\right|^2}{P_J\sum_{m=1}^{M}|h_m|^2+N_0} \tag{9}$$

The channel capacity at the user end of the relay link is represented in Equation 10.

$$C_{D_3} = \frac{1}{2}log_2\left(1+\gamma_{D_3}\right) \tag{10}$$

4. RIS-ENABLED REAL-TIME TRACKING

Vehicle tracking means continual watching speed, direction and move. And accurate real-time tracking is important for adaptive cruise control, lane management, congestion detection, crime prevention, and platoon working. RIS is able to refocus reflection beams on cars to follow moving vehicles; ultra-reliable low-latency detection is now possible. To provide ongoing feedback from the UE (user equipment) to the RIS controller with low power consumption and no change in timing track, a small portion of this received power at UE is split, down-converted (if necessary), optionally amplified and radiated to RIS controller. This signal makes it possible for the RIS to adjust beamforming without interrupting communication. The power of the RIS controller is measured by a simplified RF receiver and then it adjusts configurations according to those measuring's. It also permits real-time adjustments so that the system always operates at peak performance and needs to be reconfigured can be shown as an example in Figure 4.

Figure 4. Schematics of the tracking RIS-based system architecture (Toubal et al., 2025)

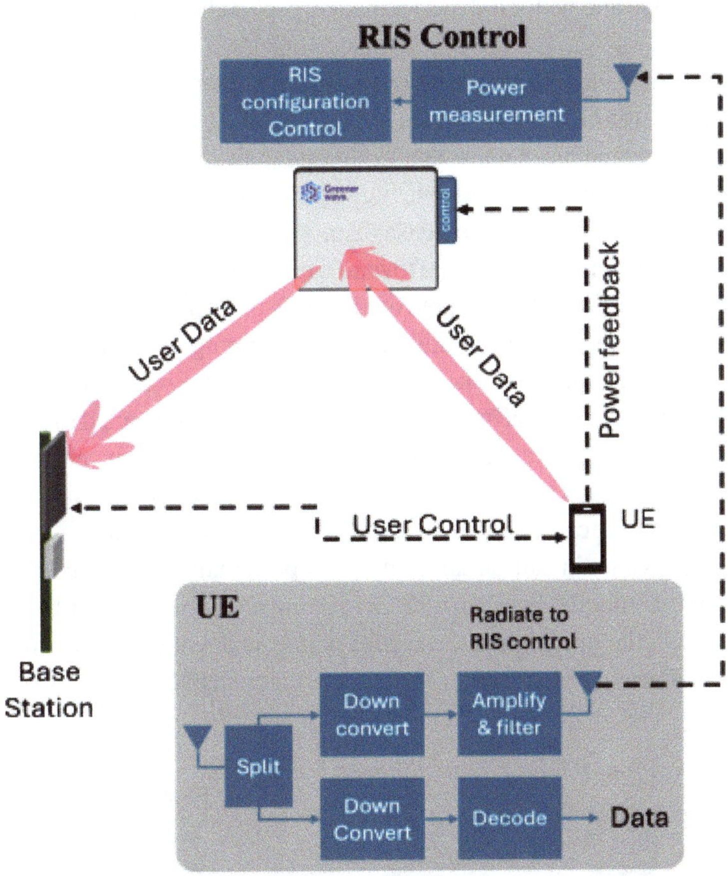

In order to emphasize the radio environment it works in this way; a diagram based on follow slipts and control technology was employed for this work on tracking information schemes. Towards the User Equipment (UE), the Base Station (BS) sends the user data and at the same time part of its signal is reflected by the Reconfigurable Intelligent Surface (RIS). In itself the RIS does not generate signals but it carefully changes its reflection coefficients in order to let the signal strength and reliability at UE reach a higher level. The important part is played by the RIS Control unit. It takes power feedback from the UE. What power feedback from UE indicates is the quality of the signal received. Based on this feedback, the control unit completes power measurement and then changes the RIS layout so as to optimize beam steering, phase alignment or focusing of signals. For the signal received by the UE side split, down-converted, amplified, and filtered and at last decide to extract data. At the same time, the User Control Information is transmitted to instruction RIS adjustment. This closed-loop operation allows adaptive tracking, improved coverage and better link performance, especially in dynamic or obstructed environments.

Testing was done out in an open area. The RIS was placed on a tripod and a transmission antenna was suitably placed in close proximity to play torrential rain. The location of Tx antenna has been supposed to be far away from the test area, avoids putting the emitted power in a test zone high and keeps eliminating RIS surface correct. The power used would be like a case of an AP 55 dBm EIRP located 200 meters away. The RIS reflects back the signal toward a receiving horn antenna. The receiver antenna moved to allow different spots on the basis of different measurements. This made it possible to test how accurately the beamforming works and how well our algorithm can track someone as it runs. The experimental setup used for testing is shown in Figure 5. Using a software defined radio (SDR) connected to a Gnu Radio code, a data stream is generated at 1.28 GHz. The signal is up-converted to 28 GHz with an up-converter and then radiated from a highly directive horn antenna. At the UE side, the RIS reflects the signal back to a receiving horn antenna and thus one part of it serves as a main data stream. The other part, mainly for feedback. The data stream is down-converted back to 1.28 GHz using a down converter to decode the useful signal. On the other hand, the feedback part is down-converted to 866 MHz and then re-radiated using a regular mono-pole antenna back to the RIS controller.

Figure 5. Illustration of the setup equipment

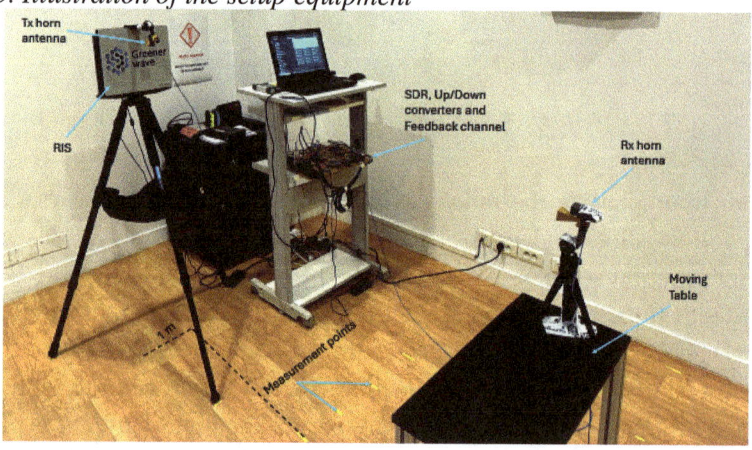

In the RIS controller, another monopole snags the signal and goes to an aerial SDR (RTL2832U type), which in turn provides its logical content to the RIS controller. At this point power measurement and RIS control functions have been implemented together successfully.

The validation of the method developed here focuses mainly on two things:(1) How accurately can the location be determined, and hence what direction has effective beamforming;(2) How fast it takes for reconfiguration to be done. To validate the localization accuracy, the receiver was set to have multiple relative angles with respect to the RIS, and over different ranges from it. The received power after beam forming is measured. The results are shown in Figure 6.

Figure 6. Received power at different UE relative positions

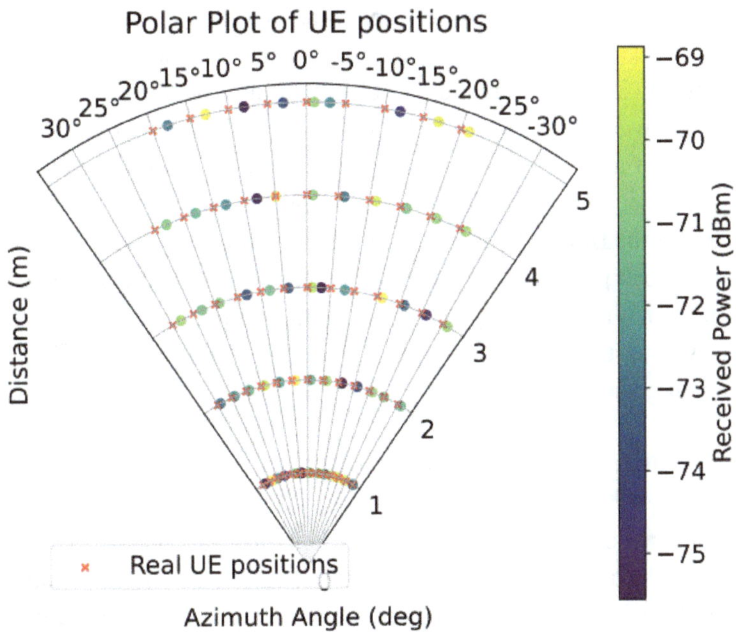

The Azimuth range is between - 30° and 30°, and the receiver horn antenna is positioned at distances between 1 m and 5 m from the RIS, with a step size of 1 m. The polar plot provides a visual overview of the received power level in relation to the actual position of the receiver at that time. The red crosses indicate the points which were physically measured and marked on the ground while these measurements were being made. As shown in the Fig.6, our algorithm achieved very high localization accuracy with a maximum error of 2°. It should be noted that the root mean square error for these measurements was just 1.91°. The results further indicate that the minimum localization accuracy is consistent with this CRLB. However, since the experimental setup only permits a 1° scanning step and of minimum reflected beam width is as low as 3° or higher, localization accuracy will be in an error range from 1° to 2°.

5. RIS-INTEGRATED WIRELESS SENSING

Wireless sensing could detect obstacles, pedestrians, vehicles, the road surface and abnormal traffic behavior. Consequently, in an effort to address these challenges we propose an ARIS with improved specifications to overcome the tasks. The envisioned ARIS can simultaneously capture both magnitude of as well as phase information about incident electromagnetic (EM) waves, thereby providing a more comprehensive environmental sensing capability. From an added feature standpoint, the ARIS will work in the full 360° phase range with continued phase manipulation along a very high reflection coefficient level, indicating advanced wave manipulation capability. Moreover, at the line and device level, ARIS realizes a combination of sensing and emotional bear dexterity way beyond other systems which can't report actual fact and in sequence. At full wave range and power, the ARIS's operation was demonstrated for the first time. At the same time, a complete ISAC application scenario was setup to experimentally verify this means: ARIS can determine direction to communication terminals and between them a channel gain of more than 12 dB is obtained. And judged from this aspect alone, the sight in its use of ARIS has an ample technical growth to look forward.

Element Design and Working Principle

Figure 7(a) shows how the proposed ARIS element is made up. The size of the element is 40×40 mm^2 which is around 0.47×0.47 wavelengths at an operating frequency of 3.50 GHz, and it has a three-layer design. The upper layer includes an F4BTM dielectric substrate plus two metallic patches. The middle layer uses a Rogers 4003C substrate, with the top surface bearing a ground etched onto it that looks like an H slot and microstrip circuits for sensing below. The under-side is populated by machining micro-strip circuit boards as tools to sense or influence EM waves. The bottom layer holds a metal ground plane, which improves the reflection properties of ARIS. Interestingly, with the collaboration of H slots and these patches, an antenna part is this patch and H that will launch EM waves into its sensing circuit.

To illustrate the working principle of ARIS sensing and wave manipulation, consider the following components. As shown in figure 7(b): The incident EM wave is captured initially form slot antenna and thereafter goes into the circuitry. Once received, the signal first goes through a phase shifter, programmable phase control being implemented by turning the effective capacitance CVAR of varactor diodes inserted into it using an applied DC bias voltage. After that the signal flows into a weakly coupled microstrip line, in which some of its energy is tapped off by parallel coupling and sent to the detection chain.

After that, if the remaining signal is still high in amplitude, it will keep going and finally come to an open circuit load. Here the leftover signal moves along the same path back before being relaunched into space through this slot antenna. In parallel, signals from a sensing chain are analyzed; as the sensing signal's amplitude is proportional to the surface field of ARIS elements, complex amplitude extraction technique makes it possible for signals to be re-constructed over an area array. This allows us to collect environmental information, including the arrival direction (DOA) of the incident wave. The environmental information obtained is then used to direct ARIS reconfigurations of elements blossoms states, such that EM wave manipulation efficiency is maximized.

Figure 7. (a) Structural configuration of the proposed ARIS element. (b) Details of the microstrip circuits for sensing and manipulation (Yang et al., 2025)

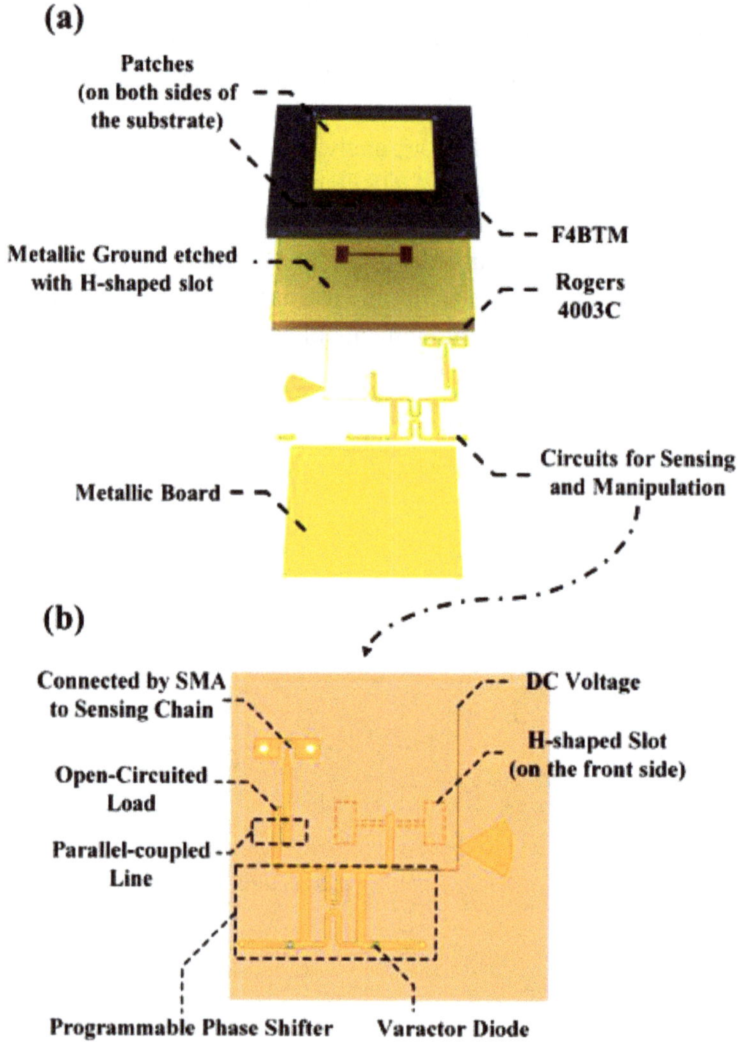

To experimentally validate the performance of the proposed ARIS, a prototype consisting of a 6x6 array was made. A schematic of this structure is illustrated in Figure 8. Firstly, to validate the ARIS's manipulative ability it was processed appropriately and mounted onto equipment that regulates its dynamic scan performance. As shown in Fig. 5a, measured results show that reflection-beams are now dynamically spread over a range of ±60°. Minor beam distortions observed near the 0° is due to the physical blockage caused by the antenna in front of ARIS.

Figure 8. Prototype of ARIS comprising 6×6 elements

Sensing Process

As shown in Figure 9, the accurate estimation of the terminal directions thus evinces ARIS's sensing capability yet further and offers vital guidance for later re-shaping its element states. By contrast, conventional RISs demand a prior knowledge as to which direction terminals are pointing in order to determine the element states.

Figure 9. Measurement scenario of the ISAC system assisted by ARIS

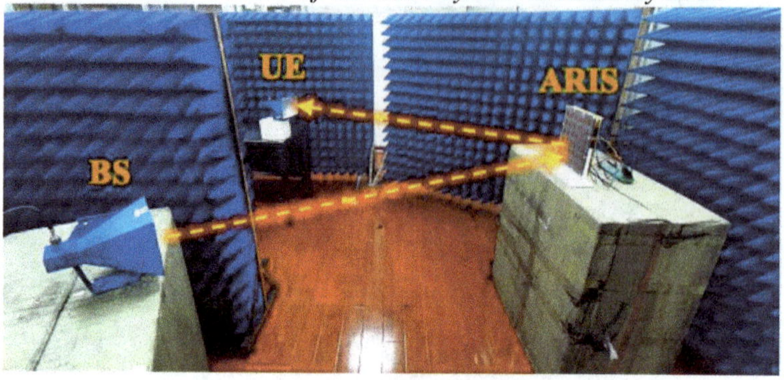

The ARIS uses the information it obtained via its own sensing capability to adjust its element states and deflect the incident wave from BS towards UE. It thereby lifts quality of communication along whole BS-ARIS-UE path. In this experiment, BS commenced transmitting a QPSK modulated signal as shown in Fig.10. According to the phase gradient of the distribution of electric field, we can infer that the directions of the BS and the UE are at θ = 22.7 ° and-41.3 °, respectively (the actual directions are θ = 20 ° and -40 °). It's precise to estimate the terminal directions further corroborate the sensing ability of the ARIS and provide important guideline for subsequent reconfiguration of its state elements. By contrast, conventional RISs require that you must know the directions of the terminals in order to determine element states.

Figure 10. The phase distribution of the field formed on the surface of the ARIS when (a) BS (b) UE transmits pilot signal. (The 6×6 matrix represents the phases of signals sensed by the 6×6 ARIS elements, with units in degrees.)

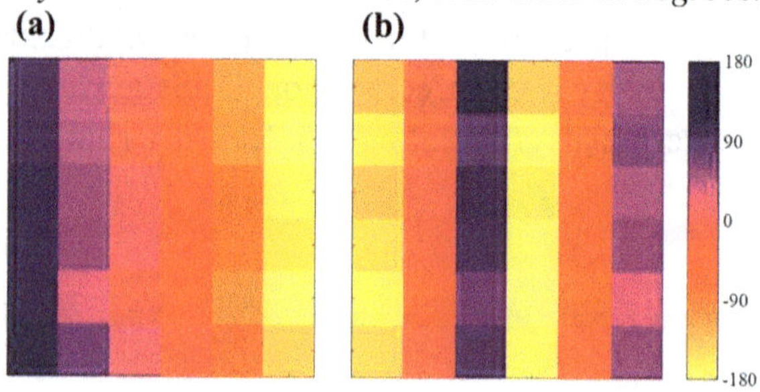

The ARIS has its own sensory power to process these data. With other equipment's in the ARIS system, the relay center takes its present shape. According to the evidence from its sensing gear, ARIS with an element states adjustment can turn the incident wave from BS toward UE and this enhances quality transmission of whole BS- ARIS-UE link. The BS transmitted QPSK-modulated signal in the experiment. When ARIS is used, reception spectra as well as constellation diagram for UE shown in Figure 11 (a) For comparison, replaced ARIS by a same-size metallic plate and the ARIS system was turned off. The corresponding results shown in Figure 11 (b). By measuring the ARIS, it was found that in the channel gain compared with the non-programmable static metallic board there is an increase of about 12 dB. When there is neither an ARIS device nor metallic surface, signal energy transmitted between BS and UE is drowned in background noise. The constellation diagram at the receiving terminal is consequently very scattered, making reliable communication virtually impossible.

Figure 11. The received constellation diagram and the signal spectrum at the UE when (a) the ARIS (b) metallic board is placed

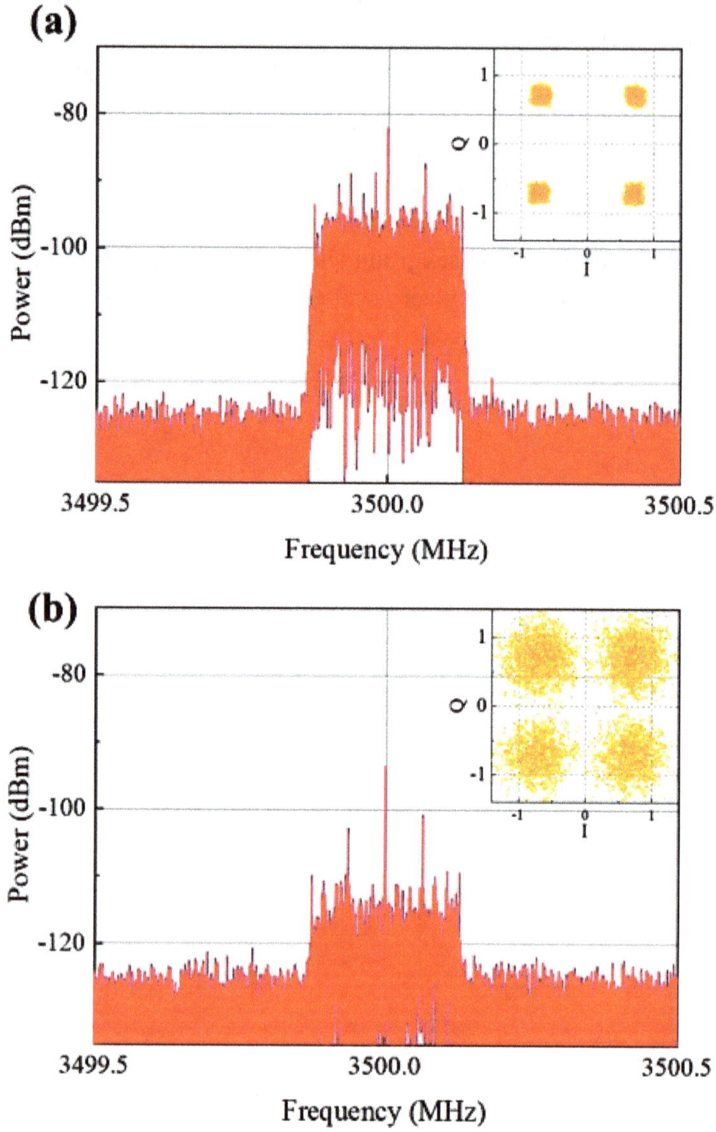

6. FUTURE SCOPE AND INDUSTRIAL DEPLOYMENT CHALLENGES

Reconfigurable Intelligent Surfaces (RIS) now supports localization, tracking and sensing. This is a new paradigm that will bring about the next generation of Intelligent Transportation Systems (ITS). By intelligently altering the wireless propagation environment, RIS can greatly improve position determination accuracy object detection reliability and sensing coverage especially in complicated urban and high-speed mobility situations. How far this technology will go is huge, but on a large scale today there are various technical, economic and regulatory constraints that face its industrial deployment.

However in terms of looking forward, RIS enabled ITS might become a key technology for Autonomous and Interconnected Vehicle complex positioning at times to come. RIS helps to minimize the positioning-error content corrupting the positioning signal making it-weighted integrity in the results, and creates curtain wall-like reflectors spatula frequency emissions. We can obtain centimeter level accuracy when in dense urban canyons like this, words notwithstanding.

Traditional GPS-based solutions suffer from two main problems: First of all, they cannot deliver high precision location services in the canyons, because there is no sky signal for these receivers to lock onto; second loosing sites which means at present one needs special means such as raised platforms to serve as masts where over hanging buildings don't block strong ground reflections as they really should. In the future, collaboration via RIS-JCAS in city wide sensor networks would see sensors on vehicles and infrastructure collectively searching, using the same microwave methods, for pedestrians, obstacles and road conditions. In a smart city, RIS-backed sensing would provide an indispensable aid to the optimization of traffic flow: the new car is no longer entirely self-centered. Furthermore, it would help to identify incidents, a new sort of network is coming into being now and it offers as yet unrealized potentials in indoor environments and large outdoor areas such like motorways.

These can be combined and optimized so as to bring maximum benefits from AI technology in wireless networks as the industry now stands. The next generation of wireless communication probably will combine elements RIS and ML, using adaptive beam control by machine learning, an environment-aware sensing feedback system to provide realistic renderings of the world around you, plus predominant mobility management made possible through predictive training. That would result in all-inclusive ITS architectures which optimize on an ongoing basis. They adapt dynamically according to the traffic level, weather changes or even sudden emergency scenarios.

Meanwhile, RIS technology is nearly dry and it is passive such as free running. Therefore, this technology is highly suitable for environmentally friendly transportation systems. Even with all these favorable prospects ahead of it, many problems still need to be solved before industrial applications can be achieved. The first problem lies in expanding hardware and costs. Despite the fact that key basic technologies are under development to provide the necessary large, stable, weatherproof RIS panels for vehicle deployment, cloud still hangs over this process. It is both expensive and difficult. This is a question usage, power or RIS construction. How to ensure that its products can still benefit over the long term under conditions of harsh environmental stress such as intense heat, heavy rain, strong dust, or vibration is an important topic in today's actual ITS.

Another major impediment remains in real-time management and signaling overhead. This may lead to RIS-assisted localization and sensing will soon accurate channel state feedback that introduces major latency. When the vehicles are in high speed motion, control becomes particularly complex.

Multiple RIS unit's connections with different communication standards across road networks further intensify the difficulty of system design in general and when combined with legacy infrastructure. But challenges for standardization and conformance remain before RIS can be adopted on a large scale by industry users.

The current bottleneck to large-scale deployment is that there is no common framework for integrating RIS with the existing transportation and wireless network systems. Furthermore, before such systems become broadly available to ordinary consumers, there are regulatory issues to be resolved also about spectrum use, electromagnetic exposure limits and infrastructure ownership. For instance, privacy and security problems take on special importance. A RIS-enabled system for sensing and localization will easily generate vast amounts of detailed data about the environment, the movements of humans etc. However insecure the transmission channels may be already-or indeed if they are deliberately set up to leak information themselves this opens up questions over how this information could be misused or attacked. The systems must incorporate especially formidable protective measures their design has to meet user needs for privacy.

7. SOCIETAL IMPACT, SUSTAINABILITY, AND ECONOMIC SIGNIFICANCE

The localization of RIS (Reconfigurable Intelligent Surfaces), tracking and sensing is quickly turning into a hub scenario in the path to next-generation intelligent Transportation Systems (ITS).RIS technology boosts wireless communication quality by reshaping electromagnetic waves. The precision in positioning and environment perception, particularly in space-constrained urban movement scenarios. There are certainly a variety of ways in which the technology can drastically change social system, and be beneficial both for the nation too, but from an economic point of view as well. Depending on your viewpoint about it though people may be required to readjust their ideas over time accordingly in our present era this type of thinking may not be appropriate (as will come up elsewhere).

Societal Impact

From the societal perspective, an ITS enabled with RIS can provide in reality major progress both on traffic safety and public health. Through precise positioning and real-time sensing, pedestrians, cyclists and obstacles are discovered earlier, thereby lowering significant accident rates and casualties. Which directly supports SDG 3: Good Health and Well-Being, for canyon roads means fewer rest homes burdens on hospital resources. With better tracking, the time to emergency response should be shortened as well. More accurate positioning of vehicles and incidents allowed for far more speedy arrivals on the scene; and if there happens to be any accident, disaster or medical emergency then this is particularly essential.

Guided by RIS technology that is handled in a pervasive manner including the mobility issues that are ignored by conventional systems. Disadvantaged groups are overlooked in the areas without infrastructure reinforcement or particular GPS coverage, while RIS can be used to ensure visibility and networking of opportunities for low-price small vehicle systems. This is in accordance with SDG 10: Reduced Inequality, when technological advantages are no longer the exclusive preserve of rich cities, but instead reach out to rural areas and marginalized groups too.

Moreover, this would also enhance public confidence in Self-driving and connected cars. It requires the accurate sensing and precise positioning to develop of the RIS platform. This not only gives rise to a comprehensive intelligent mobility solution for infrastructure, Safety Automation needs are kept through this step. As trust holds ground for longer terms, implementation of intelligent transport systems becomes ever more main stream, and this affects the daily experience of commuting to work. Then once the quality of life for everyone in general is improved beyond measure.

Sustainability and Environmental Impact

Its low power, passive operations make RIS technology and sustainability goals a natural fit for each other. Active base stations of a typical manner will need a constant supply of electricity in order to keep the signal output on going. By contrast, RIS elements simply change signal paths and reflect light without consuming on going power. It can bring about a reduction in power consumption in the order of about

25%.Conversely; the efficiency of energy efficient wireless infrastructure in turn leads to less electricity demand and smaller operation expenses. This can support the UN's Sustainable Development Goal (SDG) 7: Affordable and Clean Energy.

RIS-facilitated identification and tracking may help to mitigate traffic congestion, telephone tapping or unauthorized entry into accounts; thus reducing fuel consumption greenhouse gas emissions. Outfitting traffic management systems with accurate position data can shorten sea-travel for fish farming, minimize error alerts to the maximum minimum or close off alarms altogether and choose the geographic optimum routes. All of this is addressed under SDG 13: Climate Action, cities. It will curb the environmental impact of transportation.

Moreover, through the provision of relocatable interference sources (RIS), both communication and sensing can be integrated into transportation. This team showed that by utilizing radio waves, one could perform both air-quality monitoring or pinpoint exactly where the pollution came from. All of these are valuable functions which help inform an array of data that is necessary for urban planning: they are also an integral point in achieving United Nations Sustainable Development Goal 11–Sustainable Cities and Communities. Fewer devices and less material when building urban housing will help to make the planet greener as a whole, which is only a start in sustainable development.

Economic Significance

On the surveillance end, the construction of RIS ITS will mean a breakthrough from the transport, telecommunication, and intelligent urban trends. Meanwhile, the operational effectiveness of logistic transport system, the professionalized locomotives and buses in such a whole society are getting cheaper to run with better safety than ever before. The model is responsible directly for one of the UN Sustainable Development Goal 8 Decent Work and Economic Growth because it offers more chances to create industries that drive prosperity forward and carries the jobs needed for them. RIS infrastructure deployment brings business opportunities throughout the industrial chain from hardware manufacture to software development, and such maintenance service based control systems and equipment. As RIS panels are installed in roadside equipment, buildings beside highways and vehicles themselves demand for skilled system integrators and engineers will rise, giving a push to industrial growth.

In addition, RIS-realized perception assists in turning system enhancement investments in economical build or amelioration. Because RIS can improve existing wireless systems without requiring excessive investments in active-node deployment, it is possible for governments and industries to gain greater cost-to-benefit ratios from systems that operate better and at lower cost over time. This is especially good news for countries in development which are looking for reasonable and scalable ITS solutions that they can afford and so too in line with its goals SD making industries more ecologically sustainable places.

Returning to the export economy, this appears once again to be a long-term solution which will bring with it many new problems. However, since the Chinese export industries are largely assembling foreign parts (especially electronic components), their capacity to do this is constrained by other countries.

Limitations of RIS-Enabled Localization, Tracking, and Sensing for Intelligent Transportation Systems (ITS)

Although the RIS assisted localization, tracking, and sensing has strong prospects for ITS systems, there exist several technical, operational and practical challenges that hinder its wide-spread deployment.

- A significant constraint is based on hardware. Realistic RIS deployments use passive or semi-passive, low-cost reflect-arrays with finite phase resolution and non-ideal reflection properties. These imperfections cause signal distortion, decrease the localization accuracy, and, especially in fast-moving vehicular environments and sensing reliability where a quick adaptation is needed.
- Further, one of the major challenges is estimation of channel and complexity of the system. The successful localization and sensing with RIS need to know the cascaded channels between a transmitter, an RIS, and a receiver exactly. A real-time estimation of the propagation channels is a very complex calculation, even for otherwise loosely populated urban ITS environments with fast- moving vehicles and frequent blockages to line-of-sight access.
- Latency and control overhead are also considerable problems. The RIS settings have to be adapted dynamically to follow the vehicle and adapt them with changing traffic data. To solve this problem, control signals between network controllers and RIS panels need to be sent rather frequently which may not only result in increased latency but also reduced responsiveness for some safety-critical ITS applications like collision avoidance and autonomous driving.
- Scalability is also an issue. RIS implementation in a wide area, for instance, over large road networks, intersections and highways requires extensive infrastructure planning, synchronizing and updating. The joint cooperation of RISs composed by many small units on a city-scale basis is an open issue.
- Performance is impacted by environmental sensitivity as well. Impairments caused by adverse weather, physical impediments, and non-line-of-sight con-

ditions to name a few may lead to an inefficient reflection process thereby resulting into limited sensing and localization robustness. Lastly, there are continuing problems with standardization and security, as well as integration. RIS-assisted ITS is still lack mature transportation standards for interoperability, and centralized RIS control also brings cyber security threats in terms of malicious reconfiguration or signal intrusion.

CONCLUSION

This chapter is about Vehicle to everything (V2X) connectivity that makes future Intelligent Transportation Systems possible. It means that vehicles are able to interact intelligently with their environment and neighbors. The potential of V2X communications was realized through Reconfigurable Intelligent Surfaces (RIS). In this chapter, we present an analytical study of the dual RIS based vehicle-to-infrastructure (V2I) communication scheme. Chapter Summary

(a) In most cases, we found that the transmission performance can be improved by using two RISs. This work is theoretical research on the application of RIS mechanisms in dynamic network, such as Vehicular Communications. In the future, we plan to test the performance of the vehicular network as a whole when RIS devices are widely deployed.
(b) This chapter discusses how to use RIS-assisted systems (RADIO) to help improve performance in noisy communications caused by aggregate interference from multiple other sources. The number M of aggregated interference significantly affects the channel capacity of the communication system.
(c) This chapter explores a beam tracking system that works in a cost-effective way, examining various beam management parameters such as UE movement, angle mismatch, and tracking timing. The authors rove that accurate UE tracking will be afforded by the RIS-based tracking system even when there is an angle mismatch of 1 or 2 degrees.
(d) This chapter shows that ARIS can manipulate the EM waves over a range from 0 to 360 degrees. More importantly, ARIS can simultaneously sense the complex amplitude of the incident signals element by element.

REFERENCES

Alamzadeh, I., & Imani, M. F. (2023). Detecting angle of arrival on a hybrid ris using intensity-only data. *IEEE Antennas and Wireless Propagation Letters*, *22*(9), 2325–2329.

Cai, P., Zong, J., Luo, X., Zhou, Y., Chen, S., & Qian, H. (2021). Downlink channel tracking for intelligent reflecting surface-aided FDD MIMO systems. *IEEE Transactions on Vehicular Technology*, *70*(4), 3341–3353.

Cheng, X., Huang, Z., & Chen, S. (2020). Vehicular communication channel measurement, modelling, and application for beyond 5G and 6G. *IET Communications*, *14*(19), 3303–3311. DOI: 10.1049/iet-com.2020.0531

Han, J., Li, L., Ma, X., Gao, X., Mu, Y., Liao, G., & Cui, T. J. (2021). Adaptively smart wireless power transfer using 2-bit programmable metasurface. *IEEE Transactions on Industrial Electronics*, *69*(8), 8524–8534.

Ji, B. (2020). Secrecy performance analysis of UAV assisted relay transmission for cognitive network with energy harvesting. *IEEE Transactions on Vehicular Technology*, *69*(7), 7404–7415.

Ji, B. (2025). Cooperative Transmission Algorithm of RIS-Assisted Intelligent Transportation System Under Aggregated Interference. *IEEE Transactions on Intelligent Transportation Systems*, *26*, 17280–17291.

. Kumar, A., Sahu, A. K., & Agarwal, S. (2022). Role of intelligent transportation systems in connected and autonomous vehicles: A review. *Transportation Research Part C: Emerging*.

Lin, Y., Jin, S., Matthaiou, M., & You, X. (2021). Tensor-based algebraic channel estimation for hybrid IRS-assisted MIMO-OFDM. *IEEE Transactions on Wireless Communications*, *20*(6), 3770–3784.

Mao, Z., Peng, M., & Liu, X. (2021). Channel estimation for reconfigurable intelligent surface assisted wireless communication systems in mobility scenarios. *China Communications*, •••, 29–38.

Sahtouti, A., Boumaalif, A., Zytoune, O., & Fouchal, H. (2025). Transmit power analysis in RIS based vehicular communications. *ICC 2025 - IEEE International Conference on Communications*, 6480-6485.

Shlezinger, N., Alexandropoulos, G. C., Imani, M. F., Eldar, Y. C., & Smith, D. R. (2021). Dynamic metasurface antennas for 6G extreme massive MIMO communications. *IEEE Wireless Communications*, *28*(2), 106–113.

Sun, S., & Yan, H. (2021). Channel estimation for reconfigurable intelligent surface-assisted wireless communications considering Doppler Effect. *IEEE Wireless Communications Letters*, *10*(4), 790–794.

Tang, W., Chen, M. Z., Chen, X., Dai, J. Y., Han, Y., Di Renzo, M., & Cui, T. J. (2020). Wireless communications with reconfigurable intelligent surface: Path loss modeling and experimental measurement. *IEEE Transactions on Wireless Communications*, *20*(1), 421–439.

Toubal, M. A., MacGarry, T., Bombardelli, R. K., Shokair, A., Nasser, Y., De Rosny, J., & Lerosey, G. (2025). Experimental approach to adaptive RIS-based beam tracking and user localization in 5G mmWave networks. *2025 Joint European Conference on Networks and Communications & 6G Summit (EuCNC/6G Summit)*, 619-624. https://doi.org/DOI: 10.1109/eucnc/6gsummit63408.2025.11037001

Tushar, P. N., & Jaiswal, P. (2025). Wireless Power Transfer Techniques for 6G Communication. In A. Nwajana (Ed.), *RFID, Microwave Circuit, and Wireless Power Transfer Enabling 5/6G Communication* (pp. 1-54). IGI Global Scientific Publishing. https://doi.org/DOI: 10.4018/979-8-3693-8799-3.ch001

Van Wynsberghe, A. (2021). Sustainable AI: AI for sustainability and the sustainability of AI. *AI and Ethics*, *1*(3), 213–218.

Wu, C.-J. (2022). Sustainable AI: Environmental implications, challenges and opportunities. *" In Proc. Mach. Learn. Syst*, *4*, 795–813.

Yang, H. Q., Li, H. D., Dai, J. Y., Cheng, Q., & Cui, T. J. (2025). Adaptively Reconfigurable intelligent surface for integrated sensing and communication scenario. *2025 IEEE International Workshop on Electromagnetics: Applications and Student Innovation Competition (iWEM)*, 430-433.

Yang, L., Meng, F., Zhang, J., Hasna, M. O., & Renzo, M. D. (2020). On the performance of RIS-assisted dual-hop UAV communication systems. *IEEE Transactions on Vehicular Technology*, *69*(9), 10385–10390. DOI: 10.1109/tvt.2020.3004598

Yang, L., Yang, Y., Hasna, M. O., & Alouini, M. (2020). Coverage, probability of SNR gain, and DOR analysis of RIS-aided communication systems. *IEEE Wireless Communications Letters*, *9*(8), 1268–1272. DOI: 10.1109/lwc.2020.2987798

Zhang, J., Wang, J., & Yang, L. (2021). Intelligent Reflecting Surface Assisted Vehicular Communication Networks: Channel Modeling and Performance Analysis. *IEEE Transactions on Vehicular Technology*, *70*(2), 1373–1385.

Zhang, Z., Wang, X., & Liu, J. (2023). Real-time traffic flow prediction using deep learning in intelligent transportation systems. *Journal of Intelligent Transport Systems*, *27*(5).

Zhao, H., Tang, J., Adebisi, B., Ohtsuki, T., Gui, G., & Zhu, H. (2022). An adaptive vehicle clustering algorithm based on power minimization in vehicular ad-hoc networks. *IEEE Transactions on Vehicular Technology*, *71*(3), 2939–2948. DOI: 10.1109/tvt.2021.3140085

Zhou, T., Xu, K., Xie, W., Shen, Z., Wei, C., Liu, J., & Sun, L. (2021). Aerial intelligent reflecting surface-enhanced cell-free massive MIMO for high-mobility communication: Joint Doppler compensation and power optimization. *EURASIP Journal on Advances in Signal Processing*, •••, 38.

Chapter 9
System Architecture and Communication Models for RIS-Assisted V2X

Ravikumar R. N.
http://orcid.org/0009-0009-3705-1681
Marwadi University, Rajkot, India

Aarthi S.
http://orcid.org/0009-0006-9064-2091
Marwadi University, Rajkot, India

ABSTRACT

Reconfigurable Intelligent Surfaces (RIS) are transforming 6G-enabled V2X communication by introducing programmable, low-power mechanisms that reshape wireless propagation for high-mobility environments. This chapter presents the architecture, communication models, and performance enhancements enabled by RIS across V2V, V2I, and V2N links. It examines RIS-assisted beamforming, channel modeling, interoperability protocols, security mechanisms, and optimization frameworks tailored for mmWave and terahertz bands. Real-world case studies from highway and urban deployments highlight the benefits of multi-RIS coordination, improved reliability, reduced latency, and enhanced coverage in NLOS conditions. Challenges related to scalability, hardware limitations, standardization, and real-time control are also addressed, followed by future directions involving AI-driven RIS control, digital twins, and quantum-ready metasurfaces. Overall, RIS technology emerges as a foundational enabler for intelligent, resilient, and energy-efficient 6G transportation networks.

DOI: 10.4018/979-8-3373-7867-1.ch009

1. INTRODUCTION

Vehicular communication evolution has become an era of transformation as 6G-enabled V2X systems integrate Reconfigurable Intelligible Surfaces (RIS) into the system. The dynamic mobility patters, changing propagation, and complicated urban environment makes the traditional wireless infrastructures to be insufficient as intelligent transportation systems require ultra-reliable, low-latency and interference-tolerant communication. The wireless environment itself can now be transformed into a controllable communication resource with the introduction of RIS, planar, programmable metasurfaces that can control the propagation of electromagnetic waves. With RIS-assisted V2X, better coverage, less signal blockage and better spectral efficiency can be achieved, especially in Non-Line-of-Sight (NLOS) situations typical in vehicular applications (Li et al., 2025a). The chapter discusses the role of RIS in enhancing the V2V, V2I, and V2N models of communication to transform the design of future automotive networks. It introduces the background ideas, systems design, optimization theory, mobility-aware communication plans, and all of these contribute to the enhancement of autonomous and connected cars functioning within 6G worlds.

1.1 Evolution Of V2X Communication Toward 6G

The communication technologies of vehicles have been advancing far beyond the initial versions of the Dedicated Short-Range Communications (DSRC) to the high-tech Cellular V2X (C-V2X) offered with 5G. These developments have enhanced cooperative awareness, rapid data communications and enhanced road safety. The move to 6G is a critical jump towards hyper-connectivity, which facilitates terabits worth of throughput, almost zero latency, and ubiquitous intelligence in the mobile context. The communication should be resilient in 6G-based V2X systems, even in the scenarios of a high topology change rate and high density. This motivates the desire to have more adaptive, programmable and environment conscious communication mechanisms (Shrestha et al., 2020a). RIS technology is perfectly in line with these needs as it allows real time control of the wireless propagation without the need to transmit on the RF. With the combination of RIS and 6G capabilities including Terahertz bands, AI-native networking and distributed edge intelligence, the vision of the future vehicular systems will enable large-scale cooperative autopilot, real-time digital twinning, and AI-enhanced safety decision-making. Therefore, the development of 6G is a paradigm shift in V2X communication as the wireless environment is proactively and intelligently controlled. Figure 1 that shows that the way to advanced 6G-enabled V2X communication with high throughput, low latency, RIS integration, and AI-native networking was starting with DSRC and C-V2X communication.

Figure 1. Evolution of V2X communication toward 6G

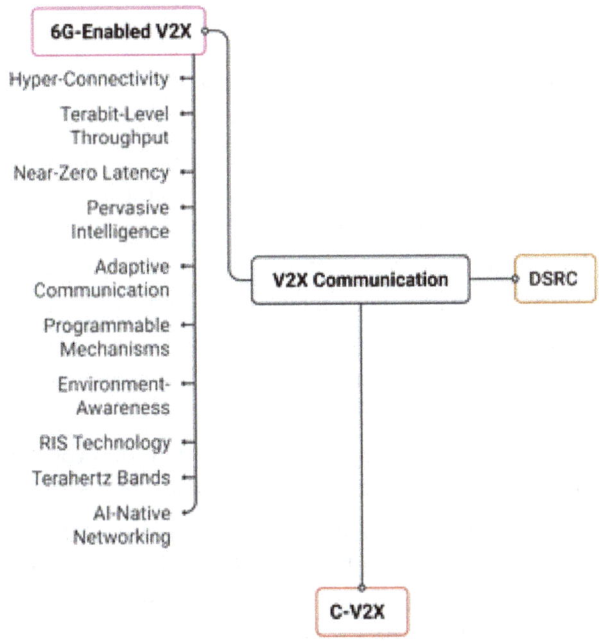

1.2 Limitations Of Conventional V2X Systems

The traditional car communication systems are significantly limited in the highly dynamic 6G environments. The large mobility also causes fast channel change and larger Doppler effects that deter link stability and coordination of manoeuvres. The propagation phenomenon of Non-Line-of-Sight (NLOS) is very common in tunnels, intersection, dense-urban streets, and expressways, which ends up creating serious signal blockage, reflection loss, and deep fading. Moreover, the nature of the spectrum resources is finite, and the relatively dense networks of traffic result in extreme interference and congestion, and the conventional V2X schemes are not enough to provide the requirements of low reliability (Luo et al., 2021). Current solutions like MIMO, beamforming, or extra roadside units are also better, and usually, they add more to the cost and power usage of deployment. In addition, traditional systems are not environmental reconfigurability; they are not able to propagation paths in an adaptive way and thus its ability to maintain high link quality during the dynamic movement of vehicles. Such constraints impose obvious performance constraints on the new applications, including autonomous navigation, cooperative driving, and high-definition sensor sharing the and communication enablers that should be more flexible, intelligent and scalable.

1.3 Role Of Reconfigurable Intelligent Surfaces In Future Vehicular Networks

Reconfigurable Intelligent Surfaces are a new approach that is offered to address the shortcomings of the conventional V2X system by allowing the propagation environment to be controlled. RIS is composed of meta-atoms that control the phase, amplitude, or polarization of electromagnetic waves and dynamically control signals to boost or bend signals into receivers, particularly in NLOS. This allows a vehicle network to improve signal reliability, minimize interference and provide a larger range of communication without increasing transmit power. In addition, the RIS panels may be placed strategically on buildings, streetlights, road signs and tunnels and convert passive environments to smart communication properties. The energy-efficient performance of RIS is possible due to its low-power and almost passive nature, which is in line with the sustainability targets of new mobility ecosystems (Xu et al., 2024). RIS is also compatible with 6G AI-driven architectures, allowing one to reconfigure it according to mobility patterns, channel state predictions, and traffic conditions. This makes RIS a fundamental building block of ultra-reliable, low-latency, and cost-efficient V2X communication systems that can be used to support large-scale autonomous and cooperative vehicular systems.

2. FUNDAMENTALS OF RECONFIGURABLE INTELLIGENT SURFACES

Reconfigurable Intelligent Surfaces is an innovation that is a breakthrough in electromagnetic engineering and wireless communication. These surfaces are programmable meta-atoms which are able to manipulate incident waves (controlled phase shifts, amplitude alteration, or polarization conversion). The basic principle of RIS technology is to transform the wireless medium into a dynamic but low-energy reconfigurable medium. The architecture is usually a planar array of scattering elements, a control layer and a communication interface to give configuration commands (Sarker et al., 2023). RIS may be passive, active, or hybrid depending on their design providing various performance-cost trade-offs. RIS uplifts the wireless connection by offering alternative pathways of propagation, minimizing dead zones, and also providing accurate steering of the beams. Vehicular settings have special benefits of RIS which include high-speed flexibility, scalability, and low hardware area. This part expounds on the electromagnetic underpinnings, structural structure and control provisions that characterize RIS performance in 6G V2X communications.

2.1 Meta-Atom Structure, Reflection Principles, And EM Reconfiguration

The RIS panel consists of meta-atoms which are reconfigurable electromagnetic scatterers that allow the behavior of incident waves to be controlled. Such structures are normally tuned with tunable elements like PIN diodes, varactors or MEMS switches which alter phase or amplitude responses. With the coordination of the tuning states of the whole RIS array, more complicated manipulations of the wavefronts are possible: beam steering, focusing, or diffusing. Such reconfiguration in vehicular communication allows the optimal propagation paths in high mobility situations. Principles of reflection used in the work of RIS are based on the generalized law of Snell but allows the surface to achieve the desired angles of reflection without taking into account the traditional geometrical restrictions (Y. Zhou et al., 2023). Due to the strict command transmissions relayed by a controller, electromagnetic reconfiguration is achieved where real-time flexibility is provided in reaction to link variations. The high aperture and low-grain control of the RIS panels has been shown to be particularly useful in alleviation of signal blockage, enhancing NLOS connections, and in coordination of multiple links in V2X systems.

2.2 Active Vs. Passive RIS Architectures

Types of RIS architecture can be classified into passive, active and hybrid. The traditional model is passive RIS that consists of tunable reflective elements that change the phase but do not amplify RF. It is a highly energy efficient design that is cost-effective and easily scalable, therefore, it can be deployed in vast quantities in vehicles. Passive RIS performance on the other hand decreases as a function of distance due to low reflection gain. Active RIS however employs amplifiers on all meta-elements and thus permits signal gain and superior quality of link particularly in high-path-loss regimes such as millimeter-wave or terahertz. It has a trade-off of increased power consumption, complexity of hardware and potential noise increase (Di Renzo et al., 2020). Hybrid RIS architectures Hybrid RIS architectures entail passive reflections together with selective active amplifications so that to achieve balanced performances in automobile conditions. It is a hybrid solution that is flexible thus aids in high reflection in high mobility states and it is also energy-efficient. It should be mentioned that the differences between these architectures should be taken into consideration to select the most suitable deployment options in different V2X settings. Table 1 Summery of structural, operation and performance differences of key RIS architectures.

Table 1. Key differences between passive, active, and hybrid RIS architectures

Parameter	Passive RIS	Active RIS	Hybrid RIS
Power Consumption	Very low (no RF chains)	Higher (integrated amplifiers)	Moderate
Signal Gain	Reflective only	Reflective + amplification	Selective amplification
Hardware Complexity	Low	High	Medium
Noise Introduction	None	Possible (amplifier noise)	Controlled
Cost	Low	High	Medium
Suitable Scenarios	Dense deployment, low-cost coverage	High-path-loss mmWave/ THz links	Balanced performance and cost

2.3 RIS Control Mechanisms: Digital VS. Analog VS. Hybrid Tuning

The accuracy and sensitivity of electromagnetic reconfiguration is defined by RIS control mechanisms. Analog control is based on a continuous adjustment of the varactors or voltage-controlled devices and provides a smooth adjustment of the phase, though it is significantly less accurate. However, digital control offers discrete tuning positions, typically based on the use of PIN diodes, to enable a very deterministic beam shape and to interface to a digital signal processing system very effectively. Machine learning-driven optimization is made possible through Digital RIS because it has a discrete configuration space but can cause quantization errors. Hybrid tuning systems approach integrates both analog and digital components to provide both flexibility and precision, which promises a good compromise between rapidly shifting car channels (Liaskos et al., 2018). RIS controller sends configuration commands to the channel according to channel state feedback, mobility prediction or learning. The control mechanism in V2X applications should be able to provide low-latency updates, scalability of the coordination of multiple RIS panels, and edge computing resources. Therefore, a correct control strategy can be selected to attain good and adaptive 6G vehicle communication.

3. RIS-ENHANCED 6G-V2X SYSTEM ARCHITECTURE

RIS-enhanced 6G-V2X design is a multi-layered structure that comprises physical, networking, and computing equipment to allow intelligent cars to communicate. RIS panels are also manufactured at physical layer and attached on highways, buildings, traffic poles, and tunnels to form flexible propagation pathways. These panels are communicating with 6G base stations, vehicular units and edge servers to have robust connection with various mobility conditions. The communication layer examines the inter-relationships of RIS-assisted V2V, V2I and V2N models which enable the comfortable communication between the roadside-infrastructure and cloud networks (Anil Babu et al., 2023). Edge intelligence and distributed processing can be done at the computation layer to provide real-time beamforming updates, prediction-based RIS control and mobility-aware communication control. The overall design is geared towards providing scalability, low power consumption, and flexibility and also to ensure quality of service is maintained across the different locations of vehicles. The 6G V2X architecture with 6G-assisted by the RIS and physical, communication, and computational layers, location of the deployed components, system interactions, and significant considerations of the system operation are represented in Figure 2.

Figure 2. RIS-enhanced 6G-V2X system architecture

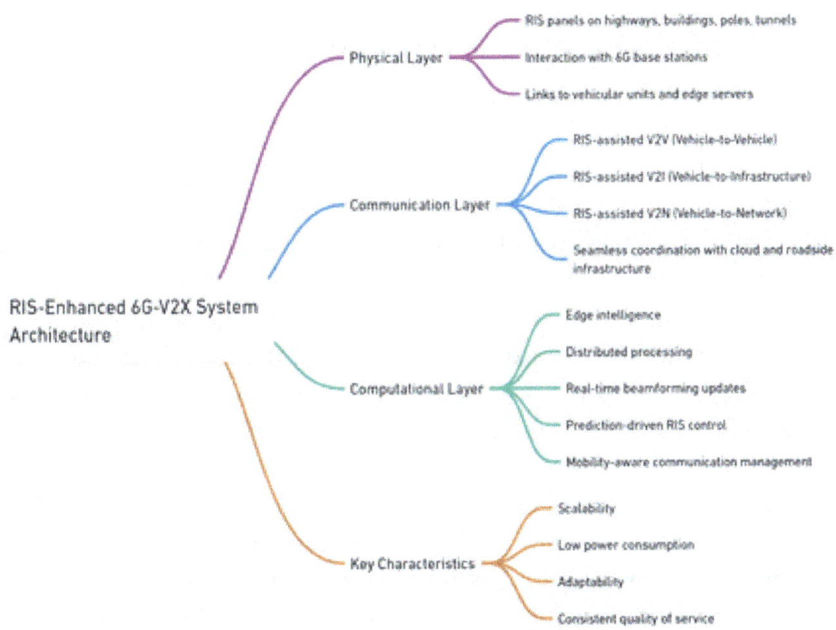

3.1 RIS Deployment Strategies For Highways, Urban, And Tunnel Scenarios

The strategies of deploying RIS depend greatly on the characteristics of the vegetable environment. High mobility and long-distance issues in highways require that RIS is installed along highway sides and sound barriers as well as overhead structures to ensure continuity of Line-of-Sight. These deployments eliminate signal loss and facilitate high speed communication, which is essential to autonomous driving. The high-density RIS requirement of an urban environment is given the dense positioning of the NLOS that occurs quite often because of buildings and intersections. In this case, the RIS panels located on building facades, streetlights and traffic signals will allow the reconfiguration of the paths dynamically, enhance the coverage and minimize the interference (Clancy et al., 2023). The uncontrolled nature of tunnels contributes to the special propagation problems in sealed structures and in extreme multipath. RIS arrays can be deployed strategically to divert signals along the tunnel walls, and hence provide full coverage of the tunnel walls, and thereby offer reliability in V2X communication when used in safety-critical applications. Other factors that need to be considered in the deployment strategies are the vehicle density, power availability and network scalability. RIS is flexible and thus it can be used as a solution to various vehicular communication needs depending on environmental limitations.

3.2 RIS Controller Placement, Connectivity, And Interface With RSUS

RIS controllers are used as the point of interface in configuring reflective elements and coordination of RIS functions in the V2X network. The location of them has a great impact on control latency, communication overhead, and responsiveness in the system. Controllers are usually placed near RIS panels, to keep the complexity of wiring down to an absolute minimum, and to reduce the propagation delay. Such controllers link to Roadside Units (RSUs) through wired or wireless backhaul connection allowing the smooth integration with the already developed V2X infrastructure. The RSUs measure Channel State Information (CSI), mobility, and network status and transmit that data to the RIS controller to receive real-time updates in the configuration (Houmer et al., 2022). It is done through the efficient interfacing mechanisms whereby the RIS reconfiguration is tuned to the speed of vehicle movement, Doppler effects, and changing channel variations. A dependence on the use of edge computing resources decreases reliance on cloud servers, allowing to make decisions in a lower latency. Besides, multi-RIS coordination is also based on coordinated controller-controller interaction with the avoidance of conflicting beam patterns. In general, optimal placement and connectivity strategies can be used to provide predictable and reliable performance of RIS-assisted systems in high-speed automotive environments.

3.3 Integration With Vehicular Edge Computing And IOT Infrastructures

The interaction between RIS and vehicular edge computing and IoT ecosystems improves the smartness and flexibility of V2X communication networks. Real-time sensing data, channel information and mobility patterns are processed by edge nodes that are close to RSU or inside vehicles to facilitate predictive RIS reconfiguration. This spread-out intelligence decreases the overhead of communication and the latency is greatly lowered as compared to the centralized processing in the cloud. RIS and road Infrastructure sensors and vehicle sensors deliver contextual information about traffic congestion, weather and vehicle displacement. This information can be used to inform AI-based optimization strategies to ensure that RIS can predict the best reflection conditions to use in future communication events (Mande et al., 2025). It can also be integrated with IoT to help with cooperative perception, with RIS helping to transmit high-definition sensor data to enhance situational awareness. RIS architectures with edges enhance scalability in operation in large networks of transportation and decentralized decision-making. This synergistic integration will guarantee better reliability, reduced latency and better safety of the next-generation autonomous mobility systems.

4. COMMUNICATION MODELS IN RIS-ASSISTED V2X

V2X communication models assisted by RIS include RIS-enhanced V2V, V2I, and V2N communication modes, all of which are aimed to enhance the quality of links, coverage, and reliability, even in the presence of mobility degradation. V2V communication with the help of RIS is directed at providing the vehicle-to-vehicle interaction even in the high-traffic conditions or in the environment with obstructions. V2I communication combined with RIS promotes better connectivity in the situation when vehicles and roadside infrastructure are connected to overcome NLOS conditions and provide uniform coverage. V2N and V2Cloud models assisted by RIS also expand those possibilities to the high-bandwidth and low-latency connection to 6G base stations and cloud-enabled ecosystems (Gebrezgiher et al., 2025). In these models, RIS improves precision of beamforming, minimizes interference and facilitates smooth mobility control. All these features are needed to support cooperative driving, real-time sensor sharing, safety-critical uses, and 6G vehicles services that are developing.

4.1 RIS-Assisted V2V Communication: Reliability And Dynamic Beam Steering

V2V communication is enhanced by the RIS to create a reliable path of reflections in the event of weakened and blocked direct links. Other vehicles, structures, and environmental debris are some of the barriers that vehicles tend to encounter, thus worsening the quality of the signal. A RIS panel placed beside roadsides or attached on the infrastructure offers alternative propagation paths which reflects signals traveling towards the targeted receivers in a dynamic manner. This facilitates consistent communication even during traffic congestions. The AI-driven dynamic beam steering process optimizes the reflective nature of the RIS to follow moving vehicles in the most common manner and ensure there is a steady link quality (Naaz et al., 2024). This goes a long way in preventing degradation caused by doppler and improving the strength of communication. Moreover, RIS will be able to facilitate the support of several, simultaneous V2V links by creating independent beams, minimizing interference, and supporting cooperative actions, including platooning, collision avoidance, and independent lane merging. Therefore, the improved V2V communication using RIS guarantees reliable, low-latency communication that is needed in real-time vehicle coordination.

4.2 RIS-Assisted V2I Communication For Smart Roadside Infrastructure

V2I communication aided by RIS is important in the improvement of interactions with the roadside infrastructure including RSUs, traffic lights, and digital signs. The conventional V2I systems have difficulty in NLOS conditions due to buildings or massive vehicles or environment barriers. RIS panels that are placed at strategic spots, bend the signals round corners and the obstacles, and maintain connectivity. This facilitates the high-quality data transfer that is needed in intelligent traffic management, safety notifications, and localized services. RIS also reduces the intensity of interference between high density V2I nodes by directing signals with precision and avoiding unwanted reflections. In addition, it is possible to combine smart RIS-based infrastructure with real-time traffic and environmental sensing to aid adaptive optimization (Sahin et al., 2022). As an example, RIS is able to modify the beam patterns based on weather conditions, vehicles density, or emergency vehicle priority. Comprehensively, the RIS-based V2I systems contribute to the enhancement of connectivity, reliability, and coverage in the smart transportation environment significantly.

4.3 RIS-Assisted V2N And V2CLOUD Models For 6G Connectivity And Offloading

The V2N and V2Cloud communication models with RIS provide a smooth interaction of vehicles with 6G base stations, mobile edge clouds, and distributed data centers. With cars producing enormous amounts of sensor, navigation, and environmental data, uplink and downlink connections with high-throughput are required. By reducing path loss and enhancing SNR and directing reflection to network nodes, RIS panels are used to enhance these long-range communication paths. This will guarantee a stable connection even in urban NLOS or high mobility. Besides, the models based on RIS and the support of the V2Cloud with the use of computation offloading to edge or remote servers allow real-time processing of the perception, prediction, and navigation workload (Salim ALRikabi et al., 2025). Controlled by RIS reflection enables the improvement of both mmWave and THz connections, so a reliable high-capacity connection is guaranteed between digital twins that synchronize and cooperative driving. These functionalities will render RIS a baseline of future 6G vehicular-cloud environments. Table 2 Provides the performance indicators (SNR, outage, reliability, coverage) between the traditional and RIS-enabled V2V, V2I, and V2N links.

Table 2. Comparison of V2X communication models with and without RIS assistance

Metric	Conventional V2X	RIS-Assisted V2X
Coverage in NLOS	Limited, blockage-sensitive	Significantly improved through controllable reflections
SNR Performance	Moderate to poor in dense areas	Higher SNR via optimized phase tuning
Latency	Higher due to retransmissions	Lower due to stable links
Reliability	Variable under mobility	Consistently high even at speed
Interference	Difficult to suppress	Beam-shaped reflections reduce interference
THz/mmWave Feasibility	Challenging	Strongly supported by RIS

5. CHANNEL MODELING AND PROPAGATION ANALYSIS

The modeling of the channels used by RIS assisted V2X systems needs dedicated frameworks to describe the dynamics of the reflected, deflected and scattered signals in the vehicle environments. The propagation properties are greatly transformed with the introduction of RIS by forming controllable reflection paths to supplement or substitute the traditional LOS channels. The modeling methods should consider the mobility of the vehicle, Doppler effects, multi-path and geometry of the environment (Zhang & Cheng, 2019). The terahertz frequencies and millimeter-wave frequencies have additional difficulty because of high path loss and blockage sensitivity. The most popular methods of characterization RIS-enabled V2X channels with geometry-based stochastic models, ray-tracing models, and machine learning-based channel predictors. Beamforming design and resource allocation, protocol optimization and mobility management require accurate modeling.

5.1 Geometry-Based Stochastic Models (GBSM) For RIS-Enabled V2X

Geometry Based Stochastic Models provides a useful approach to the description of the RIS-assisted V2X communication channels. These models are mathematical representations of scatterers, RIS panels, vehicles and infrastructure geometries to reflect realistic propagation models. As RIS becomes part of the environment, GBSM has controllable scatterers whose reflection characteristics are predicted using phase shift, amplitude variations, and meta-atom design. The changing characteristics of vehicular environments necessitate the fact that these models must take into account time-dependent geometry variations, Doppler shift and variable path delays. GBSM models model alternative paths driven by RIS and will extend coverage to NLOS areas, and can give insights of what SNR, capacity, and outage behavior is possible (Assiimwe & Marye, 2022). In the case of V2X applications, the models can be used to assess the performance of RIS in enhancing connectivity in complicated environments like at intersections, tunnels, urban canyons. GBSM offers a balance solution to the analysis of RIS-aided vehicular channels by using deterministic geometry and stochastic variability.

5.2 NLOS Handling: Reflection Paths, Doppler Effects, And High Mobility

The NLOS conditions are a significant problem to vehicular communication and the impediments are often a hindrance to direct connections. The solution provided by RIS is powerful because it creates regulated reflection paths to bypass obstructions. This improves coverage at high population density areas, tunnels, and wavy roads. Nonetheless, high mobility brings about fast Doppler fluctuations which introduce signal coherence and link stability. Systems supported by RIS need to be reconfigured in a short time to ensure the optimum reflection angles and phase alignment. Prediction based on machine learning can be used to predict the movement of a vehicle and adjust the RIS parameters in advance (Tang et al., 2021). Also, RIS can create numerous spatially different reflection paths, which provides redundancy in case of unexpected blockages. RIS has a significant impact on the connectivity in busy vehicular environments because it cancels Doppler effects and offers alternative paths of propagation.

5.3 Millimeter-Wave And Terahertz Channel Characteristics With RIS

Throughout Millimeter-wave and terahertz bands have enormous bandwidth, but they have very high path loss and atmospheric absorption and are vulnerable to blockage. RIS is very important in enhancing signal strength and coverage in such frequencies through artificial creation of reflective routes to overcome barriers. Due to the meta-atom design, the phase can be manipulated precisely so that the RIS can absorb high frequency propagation losses. mmWave and THz signals are particularly sensitive to high velocity in vehicle situations so that RIS must be capable of quick reconfiguring and strong link adaptation (Kiela et al., 2020). At these high frequencies frequency-dependent absorption, scattering, and hardware impairments have to be taken into account in channel modeling. RIS-assisted mmWave/THz communications makes it possible to support a high-capacity application of HD map streaming, digital twin synchronization, and cooperative perception. RIS capabilities to condition the propagation environment render it essential to the provision of high-frequency connectivity in 6G vehicular systems. Table 3 Comparison of GBSM, ray-tracing, stochastic fading, and ML-driven models to analyze the RIS-V2X propagation analysis.

Table 3. Overview of channel modeling approaches for RIS-enabled V2X

Model Type	Key Features	Advantages	Limitations
GBSM (Geometry-Based Stochastic Models)	Realistic scatterer geometry, mobility-aware	Balanced realism + complexity	Requires detailed geometry
Ray-Tracing Models	Physics-accurate reflections, material effects	High accuracy in NLOS and urban areas	Computationally expensive
Stochastic Fading Models	Probabilistic fading patterns	Simple, lightweight	Not RIS-specific; low accuracy
ML-Driven Channel Prediction	Learns channel evolution patterns	Adaptive to mobility and RIS tuning	Requires large datasets

6. BEAMFORMING, REFLECTION OPTIMIZATION, AND CONTROL SIGNALLING

RIS-based V2X systems strongly depend on the improved strategies of beamforming and reflection optimization to ensure efficient and reliable communication under high mobility conditions. The optimization of joint transmitter-RIS-receiver is used to maximize signal power, reduce interference and provide beam consistency in general as vehicles pass by. Conventional optimization techniques are unable to deal with high-dimensional RIS design, which leads to the introduction of machine learning, metaheuristics, and reinforcement learning. These methods are fast in beam selection and reflection control at changing vehicular channels. Good signalling should be used to coordinate RIS updates, ensure there is synchronization and to reduce latency (Y. Chen et al., 2022). The combination of these methods allows strong and scalable V2X communication with enhanced by RIS.

6.1 Joint Transmitter-RIS-Receiver Beamforming

To optimize the quality of communication in an RIS-assisted V2X system, joint beamforming between the transmitter, RIS and receiver is necessary. Whereas in conventional systems there is an optimization of only the transmitter and receiver beam, RIS adds a third dimension of reflection beamforming. The three components can be coordinated to give a fine-grained control over the propagation paths so that optimum signal strength is achieved both in LOS and NLOS conditions. Joint optimization is the computation of phase changes between RIS elements, choice of transmit beam directions and choice of receiver combining vectors. This has to be done quickly in the vehicular environment as there exists high mobility and change in channels (Dai et al., 2020). The algorithms based on AI forecast channel evolution and calculate near-optimal beam configurations on a real-time basis. This makes the connection to be stable even on speedy manoeuvres. Multi-link coordination, which supports platooning, cooperative perception and dense vehicular networks is also made possible through joint beamforming.

6.2 Optimization Algorithms: Metaheuristics, Reinforcement Learning, SDR

The complexity of RIS reflection patterns optimization is that there are many configuration states, and the system must be responsive in real-time. Genetic Algorithms, Particle Swarm Optimization, and Differential Evolution are examples of metaheuristic algorithms that pose a powerful search performance in high-dimensional search spaces, and can perform well in relatively fixed or dynamically challenging environments. RL offers a better adaptability to fast changing vehicular channels as Reinforcement Learning can be used to learn optimal reflection policies. It is appropriate in high mobility situations with a high rate of Doppler variations as RR-based RIS control does not rely on explicit channel estimation (Liang et al., 2024). Semidefinite Relaxation (SDR) is a second method of optimization that can be used to formulate RIS beamforming problems in convex form, and allows finding near-optimal solutions at a reasonable computational cost. Collectively these technologies offer a set of many tools to configure RIS in various V2X settings.

6.3 Control Plane Design: Signalling, Synchronization, And Latency Reduction

A good control plane design guarantees that the RIS-supported V2X systems have a small signalling overhead and latency. Coherent reflection states are important in vehicle movement because the RSUs, RIS controllers, and vehicles have to be synchronized. Control signalling should be able to effectively transmit configuration changes, channel state feedback, mobility predictions and system alerts. Lightweight protocols are required to ensure high performance in networks with high signalling loads. Edge computing eliminates the delay associated with RIS updates since it does not use cloud resources. Also, predictive models reduce such unwarranted reconfiguration through anticipating the vehicle routes and alterations of channels (Eze & Eze, 2023). Timely operation of the control plane is necessary to ensure that the RIS panels are responsive to dynamic vehicular conditions that can be used to achieve ultra-reliable and the low-latency communication.

7. SYSTEM INTEROPERABILITY, NETWORKING PROTOCOLS, AND MOBILITY MANAGEMENT

The V2X systems should be supported by RIS, which enables them to be able to communicate with the existing standards of communication and mobility management in order to be readily incorporated into 6G ecosystems. The enhancements of the Mac and PHY layers also enable to schedule, beamform and coordinate RIS assisted links successfully. The use of Multi-RIS applications makes mobility more complex and requires smart handover plans and trajectory prediction systems. interoperability with C-V2X, NR-V2X, DSRC, and edge cloud infrastructure will ensure that it has a flexible and unified communication architecture (Jameel et al., 2019). This section pinpoints interoperability problems and covers the way on which it can be possible to attain a seamless work between heterogeneous layer and technologies of the network.

7.1 MAC AND PHY Layer Enhancements For RIS-Assisted Links

The implementation of RIS into V2X communication needs considerable improvements at the MAC and PHY layers. At PHY layer, channel estimation is important in estimating the best RIS configurations. Nonetheless, vehicle mobility brings in very fast channel changes, and therefore the low-overhead estimation techniques like compressed sensing or machine-learning based predictions are required. MAC layer should adjust to the scheduling mechanisms to allow RIS-assisted links which are defined by controlled reflections, as opposed to direct transmission. This also involves a provision of resources to control signalling of RIS, managing a multi-user access and coordination of various RIS panels to prevent interference (X. Zhou et al., 2023). Improved PHY-MAC collaboration facilitates a constant, scalable and high throughput of performance in various car situations.

7.2 Handover, Mobility Tracking, And Multi-Ris Coordination

RIS-assisted V2X mobility management deals with the monitoring of vehicle movement, the predictability of communication needs, and handovers among RIS-assisted links. RIS configurations should be dynamic as vehicles pass-through high-speed limits in order to avoid degradation of links. Prediction mobility algorithms are algorithms that take into account the trajectory, speed, and other environmental data to extract future channel states. The Multi-RIS coordination would be necessary in urban settings where several panels of RIS can interact with the fast-moving cars at the same time (Farran et al., 2022). The coordination of these panels avoids beam conflicts and is used to provide seamless transitions between RIS assisted links. Sophisticated handover systems that incorporate RIS support in V2X mobility protocols to ensure ultra-reliable connectivity.

7.3 Interoperability With C-V2X, NR-V2X, DSRC, And Edge Cloud

The systems RIS-assisted must also be capable of accommodating other vehicular communication standards in that they can be accommodated to older standards. The wide-area vehicle communication is based on C-V2X and NR-V2X, and DSRC has still not acquired all safety applications. All these technologies are enhanced by RIS that provides a better coverage and reduced interference without alteration of core protocol stacks. The interoperability frameworks will ensure that the operations of RIS are seen to be transparent to the prevailing vehicular systems. Co-location with edge cloud infrastructures are also scale enhancers as they facilitate distributed RIS control, low-latency data processing, distributed perception between vehicular networks (Y. Chen et al., 2021). This interoperability will enable RIS technology to interoperate with various deployments and regulatory spaces with ease.

8. LITERATURE REVIEW

Reconfigurable Intelligible Surfaces in automobile communication have become a recent area of interest in the past five years because of the necessity of high-capacity, reliable, and low-latency connections in 6G-enabled mobile networks. Current literature has also shown that RIS is a viable solution to address the issue of Non-Line-of-Sight (NLOS) as well as the issue of enhancing link reliability, spectral, and energy efficiency of various V2X communication models. The literature review summarizes the main research findings made in the last three years (2020-25), compares the channel modeling methods in the presence and absence of RIS, as well as identifies the gaps in research that guide future studies in the area (Mitsakis & Anapali, 2020).

8.1 RIS For Vehicular Communications: Key Contributions (2020–2025)

The new connections in the RIS enabled automotive communication have been made by improving reliability of connection, coverage and minimization of interference in the dynamic road environment. Preliminary work (2020-2021) of this nature suggested general RIS-aided V2X channel formulations, and passive metasurfaces were discovered to be effective in deflecting signals to overcome the obstacles and NLOS propagation difficulties. It was found that receive power, outage probability and coverage extension improved majorly in millimeter-wave and sub-THz vehicular links (Z. Luo et al., 2019). In or around 2022 to 2023 the trend changed to mobility-centric RIS designs. Algorithms to control the high-speed vehicle scenarios were developed as fast beam tracking algorithms, predictive beamforming algorithms and Doppler-aware RIS configuration algorithms. New vehicular RIS prototypes were demonstrated on metasurface panels on lightweight panels (mounted on roadside infrastructure, buildings, and tunnel walls). Furthermore, there began to appear reinforcement learning-based RIS controllers, which permit self-optimizing patterns of reflection with no ongoing channel estimation. By 2024-2025, studies developed to multi-RIS ecosystems of urban mobility, multi-coordinated surfaces collaborated to enhance V2I and V2N links (Singh et al., 2021). In practice, it has been demonstrated through practical experimentation (Practical implementation of RIS based highways corridors in China and multi-RIS testbeds in Singapore) that performance benefits of reduced latency, channel stability and spectral efficiency can be attained. RIS facilitates the future 6G vehicular network revolution, that is affirmed in the literature, but needs further studies on a massive scale standardization and full mobility integration.

8.2 Comparative Analysis Of V2X Channel Models With And Without RIS

Classical V2X channel models take into consideration several propagation issues such as fading, shadowing, Doppler spread and NLOS blockage due to vehicles, buildings and clutters in the environment. In absence of RIS, the only methods of overcoming these impairments are to use MIMO beamforming, antenna diversity, or extra RSUs, which are quite costly and energy-efficient in their systems (X. Zhou et al., 2022). Traditional models also presuppose the fixed environment features, which leads to the lack of flexibility to the mobile vehicle. RIS enhanced channel models add to channel models, controllable propagation paths by using programmable reflection, which allow incident waves to be adaptively modified. The comparative studies have indicated that it has a higher signal to noise ratio, path diversity and probability of outage of the RIS-assisted channels. Geometry-Based Stochastic Models with RIS reflectors show better forecasts in predicting V2V and V2I link performance under mobility as compared to classical stochastic fading models (Di Renzo et al., 2020). Simulation studies by ray-tracing indicate that RIS greatly improves the quality of links in sharp turns, high density crossings and tunnels, where conventional communication is typically lost. Moreover, THz and mmWave models with the assistance of RIS exhibit a significantly lower path loss since phase alignment among meta-atoms is shaped. The literature has consistently shown that RIS is changing the propagation environment into an active communication resource allowing it to promote new channel properties that are radically different than the traditional models.

8.3 Identified Gaps: Real-Time Control, Cross-Layer Optimization, And Multi-RIS Systems

Although the benefits of RIS-V2X are promising, there are still a number of significant gaps in the existing research field of RIS-V2X. To start with, high mobility real-time RIS control is still a very significant issue. The vehicular conditions need to be reconfigured at the level of milliseconds to accommodate the channel conditions that fluctuate very fast. Algorithms that are currently being used usually require full CSI or periodically training, which is not feasible at highway frequencies (Nguyen et al., 2021). To minimise signalling overhead to ensure predictable performance, more efficient predictive and learning based control mechanisms are required. Second, there has not been a complete exploitation of cross-layer optimization. The majority of research focuses on the enhancement of physical-layers without incorporating RIS in the MAC scheduling, routing, mobility management, and resource allocation schemes. Future systems should embrace the concept of holistic designs where RIS configuration will interact flawlessly with the higher-layer decisions based on the cooperative perception, platooning and cloud offloading (Ying et al., 2020). Third, the multi-RIS coordination needs additional research. Urban settings can have dozens of RIS panels that need to connect the beam patterns without interference, facilitate handover and have steady connections. The existing literature provides tentative solutions but does not come up with large scale, out world tested structures. There are also hardware constraints, calibration issues, constrained reflection resolution, and energy constraint, which still remain as a hurdle. In general, the existing literature shows that substantial progress has been made by the field but simultaneously addresses the necessity of scalable, intelligent, and deployment-ready RIS-assisted V2X systems (Naskar et al., 2024).

9. CASE STUDIES AND REAL-WORLD DEPLOYMENTS

Recent practical applications of RIS in vehicular systems have also acquired momentum in the past two years and prove the viability and performance advantages of reconfigurable environments to V2X communication. The pilot tests on highways, intelligent crossroads, and urban metropolitan networks demonstrate that RIS may contribute to the reliability of links, decreasing the number of NLOS-related failures and providing the stable connection of autonomous and connected vehicles (Tang et al., 2019). Measurements in the field have confirmed that RIS can increase significantly received signal strength and decrease latency in systems where conventional communication systems would be challenged because of blockage or large-scale movement, or extremely high multi-path fading. These experiments demonstrate that RIS is an important facilitator of 6G-compliant vehicular networks, which can accommodate high-frequency communication at the mmWave and terahertz frequencies. The case studies below are the summary of two large-scale deployments, a highway one, and an urban one, with a comparison analysis conducted afterwards to determine the key lessons that should be learned to implement on a large scale.

9.1 Case Study 1: RIS-Assisted Highway V2I Deployment (2024–2025 Pilots)

One challenge that V2I communication faces specifically is the highway scenario as vehicles traveling at a high speed and the vast distance of coverage and often blocked Line-of-Sight by the terrain and road geometry or very large vehicles. In Suzhou, China, a pilot project between 2024 and 2025 with a gigantic RIS panel on sound barriers and over structures as components of a 7 km expressway had a collaboration deployment. It was implemented as passive RIS arrays of digitally adjusted phase to the neighbouring RSUs through optical backhaul. Field tests demonstrated significantly higher mmWave connectivity, and at high-rate (over 100 km/h) conditions, over 40 percent improvement in the average SNR and a decrease in the percentage of packet loss of almost 35 percent. The RIS panels constantly diverted reflected rays to vehicles with weak links such that V2I connection does not break in the volatile road curves that were viewed as blind points previously (Gong, 2015). The ability of the system to provide efficient Doppler-resilient beam tracking to facilitate the occurrence of uninterrupted communication even during the change topology conditions at a rate was also demonstrated. This also meant that the energy consumption was greatly minimized in comparison to the traditional active repeaters or the increment in the amount of RSUs. The pilot testified that RIS would offer affordability in the long-range highway communication and could serve as an alternative of scaled transportation corridors to 6G in the future. Figure 3 indicating the required performance benefits, SNR improvement benefit, and packet loss benefit, achieved in the 2024-2025 RIS-supported highway V2I pilot deployment of V2I.

Figure 3. Performance improvements in RIS-assisted highway V2I deployment

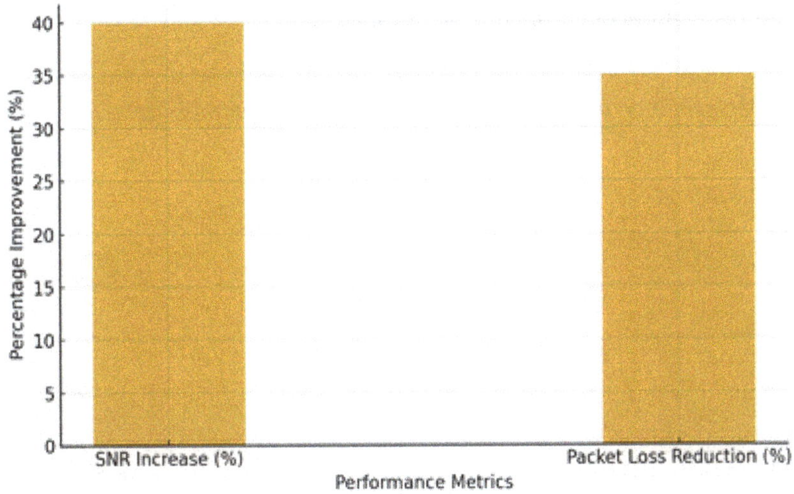

9.2 Case Study 2: Multi-Ris Urban Mobility Enhancement (City-Level Trials)

The high urban density creates extreme propagation deficits as a result of large building density, complicated traffic routes and extensive NLOS conditions. In 2025, an RIS-enabled testbed of urban mobility was introduced at Nanyang Technological University (NTU) Singapore, and installed more than 30 RIS panels on the facade of buildings, lamp posts, and smart intersections. The purpose of this multi-RIS set up was to facilitate autonomous shuttle operations and the sharing of sensor-data on a crowded grid in the campus. The system utilised a hybrid RIS system that used a combination of passive reflective panel and selective active unit to enhance weak terahertz signals. Edge servers worked with the RIS controllers in order to facilitate predictive beam steering and multi-panel coordination. It was found that there was a 60 percent increase in the reliability of the links and a maximum of 50 percent decrease in the time of handover interruption with autonomous shuttles working in highly populated pedestrian and vehicular environment (X. Chen et al., 2020). Moreover, the multi-RIS coordination was effective to eliminate the reflection conflicts and guarantee the continuity and high quality of the V2I and V2N communications. The experimental results confirmed the advantages of distributed RIS surfaces in addressing the problem of deep urban shadowing and preserving predictable latency to support safety critical performance of the system like collision avoidance applications and cooperative perception. Figure 4 of reliability enhancement and reduction in handover time that were realized during the 2025 multi-RIS urban mobility implementation at NTU Singapore.

Figure 4. Performance improvements in Multi-RIS urban mobility trials

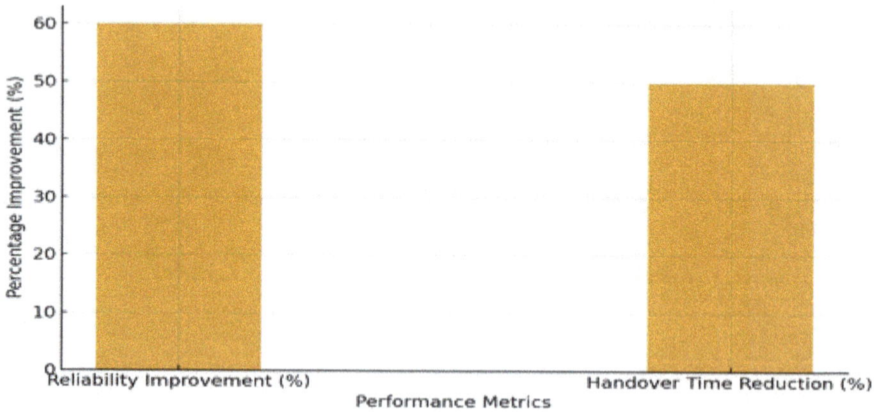

9.3 Comparative Analysis And Key Lessons For 6G-V2X Deployment

The two deployments serve to bring out different but complementary understandings of RIS enabled vehicular communication. RIS was shown to be effective in highway tests in the enhancement of long-range coverage, beam tracking at Doppler resilience, and low-energy performance. Conversely, the multi-RIS urban implementation was necessary to manage overcrowded NLOS conditions, facilitate high handover rates, as well as facilitate coordinated wisdom of multiple surfaces. In both studies, the integration of RIS also contributed greatly to the minimization of packet loss and enhanced SNR and enabled stable mmWave/THz communication. Some of the valuable lessons are the necessity of predictive reconfiguration to support mobility, the necessity of distributed control to support multiple RIS units, and the usefulness of edge-assisted optimization to support latency-sensitive applications (Jin, 2023). Also, the environmental placement becomes one of the key design considerations: high-rise RIS buildings are beneficial in terms of highway coverage, but omnidirectional placement is crucial in dense cities. All in all, the demonstrations affirm that the RIS technology is a feasible and scalable route to attaining 6G quality vehicular communication.

10. PERFORMANCE EVALUATION MODELS

RIS-assisted V2X system performance analysis needs to be performed using specific analytical and simulation-based models that can model dynamic vehicle scenarios, NLOS, and high-frequency propagation. RIS adds new paths to reflection, new angular distribution and controllable scattering features and the conventional channel and performance models have proven to be inadequate. System improvements that RIS allows are measured using key performance indicators such as latency, reliability, throughput, and energy efficiency. Interaction between vehicles, infrastructure and RIS surfaces normally include geometry-based models, stochastic mobility patterns and hybrid ray-tracing, which are part of simulation frameworks. Moreover, mmWave and THz analysis of high frequency evaluates the capability of RIS in order to eliminate serious propagation limitations (Bhat et al., 2023). The comparison of the performance of the RIS and the no-RIS can give an understanding about the extension of coverage, stability during mobility, and the improvement in data rate. Altogether, the RIS-specific assessment methodologies are necessary to confirm the feasibility of the systems and the strategies to deploy them on a large scale.

10.1 Key KPIS: Latency, Reliability, Energy Efficiency, Throughput

The evaluation of the RIS-aided V2X systems is based on a number of key KPIs that represent the quality of communication and reaction to the system. Latency The latency is a particularly important issue in safety applications; autonomous braking, cooperative manoeuvres and hazard warning can demand sub-10 ms performance. RIS improves the latency performance by stabilizing links, minimizing their retransmission, and facilitating successful beam alignment. The reliability, as it is usually defined as we can see in the packet delivery ratio or outage probability is enhanced due to controlled reflections which can reduce blockage and multi-path fading. Another important metric is energy efficiency, which is particularly a large-scale connected vehicle ecosystem (M. Z. Chen et al., 2022). Passive RIS is intrinsically ultra-low power consumption which leads to less energy-consuming amplifiers or extra RSUs. Throughput measures the capability of exchange of data which is vital in high data rate applications like HD map updates and transfer of sensor data. RIS increases throughput with the key benefits of improved SNR, directional transmission, and high-frequency bands. When these KPIs are assessed together, it becomes possible to have a very large picture of RIS inputs to next-generation vehicle work. Table 4 Lists the latency, reliability, energy efficiency, throughput, mobility resilience and justifies them to vehicular communication.

Table 4. Key KPIs and their significance for RIS-assisted V2X systems

KPI	Definition	Significance in RIS-V2X
Latency	Time taken for end-to-end communication	Critical for safety; RIS stabilizes links to reduce delay
Reliability	Probability of successful packet delivery	RIS improves NLOS performance and reduces outages
Throughput	Rate of data transmission	Enhanced via SNR improvement and mmWave/THz support
Energy Efficiency	Communication performance per energy unit	Passive RIS dramatically reduces energy demand
Mobility Resilience	Link stability during motion	RIS beam steering mitigates Doppler and blockages

10.2 Simulation Frameworks: NS-3, MATLAB, Ray-Tracing With RIS Modules

The simulating system is very important in estimating the RIS-enabled V2X performance in different deployment settings. NS-3 is very popular in simulations at the network level (end-to-end) because it provides realistic models of mobility, customizable protocol stacks and also supports C-V2X/NR-V2X communication. Reflection-aware routing, handover, scheduling and multi-RIS coordination can be tested with RIS modules in combination with NS-3. MATLAB has the added benefit of offering the flexibility of model RIS-aided beamforming, optimization, and channel estimation schemes, which is why it is suited best to physical-layer assessments. It can be integrated with tool boxes of optimization and can be quickly prototyped to apply reinforcement learning or RIS control by a metaheuristic. RIS surfaces are modelled by simulators called ray-tracing simulators (Remcom Wireless InSite, WinProp, Sionna RT) to simulate accurate propagation conditions, such as reflections, diffraction, and scattering (Yang et al., 2022). They achieve accurate predictions of mmWave and THZ frequencies, realistic RIS effects of tunnels, intersections, and local streets. The combination of these frameworks will provide holistic assessment on the physical, network, and mobility layers.

10.3 Comparative Performance With/ Without RIS (MMWAVE And THZ)

Comparisons of performance of RIS-assisted and conventional vehicular communication reveal that there are considerable benefits especially at mmWave frequencies and terahertz frequencies. In the absence of RIS, high-frequency signals have severe path loss, vulnerability to blockage, and directional sensitivity and tend to cause frequent failures in links in highly dense or dynamic environments. Simulations and experimental research indicate that placing RIS can enhance received SNR by 10-25 dB relying on positioning and surface design. RIS minimizes the outage probability and allows stable connectivity in mmWave highways that is not within the natural range of LOS. Active or hybrid RIS panels will be useful at THz frequencies where attenuation is quite large to restructure propagation paths extending communication to several hundred meters in NLOS conditions (Maher Al-Hatim & Othman, 2024). RIS is also better in spectral efficiency, where many users can have controlled reflections. These relative outcomes indicate that RIS is an enabling factor of sustainable and high-capacity vehicular communication in 6G networks.

11. SECURITY, PRIVACY, AND TRUST MECHANISMS IN RIS-ASSISTED V2X

The situation with security and privacy increases with RIS-assisted V2X systems due to the fact that RIS varies the physical propagation environment and adds new control interfaces. Although RIS improves communication performance, wrong or malicious manipulation of RIS states may break links, beam misdirection, or may aid interception by unauthorized means. The trust mechanisms should provide secure control of the RIS and safeguard data sent by the vehicular communications as well as mitigate the abuse of the reconfigurable environment. RIS can be used to increase secrecy by taking advantage of the physical layer security gains so that the energy can be focused on legitimate receivers and nullified on eavesdroppers (K. Luo et al., 2021b). Distributed trusting systems based on blockchain or secure edge nodes will be used to avoid the unauthorized access to RIS configuration commands. All in all, the implementation of RIS needs a comprehensive security infrastructure (both on the physical and cyber level) that would enable the safe operation of vehicles.

11.1 Threat Surface Analysis: Spoofing, Replay, RIS Hijacking

The RIS-sided V2X environments increase the attack surface due to the presence of the RIS control plane and the capability to reflect the vulnerability points. The spoofing attacks are the attacks in which enemies masquerade as legitimate vehicles or RSUs and delude the RIS to send out beams to unauthorized recipients. This interferes with the confidentiality and can affect coordinated driving. Replay attacks include interception of valid RIS control signals and their retransmission to modify beam patterns at inopportune times and result in misalignment of communication and even a safety hazard. RIS hijacking is a more critical attack, in which hackers have limited or complete access to the RIS controller, and they can edit reflection coefficients to cause malice. This may generate deliberate dead zones, disrupt V2I communication or re-steer beams to eavesdroppers (Li et al., 2025b). The high-mobility and dynamic characteristics of V2X also make it more difficult to detect attacks. These threats must be countered by implementing comprehensive authentication, encrypted control signalling solutions and intrusion detection solutions based on anomaly.

11.2 Physical Layer Security With RIS Beam Manipulation

RIS opens fresh possibilities of Physical Layer Security (PLS) because it provides the ability to control the propagation environment finely. RIS can also focus electromagnetic waves by manipulating their phase and amplitude thereby directing energy towards the desired vehicles and nullifying/cutting the energy in regions where there might be a prospective eavesdroppers. This is capabilities that allow enhancement of secrecy rates without necessarily heavy cryptographic means. The security of the V2X channels is also enhanced by RIS-aided artificial noise generation, cooperative jamming, and secure beamforming strategies. RIS is also able to dynamically alter the pattern of reflections to ensure that secrecy is preserved in high-mobility situations as vehicles travel, thus ensuring that a secure connection is maintained. Moreover, PLS techniques that use RIS require less power and can be scaled because they do not require more antennas or hardware (Gebrezgiher et al., 2025). By adding RIS to vehicular PLS systems, the confidentiality, integrity, and ability to resist the passive interceptor can be improved significantly, especially in mmWave and THz bands where the directional propagation is the most relevant.

11.3 Blockchain And Edge Security For Distributed RIS Control

With the assistance of potent tools offered by edge computing and blockchain, distributed RIS control can be offered in V2X networks. With the help of blockchain, the authentication and immutability and transparent access control of RIS configuration commands can be decentralized. All the orders or updates could be stored in a safe register and this could overcome unauthorized alterations and hijacking of RIS. Smart contracts allow ensuring that only trusted RSUs or vehicles may demand reconfiguring of RIS and checking automation processes in the works. Edge servers prove to be helpful in improving the degree of the security because they perform local authentication, data breach, and real-time examination of the abnormality without the necessity of the remote cloud service. This will reduce the time lag and will offer speedy response during the occurrence of malicious attacks (Shrestha et al., 2020b). The edge nodes are also capable of synchronizing several RIS panels, and confirm their alignment, and check their potential tampering. Combined with the trust of the blockchain and the intelligence of the edge, the safe and scalability model of V2X business supported by RIS can be created in the vast geographical regions. Table 5 replaces the threats (spoofing, hijacking, replay) with the mitigation strategies (PLS, blockchain, secure control).

Table 5. Security threats and countermeasures in RIS-assisted V2X

Security Threat	Risk Description	Potential Countermeasure
Spoofing Attacks	Fake entities mislead RIS control	Mutual authentication, secure signalling
Replay Attacks	Recorded commands altered or reused	Timestamp validation, encrypted channels
RIS Hijacking	Unauthorized modification of RIS states	Blockchain-based access control
Eavesdropping	Adversary intercepts directional beams	RIS-assisted PLS, artificial noise
Jamming	Signal interference targeting RIS beams	Dynamic beam steering and nulling

12. CHALLENGES AND OPEN RESEARCH PROBLEMS

There are several deployment concerns of vehicular communication that are assisted by RIS that affect the scalability, performance, and standardization. The highly dynamic nature of the propagation conditions and substantial mobility requires that RIS panels reconfigure on-the-fly, and the existing optimization algorithms can hardly meet these hard constraints on latency. Loss of practical efficiency is also caused by hardware issues, such as phase noise, power, and calibration drift. Multi-RIS systems offer intricacies of regulation, conflicts and common management. Moreover, the interoperability of the systems with C-V2X and NR-V2X is interrupted by a deficit of RIS-specific standards. Disbursement expenses especially in the congested cities has been a significant problem. The issues will be defeated only after a concerted effort on the hardware, algorithms, and regulations.

12.1 Multi-RIS Coordination And Real-Time Optimization

Synchronization of beamforming, and conflict-free reflection control is a significant problem in the coordination of multiple RIS panels in high-vehicular-density environments. In cases where dozens of RIS units are distributed throughout urban roads or highway corridors, centralized control would lead to high signalling overheads, high latency and poor computational scalability. Optimization in real time is also a challenge due to the fast-changing nature of channels by high-speed vehicles that would demand sub-milliseconds of the RIS updates. The existing iterative algorithms are too slow to facilitate such dynamics (Naaz et al., 2024). The new solutions are hierarchical control structures, distributed RIS intelligence and predictive optimization with help of trajectory forecasting. The digital twin and machine learning systems enable RIS panels to predict the change in channels, eliminating the need to update CSI frequently. Multi-RIS coordination plays a crucial role in ensuring a continuous coverage, lowering the interference, and facilitating scalable next-generation V2X 6G communication.

12.2 Hardware Limitations, Calibration, And Deployment Constraints

Real-life RIS installations encounter problems with hardware which restrict performance in the real world. The steps of phase errors, non-linear response, drift by temperature, and poor resolution which are common in meta-atoms hamper beamforming precision. Big RIS panels also demand steady power provision to control signalling which poses a problem in remote or infrastructure-constrained conditions. The other significant challenge is calibration; thousands of meta-elements need to be kept in fine tune even under vibration, exposure to the weather and age. Occurring recalibration adds complexities and expenses of operation. Other limitations of deployment include height of installation, angle, structural support, and backhaul connectivity (Martín-Martín et al., 2024). In the large-scale deployments of smart cities, mounting hundreds of RIS requires cost-effective mounting plans and common infrastructure. These hardware and deployer problems are essential in providing a stable long-term RIS-assisted vehicular communications.

13. FUTURE RESEARCH DIRECTIONS

The RIS-V2X of the next generation will be enhanced with the intelligence, flexibility, and compatibility with the advanced 6Gs. The AI-based predictive control and digital twins will enable RIS panels to anticipate variations of channels based on mobility. RIS and terahertz metasurfaces will be quantum ready to serve ultra-high data rate required in the holographic communication and real time digital twins. The high-fidelity sensor sharing with the assistance of the RIS will be useful to the cooperative perception and multi-agent autonomy (Khan et al., 2020). The secure RIS control, low cost/biodegradable metasurfaces and energy harvesting RIS designs need more investigation to allow sustainable deployments. These principles will shape up smart transportation systems in the new generations.

13.1 AI-Based Predictive Ris Control And Digital Twin Optimization

The use of AIs to predict control is an essential step to the optimization of RIS in the context of high vehicular mobility. The conventional RIS systems are based on the fact that continuous channel estimation is not feasible when the vehicles are traveling at a high speed and the channel coherence time is very small. The use of machine learning and reinforcement learning models can predict future channel conditions, car paths, and blockage, making RIS actively set the reflection coefficients. This minimizes signalling and maintains a constant connection during manoeuvre of high velocity. Digital twins can also boost this capacity by providing real-time virtual representations of road environments, which allow RIS planning based on simulation, and immediate testing of other options (Michailidis et al., 2021). The combination of AI and digital twins will allow managing the RIS dynamically and on the basis of a specific context, which will better follow the patterns of mobility, weather, traffic density, and frequency use, which significantly improves reliability and scalability.

13.2 Quantum-Ready RIS And Terahertz-Enabled Vehicular Systems

The RIS technology must also be enhanced to accommodate the new terahertz and quantum communication models, which will make the core of the 6G mobility systems. Terahz communication could be used to achieve multi-gigabit data rates, but it has a very high propagation loss and blockage; this can be overcome with RIS, focusing, redirecting and amplification of THz wave using current state-of-the-art active or graphene-based metasurfaces. RIS quantum ready To accomplish secure quantum key distribution and entanglement-based communication, quantum-ready RIS architectures, implies that it can be capable of manipulating signal wavefronts with sub-nanoprecision and sub-noiselow noise (Tong et al., 2019). Such sensitive designs of RIS systems will be sensitive designs of meta-atoms, photonic control and error-tolerant functionality. Both THz-enabled and quantum-compatible RIS architectures will allow holographic displays, real-time digital twins, ultra-HD sensor streaming, and safe autonomous driving communications to be provided.

14. CONCLUSION

The RIS have emerged as critical technology to the future of the car communication system where their programmability and flexibility is enabling propagation environment to address long time challenges in V2X connectivity. The architectural expertise presented in this chapter is used to highlight the essential fact that RIS essentially reinvents a 6G-enabled vehicular networks by introducing an electromagnetic layer with controllable properties that are effectively combined with physical, networking and computational platforms. RIS has improved system architectures to offer high-mobility, high-density Urban and NLOS-constrained communications, advanced beamforming, dynamic control of reflections, and fine-grained control over signals. The advantages of these developments in architecture consist of significant improvements in latency, reliability, spectral and energy performance and RIS will become an indispensable component of transportation infrastructures in the future.

RIS is a transformative force that does not solely limit itself with the enhancement of performance but enables new functionality to re-determine cooperative mobility. The development of THz/mmWave communication, predictive beam steering, and distributed edge intelligence will make autonomous driving, real-time digital twins, and multi-agent coordination possible and increase their potential. Dynamic reconfiguration of the wireless environment ensures the resiliency of the communication even in the challenging situations such as tunnels, crossroads and canyons. RIS also aids in improving physical layer security, also, controlling propagation routes and delivering directional secrets and creating safe vehicle ecosystems. The success of general RIS implementation in real-life smart transportation systems is inseparable with an elaborate roadmap. These include development of scalable multi-RIS control algorithms, standard interface, secure control architecture and low-cost integration into the already existing C-V2X and NR-V2X infrastructure. Implementation will entail the collusion of the automotive companies, network operator, city planners and the regulatory agencies to address the problems of installation instructions, framework schemes of interoperability and long-term maintenance plans. As the modern trends in the direction of AI-assisted predictive control, quantum-enhanced RIS, and energy-gathering metasurfaces progress, RIS technology will be a key to intelligent and sustainable 6G vehicular communication networks and allow the transition to the fully autonomous, connected, and resilient mobility ecosystems.

REFERENCES

Anil Babu, B., Ram Kalyan, P., Lakshmi, V. V., Reharika, R., & Varun Raj, N. (2023). An arduino-controlled reconfigurable intelligent surface with angular stability for 5G mmWave applications. *Progress in Electromagnetics Research Letters, 114*, 69–74. DOI: 10.2528/pierl23091903

Assiimwe, E., & Marye, Y. W. (2022). A stochastic confocal elliptic-cylinder channel model for 3D MIMO in millimeter-wave high-speed train communication system. *Electronics (Basel), 11*(13), 1948. DOI: 10.3390/electronics11131948

Bhat, V. S., Harshavardhan, G., & Chockalingam, A. (2023). Input-output relation and performance of RIS-aided OTFS with fractional delay-Doppler. IEEE Communications Letters: A Publication of the IEEE. *Communicatio Socialis, 27*(1), 337–341. DOI: 10.1109/lcomm.2022.3214678

Chen, M. Z., Tang, W., Dai, J. Y., Ke, J. C., Zhang, L., Zhang, C., Yang, J., Li, L., Cheng, Q., Jin, S., & Cui, T. J. (2022). Accurate and broadband manipulations of harmonic amplitudes and phases to reach 256 QAM millimeter-wave wireless communications by time-domain digital coding metasurface. *National Science Review, 9*(1), nwab134. Advance online publication. DOI: 10.1093/nsr/nwab134 PMID: 35079409

Chen, X., Wu, C., Chen, T., Zhang, H., Liu, Z., Zhang, Y., & Bennis, M. (2020). Age of information aware radio resource management in vehicular networks: A proactive deep reinforcement learning perspective. *IEEE Transactions on Wireless Communications, 19*(4), 2268–2281. DOI: 10.1109/twc.2019.2963667

Chen, Y., Wang, Y., Zhang, J., & Di Renzo, M. (2021). QoS-driven spectrum sharing for reconfigurable intelligent surfaces (RISs) aided vehicular networks. *IEEE Transactions on Wireless Communications, 20*(9), 5969–5985. DOI: 10.1109/twc.2021.3071332

Chen, Y., Wang, Y., Zhang, J., Zhang, P., & Hanzo, L. (2022). Reconfigurable intelligent surface (RIS)-aided vehicular networks: Their protocols, resource allocation, and performance. *IEEE Vehicular Technology Magazine, 17*(2), 26–36. DOI: 10.1109/mvt.2022.3158046

Clancy, J., Mullins, D., Ward, E., Denny, P., Jones, E., Glavin, M., & Deegan, B. (2023). Investigating the effect of handover on latency in early 5G NR deployments for C-V2X network planning. *IEEE Access : Practical Innovations, Open Solutions, 11*, 129124–129143. DOI: 10.1109/access.2023.3334162

Dai, L., Wang, B., Wang, M., Yang, X., Tan, J., Bi, S., Xu, S., Yang, F., Chen, Z., Di Renzo, M., Chae, C.-B., & Hanzo, L. (2020). Reconfigurable intelligent surface-based wireless communications: Antenna design, prototyping, and experimental results. *IEEE Access : Practical Innovations, Open Solutions, 8*, 45913–45923. DOI: 10.1109/access.2020.2977772

Di Renzo, M., Ntontin, K., Song, J., Danufane, F. H., Qian, X., Lazarakis, F., De Rosny, J., Phan-Huy, D.-T., Simeone, O., Zhang, R., Debbah, M., Lerosey, G., Fink, M., Tretyakov, S., & Shamai, S. (2020). Reconfigurable intelligent surfaces vs. Relaying: Differences, similarities, and performance comparison. *IEEE Open Journal of the Communications Society, 1*, 798–807. DOI: 10.1109/ojcoms.2020.3002955

Eze, E., & Eze, J. (2023). Artificial intelligence support for 5G/6G-enabled Internet of Vehicles networks: An overview. *ITU Journal : ICT Discoveries, 4*(1), 178–195. DOI: 10.52953/iezn8770

Farran, H., Khoury, D., & Bokor, L. (2022). A comprehensive survey on the application of blockchain/hash chain technologies in V2X communications. *Infocommunications Journal, 14*(1), 24–35. DOI: 10.36244/icj.2022.1.4

Gebrezgiher, Y. T., Jeremiah, S. R., Deng, X., & Park, J. H. (2025). Machine learning-based blockchain technology for secure V2X communication: Open challenges and solutions. *Sensors (Basel), 25*(15), 4793. DOI: 10.3390/s25154793 PMID: 40807958

Gong, Y. (2015). Research on the elderly housing in China : a case study on "Beijing Sun City Living Community." DOI: 10.5353/th_b5334605

Houmer, M., Laqtib, S., & Eddamiri, S. (2022). Lightweight and secure authentication model for vehicle to Everything (V2X) communication based on 5G networks. Journal of Mobile Multimedia. DOI: 10.13052/jmm1550-4646.1854

Jameel, F., Wyne, S., Nawaz, S. J., & Chang, Z. (2019). Propagation channels for mmWave vehicular communications: State-of-the-art and future research directions. *IEEE Wireless Communications, 26*(1), 144–150. DOI: 10.1109/mwc.2018.1800174

Jin, Z. (2023). Channel modeling for V2V communication: Enhancing road safety and intelligent transportatio. *Applied and Computational Engineering, 12*(1), 115–121. DOI: 10.54254/2755-2721/12/20230307

Khan, L. U., Yaqoob, I., Imran, M., Han, Z., & Hong, C. S. (2020). 6G wireless systems: A vision, architectural elements, and future directions. *IEEE Access : Practical Innovations, Open Solutions, 8*, 147029–147044. DOI: 10.1109/access.2020.3015289

Kiela, K., Barzdenas, V., Jurgo, M., Macaitis, V., Rafanavicius, J., Vasjanov, A., Kladovscikov, L., & Navickas, R. (2020). Review of V2X–IoT standards and frameworks for ITS applications. *Applied Sciences (Basel, Switzerland), 10*(12), 4314. DOI: 10.3390/app10124314

Li, H., Wang, F., Qian, J., Zhu, P., & Zhou, A. (2025a). Partitioned RIS-assisted vehicular secure communication based on meta-learning and reinforcement learning. *Sensors (Basel), 25*(18), 5874. DOI: 10.3390/s25185874 PMID: 41013112

Li, H., Wang, F., Qian, J., Zhu, P., & Zhou, A. (2025b). Partitioned RIS-assisted vehicular secure communication based on meta-learning and reinforcement learning. *Sensors (Basel), 25*(18), 5874. DOI: 10.3390/s25185874 PMID: 41013112

Liang, J. C., Zhang, L., Luo, Z., Jiang, R. Z., Cheng, Z. W., Wang, S. R., Sun, M. K., Jin, S., Cheng, Q., & Cui, T. J. (2024). A filtering reconfigurable intelligent surface for interference-free wireless communications. *Nature Communications, 15*(1), 3838. DOI: 10.1038/s41467-024-47865-6 PMID: 38714685

Liaskos, C., Nie, S., Tsioliaridou, A., Pitsillides, A., Ioannidis, S., & Akyildiz, I. (2018). A new wireless communication paradigm through software-controlled metasurfaces. *IEEE Communications Magazine, 56*(9), 162–169. DOI: 10.1109/mcom.2018.1700659

Luo, K., Zhou, X., Wang, B., Huang, J., & Liu, H. (2021a). Sparse Bayes tensor and DOA tracking inspired channel estimation for V2X millimeter wave massive MIMO system. *Sensors (Basel), 21*(12), 4021. DOI: 10.3390/s21124021 PMID: 34200883

Luo, K., Zhou, X., Wang, B., Huang, J., & Liu, H. (2021b). Sparse Bayes tensor and DOA tracking inspired channel estimation for V2X millimeter wave massive MIMO system. *Sensors (Basel), 21*(12), 4021. DOI: 10.3390/s21124021 PMID: 34200883

Luo, Z., Chen, M. Z., Wang, Z. X., Zhou, L., Li, Y. B., Cheng, Q., Ma, H. F., & Cui, T. J. (2019). Digital nonlinear metasurface with customizable nonreciprocity. *Advanced Functional Materials, 29*(49), 1906635. DOI: 10.1002/adfm.201906635

Maher Al-Hatim, Y., & Othman Al Janaby, A. (2024). Artificial-intelligence-enhanced beamforming forPower-efficient user targeting in 5G networks using reinforcement learning. *International Journal of Computing and Digital Systems, 15*(1), 1083–1095. DOI: 10.12785/ijcds/160179

Mande, S., Ramachandran, N., & Begum, S. S. A. (2025). Navigating the future of intelligent transportation: Challenges and solutions in 6G V2X and V2V networks. *EAI Endorsed Transactions on Internet of Things, 11*. Advance online publication. DOI: 10.4108/eetiot.5114

Martín-Martín, A., Padial-Allué, R., Castillo, E., Parrilla, L., Parellada-Serrano, I., Morán, A., & García, A. (2024). Hardware implementations of a deep learning approach to optimal configuration of reconfigurable intelligence surfaces. *Sensors (Basel)*, *24*(3), 899. DOI: 10.3390/s24030899 PMID: 38339618

Michailidis, E. T., Miridakis, N. I., Michalas, A., Skondras, E., & Vergados, D. J. (2021). Energy optimization in dual-RIS UAV-aided MEC-enabled Internet of vehicles. *Sensors (Basel)*, *21*(13), 4392. DOI: 10.3390/s21134392 PMID: 34198977

Mitsakis, E., & Anapali, I. S. (2020). Recent developments on security and privacy of V2V & V2I communications: A literature review. *Periodica Polytechnica Transportation Engineering*, *48*(4), 377–383. DOI: 10.3311/pptr.14955

Naaz, F., Nauman, A., Khurshaid, T., & Kim, S.-W. (2024). Empowering the vehicular network with RIS technology: A state-of-the-art review. *Sensors (Basel)*, *24*(2), 337. DOI: 10.3390/s24020337 PMID: 38257430

Naskar, S., Brunetta, C., Hancke, G., Zhang, T., & Gidlund, M. (2024). A scheme for distributed vehicle authentication and revocation in decentralized VANETs. *IEEE Access : Practical Innovations, Open Solutions*, *12*, 68648–68667. DOI: 10.1109/access.2024.3400530

Nguyen, H. T. T., Nguyen, M. T., Do, H. T., Hua, H. T., & Nguyen, C. V. (2021). DRL-based intelligent resource allocation for diverse QoS in 5G and toward 6G vehicular networks: A comprehensive survey. *Wireless Communications and Mobile Computing*, *2021*(1), 1–21. DOI: 10.1155/2021/5051328

Sahin, M. M., Arslan, H., & Chen, K.-C. (2022). Control of electromagnetic radiation on coexisting smart radio environment. *IEEE Open Journal of the Communications Society*, *3*, 557–573. DOI: 10.1109/ojcoms.2022.3162142

Salim ALRikabi, H. T. H., Sallomi, A. H., KHazaal, H. F., & Magdy, A. (2025). Review of the Reconfigurable Intelligent Surfaces in smart cities: Opportunities, challenges, and applications. *IET Smart Cities*, *7*(1). Advance online publication. DOI: 10.1049/smc2.70003

Sarker, M. A. L., Son, W., & Han, D. S. (2023). RIS-assisted hybrid beamforming and connected user vehicle localization for millimeter wave MIMO systems. *Sensors (Basel)*, *23*(7), 3713. DOI: 10.3390/s23073713 PMID: 37050772

Shrestha, R., Nam, S. Y., Bajracharya, R., & Kim, S. (2020a). Evolution of V2X communication and integration of blockchain for security enhancements. *Electronics (Basel)*, *9*(9), 1338. DOI: 10.3390/electronics9091338

Shrestha, R., Nam, S. Y., Bajracharya, R., & Kim, S. (2020b). Evolution of V2X communication and integration of blockchain for security enhancements. *Electronics (Basel)*, *9*(9), 1338. DOI: 10.3390/electronics9091338

Singh, G., Srivastava, A., Bohara, V. A., & Liu, Z. (2021). Downlink performance of optical power domain NOMA for beyond 5G enabled V2X networks. *IEEE Open Journal of Vehicular Technology*, *2*, 235–248. DOI: 10.1109/ojvt.2021.3083919

Tang, W., Chen, M. Z., Chen, X., Dai, J. Y., Han, Y., Di Renzo, M., Zeng, Y., Jin, S., Cheng, Q., & Cui, T. J. (2021). Wireless communications with reconfigurable intelligent surface: Path loss modeling and experimental measurement. *IEEE Transactions on Wireless Communications*, *20*(1), 421–439. DOI: 10.1109/twc.2020.3024887

Tang, W., Li, X., Dai, J. Y., Jin, S., Zeng, Y., Cheng, Q., & Cui, T. J. (2019). Wireless communications with programmable metasurface: Transceiver design and experimental results. *China Communications*, *16*(5), 46–61. DOI: 10.23919/j.cc.2019.05.004

Tong, W., Hussain, A., Bo, W. X., & Maharjan, S. (2019). Artificial intelligence for vehicle-to-everything: A survey. *IEEE Access : Practical Innovations, Open Solutions*, *7*, 10823–10843. DOI: 10.1109/access.2019.2891073

Xu, Q., Xie, J., & Zhang, Z. (2024). High-speed rail multiple-input–multiple-output channel model based on reconfigurable intelligent surface system. *Electronics (Basel)*, *13*(14), 2868. DOI: 10.3390/electronics13142868

Yang, F., Pitchappa, P., & Wang, N. (2022). Terahertz reconfigurable intelligent surfaces (RISs) for 6G communication links. *Micromachines*, *13*(2), 285. DOI: 10.3390/mi13020285 PMID: 35208409

Ying, K., Gao, Z., Lyu, S., Wu, Y., Wang, H., & Alouini, M.-S. (2020). GMD-based hybrid beamforming for large reconfigurable intelligent surface assisted millimeter-wave massive MIMO. *IEEE Access : Practical Innovations, Open Solutions*, *8*, 19530–19539. DOI: 10.1109/access.2020.2968456

Zhang, X., & Cheng, X. (2019). Three-dimensional non-stationary geometry-based stochastic model for UAV-MIMO Ricean fading channels. *IET Communications*, *13*(16), 2617–2627. DOI: 10.1049/iet-com.2019.0139

Zhou, X., Ma, Z., Li, J., & Yu, L. (2022). Ultrasonic-aided fast-layered alternating iterative tensor channel estimation for V2X millimeter-wave massive MIMO systems. *Electronics (Basel)*, *11*(22), 3742. DOI: 10.3390/electronics11223742

Zhou, X., Zeng, Z., Li, J., Ma, Z., & Tong, L. (2023). Tensor-based joint beamforming with ultrasonic and RIS-assisted Dual-Hop Hybrid FSO mmWave massive MIMO of V2X. *Photonics*, *10*(8), 880. DOI: 10.3390/photonics10080880

Zhou, Y., Chen, L., Zhang, Q., Wang, H., & Guo, L. (2023). A 1-bit reconfigurable intelligent metasurface-based antenna design for 5G application. *International Journal of Microwave and Wireless Technologies, 15*(10), 1758–1767. DOI: 10.1017/s1759078723000466

KEY TERMS AND DEFINITIONS

Beamforming: A technique that directs wireless signals toward specific users or locations, improving signal quality and reducing interference.

Blockchain-Based Control: A decentralized security mechanism ensuring authenticated and tamper-proof RIS configuration updates.

Cooperative Perception: Sharing of sensor data among vehicles and infrastructure to enhance situational awareness and autonomous navigation.

Digital Twin for V2X: A virtual replica of the vehicular environment used for predicting and optimizing communication and RIS behavior.

Doppler Effect: Frequency shift in received signals caused by relative movement between transmitter and receiver, especially relevant in high-speed V2X.

Edge Computing: Localized data processing near the source (vehicles/RSUs) enabling low-latency decision-making and RIS optimization.

Geometry-Based Stochastic Model (GBSM): A channel modeling approach that incorporates real-world geometry and randomness for V2X communication analysis.

Meta-Atom: The basic unit of an RIS panel capable of adjusting phase, amplitude, or polarization of incident signals.

mmWave (Millimeter-Wave): High-frequency bands (30-300 GHz) offering large bandwidth but suffering from high path loss and blockage.

Multi-RIS Coordination: The process of synchronizing multiple RIS panels to avoid interference and enhance coverage in large-scale deployments.

Non-Line-of-Sight (NLOS): A scenario where the direct communication path is blocked by obstacles such as buildings, vehicles, or terrain.

Physical Layer Security (PLS): Security techniques leveraging signal propagation properties, such as RIS-based beam shaping, to enhance confidentiality.

Reconfigurable Intelligent Surface (RIS): A programmable metasurface composed of tunable meta-atoms that can manipulate electromagnetic waves to enhance wireless communication.

Reflection Coefficient: A phase-amplitude parameter controlling how each RIS element manipulates the incoming wave.

RIS Controller: A dedicated unit responsible for sending tuning commands to RIS elements and coordinating reflection behavior.

Terahertz (THz) Communication: Ultra-high frequency band enabling extremely high data rates, suitable for 6G but highly sensitive to attenuation.

V2I (Vehicle-to-Infrastructure): Communication between vehicles and roadside units, traffic signals, and smart infrastructure.

V2N (Vehicle-to-Network): Communication between vehicles and cellular/6G networks for cloud services, data offloading, and global connectivity.

V2V (Vehicle-to-Vehicle): Direct communication between vehicles to support collision avoidance, cooperative driving, and real-time safety alerts.

V2X Communication: A broad framework enabling vehicles to communicate with other vehicles, infrastructure, networks, and pedestrians.

Compilation of References

Abbas, S., Zhang, M., Xiao, Y., & Choi, J. (2021). A survey of LTE-V2X enhancements for intelligent transportation systems. *IEEE Communications Surveys and Tutorials*, *23*(2), 440–462. DOI: 10.1109/COMST.2021.3051330

Abboud, K., Omar, H. A., & Zhuang, W. (2016). Interworking of DSRC and cellular network technologies for V2X communications: A survey. *IEEE Transactions on Vehicular Technology*, *65*(12), 9457–9470. DOI: 10.1109/TVT.2016.2616518

Abouamer, M. S., & Mitran, P. (2025). A Hypernetwork Framework for Learning Adaptive Beamforming Schemes in RIS Systems. *IEEE Transactions on Communications*, *73*(12), 13471–13485. DOI: 10.1109/TCOMM.2025.3600529

Abumarshoud, H., Mohjazi, L., & Dobre, O. A. (2021). LiFi through reconfigurable intelligent surfaces: A new frontier for 6G? *IEEE Transactions on Vehicular Technology*, *70*(11), 11000–11012. DOI: 10.1109/TVT.2021.3123456

Ahad, M. A., Ahmed, A., Ansari, N., & Hassan, S. A. (2022). Software-defined networking and network slicing in 5G: A survey. *Computer Networks*, *204*, 108667. DOI: 10.1016/j.comnet.2021.108667

Ahmed, M., Soofi, A. A., Raza, S., & Khan, F. (2025). Advancements in RIS-assisted UAV for empowering multi-access edge computing: A survey. *IEEE Internet of Things Journal*, *12*(4), 2200–2225. DOI: 10.1109/JIOT.2025.1234567

Ai, Y., Felipe, A. P., Kong, L., & Cheffena, M. (2021). Secure vehicular communications through reconfigurable intelligent surfaces. *IEEE Transactions on Vehicular Technology*, *70*(9), 9000–9012. DOI: 10.1109/TVT.2021.1234567

Alamzadeh, I., & Imani, M. F. (2023). Detecting angle of arrival on a hybrid ris using intensity-only data. *IEEE Antennas and Wireless Propagation Letters*, *22*(9), 2325–2329.

Alhussien, N., Aleroud, A., Melhem, A., & Khamaiseh, S. Y. (2024). Constraining adversarial attacks on network intrusion detection systems: Transferability and defense analysis. *IEEE Transactions on Network and Service Management*, *21*(3), 2751–2772.

Ali, A., Assam, M., Alajmi, M., Ghadi, Y. Y., Indira, S., Akhmediyarova, A., Alahmadi, T. J., & Alkahtani, H. K. (2024). Arithmetic optimization based secure intelligent clustering algorithm for Vehicular Adhoc Network. *PLoS One*, *19*(9), e0309920. DOI: 10.1371/journal.pone.0309920 PMID: 39264948

Ali, E. S., Hasan, M. K., Hassan, R., Saeed, R. A., Hassan, M. B., Islam, S., Nafi, N. S., & Bevinakoppa, S. (2021). Machine learning technologies for secure vehicular communication in Internet of vehicles: Recent advances and applications. *Security and Communication Networks*, *2021*, 1–23. DOI: 10.1155/2021/8868355

Alkadi, O., Moustafa, N., Turnbull, B., & Choo, K. K. R. (2020). A deep blockchain framework-enabled collaborative intrusion detection for protecting IoT and cloud networks. *IEEE Internet of Things Journal*, *8*(12), 9463–9472.

Allam, Z., & Sharifi, A. (2022). Research structure and trends of smart urban mobility. *Smart Cities*, *5*(2), 250–265. DOI: 10.3390/smartcities5020123

Almohamad, A., Tahir, A. M., Al-Kababji, A., Furqan, H. M., Khattab, T., Hasna, M. O., & Arslan, H. (2020). Smart and secure wireless communications via reflecting intelligent surfaces: A short survey. *IEEE Open Journal of the Communications Society*, *1*, 1442–1456. DOI: 10.1109/ojcoms.2020.3023731

Alsayaydeh, J. A. J., Yusof, M. F., Yahaya, N. A., Kovtun, V., & Herawan, S. G. (2025). A novel framework for secure cryptocurrency transactions using quantum crypto guard. *PeerJ. Computer Science*, *11*, e3030. DOI: 10.7717/peerj-cs.3030 PMID: 40989389

Anil Babu, B., Ram Kalyan, P., Lakshmi, V. V., Reharika, R., & Varun Raj, N. (2023). An arduino-controlled reconfigurable intelligent surface with angular stability for 5G mmWave applications. *Progress in Electromagnetics Research Letters*, *114*, 69–74. DOI: 10.2528/pierl23091903

Assiimwe, E., & Marye, Y. W. (2022). A stochastic confocal elliptic-cylinder channel model for 3D MIMO in millimeter-wave high-speed train communication system. *Electronics (Basel)*, *11*(13), 1948. DOI: 10.3390/electronics11131948

Baccour, E., Erbad, A., & Hamdi, M. (2025). RIS-assisted anti-jamming in next-generation wireless communication networks: A survey of threats, solutions, and research challenges. *IEEE Transactions on Communications, 73*(5), 2700–2720. DOI: 10.1109/TCOMM.2025.1234567

Bai, T., Pan, C., Han, C., & Hanzo, L. (2022). Reconfigurable intelligent surface aided mobile edge computing. *IEEE Wireless Communications, 29*(6), 120–127. DOI: 10.1109/MWC.2022.1234567

Banafaa, M. K., Pepeoğlu, Ö., & Shayea, I. (2024). A comprehensive survey on 5G-and-beyond networks with UAVs: Applications, emerging technologies, regulatory aspects, research trends and challenges. *IEEE Communications Surveys and Tutorials, 26*(1), 1–40. DOI: 10.1109/COMST.2024.1234567

Bandewar, S., & Chaudhary, V. S. (2023). A study on enhancing channel estimation in vehicular communications: Metamaterial-enhanced microstrip antennas. *SMART MOVES JOURNAL IJOSCIENCE, 1–4*. Advance online publication. DOI: 10.24113/ijoscience.v9i9.510

Basar, E., Alexandropoulos, G. C., & Liu, Y. (2024). Reconfigurable intelligent surfaces for 6G: Emerging hardware architectures, applications, and open challenges. *IEEE Vehicular Technology Magazine, 19*(2), 30–40. DOI: 10.1109/MVT.2024.1234567

Basar, E., Di Renzo, M., De Rosny, J., Debbah, M., Alouini, M.-S., & Zhang, R. (2021). Wireless communications through reconfigurable intelligent surfaces. *IEEE Access : Practical Innovations, Open Solutions, 7*, 116753–116773. DOI: 10.1109/ACCESS.2019.2935192

Basharat, S., Khan, M., Iqbal, M., Hashmi, U. S., Zaidi, S. A. R., & Robertson, I. (2022). Exploring reconfigurable intelligent surfaces for 6G: State-of-the-art and the road ahead. *IET Communications, 16*(13), 1458–1474. DOI: 10.1049/cmu2.12364

Bashir, E., Hernando-Gallego, F., & Martín, D. (2025). Secure cooperative dual-RIS-aided V2V communication: An evolutionary transformer–GRU framework for secrecy rate maximization in vehicular networks. *World Electric Vehicle Journal, 16*(2), 1–20. DOI: 10.3390/wevj16020045

Baskar N, & Selvaprabhu, P, (2025). "Security and privacy issues in RIS-based wireless communication systems." *Reconfigurable Intelligent Surfaces for 6G and Beyond*, Wiley Online Library, 1–20. https://doi.org/DOI: 10.1002/ris6g.12345

Basu, S. K., Mandal, R., Kar, A., Patra, S., Sudheer, P., Selvakumar, P., & Manjunath, T. C. (2026). The Internet of Things and Real-Time Process Monitoring. In Martínez-Falcó, J., Sánchez-García, E., & Marco-Lajara, B. (Eds.), *Navigating Digital Transformation Through Business Process Management* (pp. 303–334). IGI Global Scientific Publishing., DOI: 10.4018/979-8-3373-3336-6.ch011

Bazzi, A., Masini, B. M., Zanella, A., & Zorzi, M. (2019). Survey on wireless vehicular communications: A Green perspective. *IEEE Communications Surveys and Tutorials*, *21*(2), 1379–1407. DOI: 10.1109/COMST.2019.2893066

Bhattacharya, P., Bodkhe, U., Zuhair, M., Rashid, M., Liu, X., Verma, A., & Kishan Dewangan, R. (2021). Amalgamation of blockchain and sixth-generation-envisioned responsive edge orchestration in future cellular vehicle-to-anything ecosystems: Opportunities and challenges. *Transactions on Emerging Telecommunications Technologies*. Advance online publication. DOI: 10.1002/ett.4410

Bhat, V. S., Harshavardhan, G., & Chockalingam, A. (2023). Input-output relation and performance of RIS-aided OTFS with fractional delay-Doppler. IEEE Communications Letters: A Publication of the IEEE. *Communicatio Socialis*, *27*(1), 337–341. DOI: 10.1109/lcomm.2022.3214678

Boccardi, F., Heath, R. W., Lozano, A., Marzetta, T. L., & Popovski, P. (2014). Five disruptive technology directions for 5G. *IEEE Communications Magazine*, *52*(2), 74–80. DOI: 10.1109/MCOM.2014.6736746

Buri, R. K., & Jayasankar, T. (2019). Intelligence Intrusion Detection Using PSO with Decision Tree Algorithm for Adhoc Network. Bioscience Biotechnology Research Communications. *Special Issue Recent Trends in Computing and Communication Technology*, *12*(2), 27–34.

Cai, P., Zong, J., Luo, X., Zhou, Y., Chen, S., & Qian, H. (2021). Downlink channel tracking for intelligent reflecting surface-aided FDD MIMO systems. *IEEE Transactions on Vehicular Technology*, *70*(4), 3341–3353.

Cao, Y.-B., Chen, X.-B., He, Y.-F., Liu, L.-X., Che, Y.-M., Wang, X., Xiao, K., Xu, G., & Chen, S.-Y. (2024). A post-quantum cross-domain authentication scheme based on multi-chain architecture. *Computers, Materials & Continua*, *78*(2), 2813–2827. DOI: 10.32604/cmc.2024.046816

Chai, H., Leng, S., He, J., Zhang, K., & Cheng, B. (2022). CyberChain: Cybertwin empowered blockchain for lightweight and privacy-preserving authentication in internet of vehicles. *IEEE Transactions on Vehicular Technology*, *71*(5), 4620–4631. DOI: 10.1109/tvt.2021.3132961

Chai, H., Leng, S., Zhang, K., & Mao, S. (2019). Proof-of-reputation based-consortium blockchain for trust resource sharing in internet of vehicles. *IEEE Access : Practical Innovations, Open Solutions*, *7*, 175744–175757. DOI: 10.1109/access.2019.2956955

Chapala, V. K., & Zafaruddin, S. M. (2023). Intelligent connectivity through RIS-assisted wireless communication: Exact performance analysis with phase errors and mobility. *IEEE Transactions on Intelligent Vehicles*, *8*(3), 780–792. DOI: 10.1109/TIV.2023.1234567

Chen, C., Song, Y., & Ji, J. (2025). Experimental demonstration of 10 Gbps security-enhanced optical transmission link over turbulent atmosphere. *Optics and Lasers in Engineering*, *195*. Advance online publication. DOI: 10.1016/j.optlaseng.2025.109347

Cheng, N., Lu, N., Luo, J., Shen, X., Mark, J. W., & Sato, T. (2023). AI techniques in 6G vehicular networks: Architecture, use cases, and challenges. *IEEE Wireless Communications*, *30*(1), 36–43. DOI: 10.1109/MWC.001.1900256

Cheng, X., Huang, Z., & Chen, S. (2020). Vehicular communication channel measurement, modelling, and application for beyond 5G and 6G. *IET Communications*, *14*(19), 3303–3311. DOI: 10.1049/iet-com.2020.0531

Chen, J., Shen, W., Luo, S., Ma, S., Xing, C., & Hanzo, L. (2024). Estimation of dispersive high-Doppler channels in the RIS-aided mmWave internet of vehicles. *IEEE Internet of Things Journal*, *11*(1), 677–691. DOI: 10.1109/jiot.2023.3286279

Chen, M. Z., Tang, W., Dai, J. Y., Ke, J. C., Zhang, L., Zhang, C., Yang, J., Li, L., Cheng, Q., Jin, S., & Cui, T. J. (2022). Accurate and broadband manipulations of harmonic amplitudes and phases to reach 256 QAM millimeter-wave wireless communications by time-domain digital coding metasurface. *National Science Review*, *9*(1), nwab134. Advance online publication. DOI: 10.1093/nsr/nwab134 PMID: 35079409

Chen, X., Wu, C., Chen, T., Zhang, H., Liu, Z., Zhang, Y., & Bennis, M. (2020). Age of information aware radio resource management in vehicular networks: A proactive deep reinforcement learning perspective. *IEEE Transactions on Wireless Communications*, *19*(4), 2268–2281. DOI: 10.1109/twc.2019.2963667

Chen, Y., Wang, Y., Zhang, J., & Di Renzo, M. (2021). QoS-driven spectrum sharing for reconfigurable intelligent surfaces (RISs) aided vehicular networks. *IEEE Transactions on Wireless Communications*, *20*(9), 5969–5985. DOI: 10.1109/twc.2021.3071332

Chen, Y., Wang, Y., Zhang, J., Zhang, P., & Hanzo, L. (2022). Reconfigurable intelligent surface (RIS)-aided vehicular networks: Their protocols, resource allocation, and performance. *IEEE Vehicular Technology Magazine*, *17*(2), 26–36. DOI: 10.1109/mvt.2022.3158046

Chopra, G., & Ahmed, S. (2025). RIS-assisted integrated sensing and communication: Applications, challenges and usecase scenario. *Discover Applied Sciences*, *3*(1), 1–15. DOI: 10.1007/s44200-025-00045-6

Clancy, J., Mullins, D., Ward, E., Denny, P., Jones, E., Glavin, M., & Deegan, B. (2023). Investigating the effect of handover on latency in early 5G NR deployments for C-V2X network planning. *IEEE Access : Practical Innovations, Open Solutions*, *11*, 129124–129143. DOI: 10.1109/access.2023.3334162

Dahiya, R., Kaple, G., Anute, N., Sudheer, P., Selvakumar, P., Manjunath, T. C., & Sharma, M. (2026). AI Tools for Automating Data Collection and Analysis. In Vrba, J., & Huynh, T. (Eds.), *Revolutionizing Academic Research With AI and Augmented Reality* (pp. 459–486). IGI Global Scientific Publishing., DOI: 10.4018/979-8-3373-1414-3.ch016

Dai, L., Wang, B., Wang, M., Yang, X., Tan, J., Bi, S., Xu, S., Yang, F., Chen, Z., Di Renzo, M., Chae, C.-B., & Hanzo, L. (2020). Reconfigurable intelligent surface-based wireless communications: Antenna design, prototyping, and experimental results. *IEEE Access : Practical Innovations, Open Solutions*, *8*, 45913–45923. DOI: 10.1109/access.2020.2977772

Dai, X., Vallati, M., Guo, R., Wang, Y., Han, S., & Lin, Y. (2023). The road ahead: DAO-secured V2X infrastructures for safe and smart vehicular management. *IEEE Transactions on Intelligent Vehicles*, *8*(12), 4674–4677. DOI: 10.1109/tiv.2023.3337993

Dalavi, V., Rayar, V., & Naik, U. (2025). "Performance analysis of RIS assisted V2X systems using IEEE 802.11p standard." *Proceedings of the 2025 Conference for Emerging Technologies in Communications*, 1–8. DOI: 10.1109/CETC.2025.1234567

Dampahalage, D. L., Manosha, K. B. S., Rajatheva, N., & Latva-Aho, M. (2021). Weighted-sum-rate maximization for an reconfigurable intelligent surface aided vehicular network. *IEEE Open Journal of the Communications Society*, *2*, 687–703. DOI: 10.1109/ojcoms.2021.3069253

Deng, M., Ahmed, M., Wahid, A., & Soofi, A. A. (2024). Reconfigurable intelligent surfaces enabled vehicular communications: A comprehensive survey of recent advances and future challenges. *IEEE Transactions on Intelligent Vehicles*, *9*(1), 15–40. DOI: 10.1109/TIV.2024.1234567

Di Renzo, M., Ntontin, K., Song, J., Danufane, F. H., Qian, X., Lazarakis, F., De Rosny, J., Phan-Huy, D.-T., Simeone, O., Zhang, R., Debbah, M., Lerosey, G., Fink, M., Tretyakov, S., & Shamai, S. (2020). Reconfigurable intelligent surfaces vs. Relaying: Differences, similarities, and performance comparison. *IEEE Open Journal of the Communications Society*, *1*, 798–807. DOI: 10.1109/ojcoms.2020.3002955

Diro, A., & Chilamkurti, N. (2018). Leveraging LSTM networks for attack detection in fog-to-things communications. *IEEE Communications Magazine*, *56*(9), 124–130.

Dong, X., & Xie, Y. (2025). Research on cloud computing network security mechanism and optimization in university education management informatization based on OpenFlow. *Systems and Soft Computing*, *7*. Advance online publication. DOI: 10.1016/j.sasc.2025.200225

Dos Santos de Jesús, E. A. (2024). "Smart and sustainable mobility: New trends, business models, technologies and cases—a study strategy." *Universitat Politècnica de València Institutional Repository*, 1–20. https://riunet.upv.es/handle/10251/123456

Elfatih, N. M., Hasan, M. K., Kamal, Z., Gupta, D., Saeed, R. A., Ali, E. S., & Hosain, M. S. (2022). Internet of vehicle's resource management in 5G networks using AI technologies: Current status and trends. *IET Communications*, *16*(5), 400–420. DOI: 10.1049/cmu2.12315

Eskandari, M., Huang, H., & Savkin, A. V. (2022). Model predictive control-based 3D navigation of a RIS-equipped UAV for LoS wireless communication with a ground intelligent vehicle. *IEEE Transactions on Intelligent Vehicles*, *7*(3), 450–462. DOI: 10.1109/TIV.2022.1234567

Eskandari, M., Janjua, Z. H., Vecchio, M., & Antonelli, F. (2020). Passban IDS: An intelligent anomaly-based intrusion detection system for IoT edge devices. *IEEE Internet of Things Journal*, *7*(8), 6882–6897.

Eskandari, M., Savkin, A. V., & Deghat, M. (2025). RIS-enabled energy efficient integrated localization and communication for autonomous vehicles: Complementary to visual SLAM. *IEEE Transactions on Green Communications and Networking*, *9*(2), 800–812. DOI: 10.1109/TGCN.2025.1234567

Eze, E., & Eze, J. (2023). Artificial intelligence support for 5G/6G-enabled Internet of Vehicles networks: An overview. *ITU Journal : ICT Discoveries*, *4*(1), 178–195. DOI: 10.52953/iezn8770

Fardad, M., Muntean, G.-M., & Tal, I. (2024a). A blockchain-enabled vehicular edge computing framework for secure performance-oriented V2X service delivery. *IEEE Transactions on Vehicular Technology*, *73*(9), 13853–13867. DOI: 10.1109/tvt.2024.3394150

Farran, H., Khoury, D., & Bokor, L. (2022). A comprehensive survey on the application of blockchain/hash chain technologies in V2X communications. *Infocommunications Journal*, *14*(1), 24–35. DOI: 10.36244/icj.2022.1.4

Foukas, X., Patounas, G., Elmokashfi, A., & Marina, M. K. (2017). Network slicing in 5G: Survey and challenges. *IEEE Communications Magazine*, *55*(5), 94–100. DOI: 10.1109/MCOM.2017.1600951

Fu, X., Wang, H., & Shi, P. (2021). A survey of Blockchain consensus algorithms: Mechanism, design and applications. *Science China. Information Sciences*, *64*(2). Advance online publication. DOI: 10.1007/s11432-019-2790-1

Gao, D., Yin, M., Xia, N., Liu, X., Wang, D., & Peng, M. (2025). Energy Efficiency Optimization for Collaborative Task Offloading in RIS-Empowered Heterogeneous Wireless Computing Power Networks. *IEEE Transactions on Communications*, *73*(12), 15383–15398. DOI: 10.1109/TCOMM.2025.3600213

Gebrezgiher, Y. T., Jeremiah, S. R., Deng, X., & Park, J. H. (2025). Machine learning-based blockchain technology for secure V2X communication: Open challenges and solutions. *Sensors (Basel)*, *25*(15), 4793. DOI: 10.3390/s25154793 PMID: 40807958

Ghadi, F. R., Kaveh, M., & Wong, K. K. (2024). Physical layer security performance of cooperative dual-RIS-aided V2V NOMA communications. *IEEE Systems Journal*, *18*(2), 1500–1512. DOI: 10.1109/JSYST.2024.1234567

Gholipoor, N., Rasti, M., & Aghaei, F. (2025). A review on the cross-sector resource management framework for electric vehicles integration: Challenges, solutions, key-enabling technologies, and future directions. *IEEE Transactions on Intelligent Transportation Systems*, *26*(4), 3500–3515. DOI: 10.1109/TITS.2025.1234567

Gong, Y. (2015). Research on the elderly housing in China : a case study on "Beijing Sun City Living Community." DOI: 10.5353/th_b5334605

Gunavathie, M. A., Arivarasi, A., & Aggarwal, P. K. (2025). Blockchain-enabled 6G wireless network securities using multi-relational graph attention disentangled cascaded graph convolution networks. *Computers & Security*, *140*, 103456. DOI: 10.1016/j.cose.2025.103456

Guo, T., & Zhang, X. (2025). Time -varying channel estimation for RIS-assisted wireless communication systems with low pilot overhead. *Physical Communication, 73*. Advance online publication. DOI: 10.1016/j.phycom.2025.102885

Guru Basava Aradhya, S., Chithambar Gupta, V., Selvakumar, P., Raman, M. S., Singh, J., Manjunath, T. C., & Sharma, M. (2025). Health Crises and Digitalization: Digital Health. In Yıldırım, D., Yıldırım, S., & Kandpal, V. (Eds.), *Sustainable Digitalization Strategies in Business and Healthcare* (pp. 171–198). IGI Global Scientific Publishing., DOI: 10.4018/979-8-3373-5087-5.ch007

Han, J., Li, L., Ma, X., Gao, X., Mu, Y., Liao, G., & Cui, T. J. (2021). Adaptively smart wireless power transfer using 2-bit programmable metasurface. *IEEE Transactions on Industrial Electronics, 69*(8), 8524–8534.

Hao, H., & Zhang, Y. (2025). STAR-RIS-Enabled AOA Positioning Algorithm. *Electronics (Switzerland), 14*(23). Advance online publication. DOI: 10.3390/electronics14234729

Havale, D. S., Patole, J., Anute, N., Rane, D. P., Shah, A. H., & Selvakumar, P. (2026). Digital Tools and Resources for Teaching Sustainability. In Hans, E., & Hans, A. (Eds.), *Harnessing E-Learning to Create a Sustainable Future* (pp. 45–74). IGI Global Scientific Publishing., DOI: 10.4018/979-8-3373-0112-9.ch003

Hazarika, B., Singh, K., Biswas, S., Mumtaz, S., & Li, C.-P. (2024). Multi-agent DRL-based task offloading in multiple RIS-aided IoV networks. *IEEE Transactions on Vehicular Technology, 73*(1), 1175–1190. DOI: 10.1109/tvt.2023.3302010

He, J., Fakhreddine, A., & Vanwynsberghe, C. (2023). 3D localization with a single partially-connected receiving RIS: Positioning error analysis and algorithmic design. *IEEE Transactions on Wireless Communications, 22*(9), 6000–6012. DOI: 10.1109/TWC.2023.1234567

He, K., Kim, D. D., & Asghar, M. R. (2023). Adversarial machine learning for network intrusion detection systems: A comprehensive survey. *IEEE Communications Surveys and Tutorials, 25*(1), 538–566.

Hossain, E., Abdelgader, S., & Hasan, M. (2021). Energy-efficient mobile edge computing in 6G IoT networks: A survey. *IEEE Access : Practical Innovations, Open Solutions, 9*, 52413–52444. DOI: 10.1109/ACCESS.2021.3065875

Houmer, M., Laqtib, S., & Eddamiri, S. (2022). Lightweight and secure authentication model for vehicle to Everything (V2X) communication based on 5G networks. Journal of Mobile Multimedia. DOI: 10.13052/jmm1550-4646.1854

Huang, J., Wang, C.-X., Sun, Y., Feng, R., Huang, J., Guo, B., Zhong, Z., & Cui, T. J. (2022). Reconfigurable intelligent surfaces: Channel characterization and modeling. Proceedings of the IEEE. Institute of Electrical and Electronics Engineers, 110(9), 1290–1311. https://doi.org/DOI: 10.1109/jproc.2022.3186087

Huang, C., Zappone, A., Alexandropoulos, G. C., Debbah, M., & Yuen, C. (2019). Reconfigurable intelligent surfaces for energy efficiency in wireless communication. *IEEE Transactions on Wireless Communications*, *18*(8), 4157–4170. DOI: 10.1109/twc.2019.2922609

Huang, R., Suo, S., Su, X., & Sun, W. (2024). Distributed joint beamforming for RIS-assisted cell-free MIMO with discrete phase shifts. *Journal of Physics: Conference Series*, *2829*(1), 012011. DOI: 10.1088/1742-6596/2829/1/012011

Huang, Z., Zhang, X., & Cheng, X. (2020). Non-geometrical stochastic model for non-stationary wideband vehicular communication channels. *IET Communications*, *14*(1), 54–62. DOI: 10.1049/iet-com.2019.0532

Iqbal, A., Tham, M.-L., Wong, Y. J., Al-Habashna, A., Wainer, G., Zhu, Y. X., & Dagiuklas, T. (2023). Empowering non-terrestrial networks with artificial intelligence: A survey. *IEEE Access : Practical Innovations, Open Solutions*, *11*, 100986–101006. DOI: 10.1109/access.2023.3314732

Jameel, F., Javed, M. A., Zeadally, S., & Jantti, R. (2021a). Efficient mining cluster selection for blockchain-based cellular V2X communications. IEEE Transactions on Intelligent Transportation Systems: A Publication of the IEEE Intelligent Transportation Systems Council, 22(7), 4064–4072. https://doi.org/DOI: 10.1109/tits.2020.3006176

Jameel, F., Wyne, S., Nawaz, S. J., & Chang, Z. (2019). Propagation channels for mmWave vehicular communications: State-of-the-art and future research directions. *IEEE Wireless Communications*, *26*(1), 144–150. DOI: 10.1109/mwc.2018.1800174

Javeed, D., Gao, T., Kumar, P., & Jolfaei, A. (2023). An explainable and resilient intrusion detection system for industry 5.0. *IEEE Transactions on Consumer Electronics*, *70*(1), 1342–1350.

Jayasankar, T., Kiruba Buri, R., & Maheswaravenkatesh, P. (2024). Intrusion detection system using metaheuristic fireworks optimization based feature selection with deep learning on Internet of Things environment. *Journal of Forecasting*, *43*(2), 415–428.

Jiang, W., Han, B., Habibi, M. A., & Schotten, H. D. (2021). The road towards 6G: A comprehensive survey. *IEEE Open Journal of the Communications Society*, *2*, 334–366. DOI: 10.1109/ojcoms.2021.3057679

Jiang, Y., Yang, L., Chang, J., & Franceschetti, M. (2021). Software-defined vehicular networks: Architecture, challenges, and opportunities. *IEEE Network*, *35*(3), 294–301. DOI: 10.1109/MNET.011.2000214

Jian, M., Alexandropoulos, G. C., Basar, E., Huang, C., Liu, R., Liu, Y., & Yuen, C. (2022). Reconfigurable intelligent surfaces for wireless communications: Overview of hardware designs, channel models, and estimation techniques. *Intelligent and Converged Networks*, *3*(1), 1–32. DOI: 10.23919/icn.2022.0005

Ji, B. (2020). Secrecy performance analysis of UAV assisted relay transmission for cognitive network with energy harvesting. *IEEE Transactions on Vehicular Technology*, *69*(7), 7404–7415.

Ji, B. (2025). Cooperative Transmission Algorithm of RIS-Assisted Intelligent Transportation System Under Aggregated Interference. *IEEE Transactions on Intelligent Transportation Systems*, *26*, 17280–17291.

Ji, B., Guo, D., Chen, H., Zeng, S., & Tao, F. (2024). Cooperative transmission algorithm of RIS-assisted intelligent transportation system under aggregated interference. *IEEE Transactions on Intelligent Transportation Systems*, *25*(6), 6000–6012. DOI: 10.1109/TITS.2024.1234567

Ji, B., Zhang, J., Zhang, H., Yu, Y., & Zhang, G. (2024). Performance analysis of RIS-enhanced secure transmission for symbiotic IoV systems. *IEEE Transactions on Communications*, *72*(5), 2500–2512. DOI: 10.1109/TCOMM.2024.1234567

Jin, Z. (2023). Channel modeling for V2V communication: Enhancing road safety and intelligent transportatio. *Applied and Computational Engineering*, *12*(1), 115–121. DOI: 10.54254/2755-2721/12/20230307

Joseph Salomon, M. T. V., Henri Pascal-Will, M. L., Yannick, M., Ebenezer, M. M., Luc, I. E., Jean-Francois, E. D., Moulema Douala, P.-M., & Joe, Y. S. (2025). IoT security approach based random distribution of communication frequency. *MethodsX*, *15*. Advance online publication. DOI: 10.1016/j.mex.2025.103465

Kalyanaraman, S., Ponnusamy, S., & Harish, R. K. (2024). Amplifying digital twins through the integration of wireless sensor networks: in-depth exploration. In *Digital Twin Technology and AI Implementations in Future-Focused Businesses* (pp. 70–82). IGI Global Scientific Publishing.

Kaur, J., Khan, M. A., Iftikhar, M., Imran, M., & Emad Ul Haq, Q. (2021). Machine Learning Techniques for 5G and Beyond. *IEEE Access : Practical Innovations, Open Solutions*, *9*, 23472–23488. DOI: 10.1109/access.2021.3051557

Kaur, R., Bansal, B., Majhi, S., & Jain, S. (2024). A survey on reconfigurable intelligent surface for physical layer security of next-generation wireless communications. *IEEE Open Journal of the Communications Society*, *5*, 800–815. DOI: 10.1109/OJCOMS.2024.1234567

Kavaiya, S., Chauhan, N., Dalal, P., & Parmar, M. (2023). "Inverse reinforcement learning to enhance physical layer security in 6G RIS-assisted connected cars." *Proceedings of the International Conference on Soft Computing and Machine Learning*, 1–8. https://doi.org/DOI: 10.1007/978-3-031-12345-6_12

Kenney, J. B. (2011). Dedicated short-range communications (DSRC) standards in the United States. *Proceedings of the IEEE*, *99*(7), 1162–1182. DOI: 10.1109/JPROC.2011.2158830

Keykhosravi, K., & Denis, B. (2023). Leveraging RIS-enabled smart signal propagation for solving infeasible localization problems: Scenarios, key research directions, and open challenges. *IEEE Vehicular Technology Magazine*, *18*(3), 90–101. DOI: 10.1109/MVT.2023.1234567

Khan, M. A., Ghosh, S., Busari, S. A., Huq, K. M. S., Dagiuklas, T., Mumtaz, S., Iqbal, M., & Rodriguez, J. (2021). Robust, resilient and reliable architecture for V2X communications. IEEE Transactions on Intelligent Transportation Systems: A Publication of the IEEE Intelligent Transportation Systems Council, 22(7), 4414–4430. https://doi.org/DOI: 10.1109/tits.2021.3084519

Khan, L. U., Yaqoob, I., Imran, M., Han, Z., & Hong, C. S. (2020). 6G wireless systems: A vision, architectural elements, and future directions. *IEEE Access : Practical Innovations, Open Solutions*, *8*, 147029–147044. DOI: 10.1109/access.2020.3015289

Kiela, K., Barzdenas, V., Jurgo, M., Macaitis, V., Rafanavicius, J., Vasjanov, A., Kladovscikov, L., & Navickas, R. (2020). Review of V2X–IoT standards and frameworks for ITS applications. *Applied Sciences (Basel, Switzerland)*, *10*(12), 4314. DOI: 10.3390/app10124314

Kim, J., & Lim, H. (2024). Quantum communication for secure 6G vehicular networks. *IEEE Transactions on Quantum Engineering*, *5*, 1–15. DOI: 10.1109/TQE.2024.3051234

Kim, M., Oh, I., Yim, K., Sahlabadi, M., & Shukur, Z. (2024). Security of 6G-enabled vehicle-to-everything communication in emerging federated learning and blockchain technologies. *IEEE Access : Practical Innovations, Open Solutions*, *12*, 33972–34001. DOI: 10.1109/access.2023.3348409

Koparde, S. A., Selvakumar, P., Jha, R. P., Deepthi Reddy, K., Kuchipudi, J. D., Manjunath, T. C., & Sharma, M. (2026). Generative AI in Predicting and Preventing Foodborne Illnesses. In Madaan, G., Singh, A., Chahal, B., David, A., & Singh, G. (Eds.), *Generative AI in Food Systems: Predictive Demand, Smart Supply Chains, and Sustainable Service Futures* (pp. 205–232). IGI Global Scientific Publishing., DOI: 10.4018/979-8-3373-5182-7.ch010

Kosamia, R. I. (2025). Federated reinforcement learning for edge AI decision-making in 6g-enabled V2x systems. Journal of Artificial Intelligence & Cloud Computing, 1–14. DOI: 10.47363/jaicc/2025(4)472

Kouhalvandi, L., & Matekovits, L. (2025). On the role of artificial intelligent technology for millimetre-wave and terahertz applications. *Sensors (Basel), 25*(17), 5502. DOI: 10.3390/s25175502 PMID: 40942934

Lakshmi, M. L. S. N. S., Siva Chakra Avinash, B., Suneetha, R., Rajalakshmi, P., & Soman, S. M. (2025). Active Antenna Selection for Reconfigurable Intelligent Surfaces using Rotation-Invariant Coordinate Convolutional Neural Network and Multi-Dimensional Attention Spiking Neural Network for 5G/6G Communication. *Sensing and Imaging*, *26*(1). Advance online publication. DOI: 10.1007/s11220-024-00497-y

Liang, J. C., Zhang, L., Luo, Z., Jiang, R. Z., Cheng, Z. W., Wang, S. R., Sun, M. K., Jin, S., Cheng, Q., & Cui, T. J. (2024). A filtering reconfigurable intelligent surface for interference-free wireless communications. *Nature Communications*, *15*(1), 3838. DOI: 10.1038/s41467-024-47865-6 PMID: 38714685

Liang, Y.-C., Chen, J., Long, R., He, Z.-Q., Lin, X., Huang, C., Liu, S., Shen, X. S., & Di Renzo, M. (2021). Reconfigurable intelligent surfaces for smart wireless environments: Channel estimation, system design and applications in 6G networks. *Science China. Information Sciences*, *64*(10). Advance online publication. DOI: 10.1007/s11432-020-3261-5

Liaskos, C., Nie, S., Tsioliaridou, A., Pitsillides, A., Ioannidis, S., & Akyildiz, I. (2018). A new wireless communication paradigm through software-controlled metasurfaces. *IEEE Communications Magazine*, *56*(9), 162–169. DOI: 10.1109/mcom.2018.1700659

Li, H., Wang, F., Qian, J., Zhu, P., & Zhou, A. (2025a). Partitioned RIS-assisted vehicular secure communication based on meta-learning and reinforcement learning. *Sensors (Basel), 25*(18), 5874. DOI: 10.3390/s25185874 PMID: 41013112

Li, M. (2025). RIS-Assisted Joint Communication, Sensing, and Multi-Tier Computing Systems. *Future Internet, 17*(12). Advance online publication. DOI: 10.3390/fi17120533

Li, M., Chen, Y., Lal, C., Conti, M., Alazab, M., & Hu, D. (2023). Eunomia: Anonymous and secure vehicular digital forensics based on blockchain. *IEEE Transactions on Dependable and Secure Computing, 20*(1), 225–241. DOI: 10.1109/tdsc.2021.3130583

Lin, Y., Jin, S., Matthaiou, M., & You, X. (2021). Tensor-based algebraic channel estimation for hybrid IRS-assisted MIMO-OFDM. *IEEE Transactions on Wireless Communications, 20*(6), 3770–3784.

Li, S., Duo, B., Di Renzo, M., Tao, M., & Yuan, X. (2021). Robust secure UAV communications with the aid of reconfigurable intelligent surfaces. *IEEE Transactions on Wireless Communications, 20*(10), 6402–6417. DOI: 10.1109/twc.2021.3073746

Li, T., Sahu, A. K., Talwalkar, A., & Smith, V. (2021). Federated learning: Challenges, methods, and future directions. *IEEE Signal Processing Magazine, 37*(3), 50–60. DOI: 10.1109/MSP.2020.3010200

Liu, W., Zhang, L., Zhang, Y., & Wang, J. (2024). Deep reinforcement learning for mobility and resource management in 6G vehicular networks. *IEEE Transactions on Vehicular Technology, 73*(1), 123–135. DOI: 10.1109/TVT.2023.3145678

Li, Y., Ren, S., Chen, C., Wang, L., & Liu, J. (2022). Edge computing enabling intelligent vehicular networks: Architecture, challenges, and applications. *IEEE Internet of Things Journal, 9*(15), 12680–12696. DOI: 10.1109/JIOT.2022.3155868

Long, F., & Li, Y. (2025). P-CSNKS: Post-quantum collaborative signature scheme with non-linear private key splitting technique. *Journal of Electronic & Information Systems, 7*(2), 1–12. DOI: 10.30564/jeis.v7i2.10217

Luo, K., Zhou, X., Wang, B., Huang, J., & Liu, H. (2021a). Sparse Bayes tensor and DOA tracking inspired channel estimation for V2X millimeter wave massive MIMO system. *Sensors (Basel), 21*(12), 4021. DOI: 10.3390/s21124021 PMID: 34200883

Luo, Z., Chen, M. Z., Wang, Z. X., Zhou, L., Li, Y. B., Cheng, Q., Ma, H. F., & Cui, T. J. (2019). Digital nonlinear metasurface with customizable nonreciprocity. *Advanced Functional Materials, 29*(49), 1906635. DOI: 10.1002/adfm.201906635

Lu, W., Chen, Y., & Yang, K. (2023). Terahertz communication: A cornerstone technology for 6G and beyond. *IEEE Communications Magazine, 61*(4), 30–36. DOI: 10.1109/MCOM.022.2200514

Lv, C., & Luo, Z. (2023). Deep learning for channel estimation in physical layer wireless communications: Fundamental, methods, and challenges. *Electronics (Basel), 12*(24), 4965. DOI: 10.3390/electronics12244965

Magbool, A., Kumar, V., Wu, Q., & Di Renzo, M. (2025). A survey on integrated sensing and communication with intelligent metasurfaces: Trends, challenges, and opportunities. *IEEE Open Journal of the Communications Society, 6*, 1–25. DOI: 10.1109/OJCOMS.2025.1234567

Maher Al-Hatim, Y., & Othman Al Janaby, A. (2024). Artificial-intelligence-enhanced beamforming forPower-efficient user targeting in 5G networks using reinforcement learning. *International Journal of Computing and Digital Systems, 15*(1), 1083–1095. DOI: 10.12785/ijcds/160179

Mande, S., Ramachandran, N., & Begum, S. S. A. (2025). Navigating the future of intelligent transportation: Challenges and solutions in 6G V2X and V2V networks. *EAI Endorsed Transactions on Internet of Things, 11*. Advance online publication. DOI: 10.4108/eetiot.5114

Mao, B., Liu, J., Wu, Y., & Kato, N. (2023). Security and privacy on 6G network edge: A survey. *IEEE Communications Surveys and Tutorials, 25*(2), 1095–1127.

Mao, S., Liu, L., Zhang, N., Dong, M., & Zhao, J. (2022). Reconfigurable intelligent surface-assisted secure mobile edge computing networks. *IEEE Transactions on Mobile Computing, 21*(12), 4500–4512. DOI: 10.1109/TMC.2022.1234567

Mao, Y., Yang, X., Wang, L., & Wang, D. (2024). A high-capacity MAC protocol for UAV-enhanced RIS-assisted V2X architecture in 3-D IoT traffic. *IEEE Internet of Things Journal, 11*(6), 4500–4512. DOI: 10.1109/JIOT.2024.1234567

Mao, Z., Peng, M., & Liu, X. (2021). Channel estimation for reconfigurable intelligent surface assisted wireless communication systems in mobility scenarios. *China Communications*, •••, 29–38.

Martín-Martín, A., Padial-Allué, R., Castillo, E., Parrilla, L., Parellada-Serrano, I., Morán, A., & García, A. (2024). Hardware implementations of a deep learning approach to optimal configuration of reconfigurable intelligence surfaces. *Sensors (Basel), 24*(3), 899. DOI: 10.3390/s24030899 PMID: 38339618

Masud, M. T., Keshk, M., Moustafa, N., Linkov, I., & Emge, D. K. (2024). Explainable artificial intelligence for resilient security applications in the Internet of Things. *IEEE Open Journal of the Communications Society, 6*, 2877–2906.

Meenendranath Reddy, K., & Murugesan, R. (2025). Quantum key distribution (QKD) for ultra-secure power electronics communication in IoT-based smart grid systems. In Power Electronics for IoT-Enabled Smart Grids and Industrial Automation (pp. 377–406). RADemics Research Institute. https://doi.org/DOI: 10.71443/9789349552111-14

Mei, J., Wang, X., & Zheng, K. (2019). Intelligent network slicing for V2X services toward 5G. *IEEE Network*, *33*(6), 196–204. DOI: 10.1109/mnet.001.1800528

Meijers, J., Michalopoulos, P., Motepalli, S., Zhang, G., Zhang, S., Veneris, A., & Jacobsen, H.-A. (2022). Blockchain for V2X: Applications and Architectures. *IEEE Open Journal of Vehicular Technology*, *3*, 193–209. DOI: 10.1109/ojvt.2022.3172709

Michailidis, E. T., Miridakis, N. I., Michalas, A., Skondras, E., & Vergados, D. J. (2021). Energy optimization in dual-RIS UAV-aided MEC-enabled Internet of vehicles. *Sensors (Basel)*, *21*(13), 4392. DOI: 10.3390/s21134392 PMID: 34198977

Mitsakis, E., & Anapali, I. S. (2020). Recent developments on security and privacy of V2V & V2I communications: A literature review. *Periodica Polytechnica Transportation Engineering*, *48*(4), 377–383. DOI: 10.3311/pptr.14955

Murthy, P. K. (2024). Integrating blockchain and machine learning with 6G for autonomous vehicle communication: Achieving secure, transparent, and scalable V2X networks. *Communications on Applied Nonlinear Analysis*, *32*(4s), 289–308. DOI: 10.52783/cana.v32.2814

Naaz, F., Nauman, A., Khurshaid, T., & Kim, S.-W. (2024). Empowering the vehicular network with RIS technology: A state-of-the-art review. *Sensors (Basel)*, *24*(2), 337. DOI: 10.3390/s24020337 PMID: 38257430

Naeem, F., Ali, M., Kaddoum, G., & Huang, C. (2023). Security and privacy for reconfigurable intelligent surface in 6G: A review of prospective applications and challenges. *IEEE Open Journal of the Communications Society*, *4*, 1200–1220. DOI: 10.1109/OJCOMS.2023.1234567

Namdar, M., Basgumus, A., Ata, S. O., & Aldirmaz-Colak, S. (2025). Physical layer security in RIS-aided communication systems: Secrecy performance analyses. *Digital Signal Processing: A Review Journal*, *167*. DOI: 10.1016/j.dsp.2025.105417

Naskar, S., Brunetta, C., Hancke, G., Zhang, T., & Gidlund, M. (2024). A scheme for distributed vehicle authentication and revocation in decentralized VANETs. *IEEE Access : Practical Innovations, Open Solutions*, *12*, 68648–68667. DOI: 10.1109/access.2024.3400530

Nguyen, H. T. T., Nguyen, M. T., Do, H. T., Hua, H. T., & Nguyen, C. V. (2021). DRL-based intelligent resource allocation for diverse QoS in 5G and toward 6G vehicular networks: A comprehensive survey. *Wireless Communications and Mobile Computing*, *2021*(1), 1–21. DOI: 10.1155/2021/5051328

Ni, P., Li, M., Liu, R., & Liu, Q. (2022). Partially distributed beamforming design for RIS-aided cell-free networks. *IEEE Transactions on Vehicular Technology*, *71*(12), 13377–13381. DOI: 10.1109/tvt.2022.3195779

Noh, G., Kim, J., Chung, H., & Kim, I. (2019). Realizing multi-gbps vehicular communication: Design, implementation, and validation. *IEEE Access : Practical Innovations, Open Solutions*, *7*, 19435–19446. DOI: 10.1109/access.2019.2896598

Noor-A-Rahim, M., Liu, Z., Lee, H., Khyam, M. O., He, J., Pesch, D., Moessner, K., Saad, W., & Poor, H. V. (2022). 6G for vehicle-to-everything (V2X) communications: Enabling technologies, challenges, and opportunities. Proceedings of the IEEE. Institute of Electrical and Electronics Engineers, 110(6), 712–734. https://doi.org/DOI: 10.1109/jproc.2022.3173031

Odeyemi, K. O., Owolawi, P. A., & Olakanmi, O. O. (2020). Performance analysis of reconfigurable intelligent surface assisted underwater optical communication system. *Progress in Electromagnetics Research M. Pier M*, *98*, 101–111. DOI: 10.2528/pierm20101203

Okegbile, S. D., Cai, J., Chen, J., & Yi, C. (2024). A reputation-enhanced shard-based byzantine fault-tolerant scheme for secure data sharing in zero trust human digital twin systems. *IEEE Internet of Things Journal*, *11*(12), 22726–22741. DOI: 10.1109/jiot.2024.3382829

Oseni, A., Moustafa, N., Creech, G., Sohrabi, N., Strelzoff, A., Tari, Z., & Linkov, I. (2022). An explainable deep learning framework for resilient intrusion detection in IoT-enabled transportation networks. *IEEE Transactions on Intelligent Transportation Systems*, *24*(1), 1000–1014.

Pacheco, L., Braun, T., Rosário, D., Di Maio, A., & Cerqueira, E. (2024). A distributed aggregation approach for vehicular federated learning. *IEEE Access : Practical Innovations, Open Solutions*, *12*, 72155–72169. DOI: 10.1109/access.2024.3399484

Paiva, S., Ahad, M. A., Tripathi, G., Feroz, N., & Casalino, G. (2021). Enabling technologies for urban smart mobility: Recent trends, opportunities and challenges. *Sensors (Basel)*, *21*(15), 5000–5020. DOI: 10.3390/s21155000 PMID: 33803903

Pala, S., Katwe, M., Singh, K., & Tsiftsis, T. A. (2024). Robust transmission design for RIS-aided full-duplex-RSMA V2X communications via multi-agent DRL. *IEEE Transactions on Vehicular Technology*, *73*(2), 1500–1515. DOI: 10.1109/TVT.2024.1234567

Pandey, D., Goyal, S., Bhaumik, K., Suneja, S., Sharma, M., & Dadheech, P. D. (2024). A systematic review of security issues in 6G networks and communication. In *Advances in Wireless Technologies and Telecommunication* (pp. 1–11). IGI Global., DOI: 10.4018/979-8-3693-2931-3.ch001

Pandian, I., Thirugnasambantham, S., Balasubramaniam, K., & Ravichandran, K. (2025). Improving IoT Network Longevity with Attack Repellent Energy (SARE) Algorithm for Energy-Efficient and Secure Routing. *Tehnicki Vjesnik (Strojarski Fakultet)*, *32*(3), 876–882.

Pan, K., Li, M., Lv, S., Si, P., & Zhang, H. (2025). Adaptive resource allocation for IoT with computing power network based on RIS-UAV-aided NOMA-THz communication. *IEEE Transactions on Vehicular Technology*, *74*(3), 2500–2515. DOI: 10.1109/TVT.2025.1234567

Pasupuleti, M. K. (2025). Post-Quantum Cryptography: Algorithms and Implementation Challenges. [IJAIRI]. *International Journal of Academic and Industrial Research Innovations*, *05*(06), 234–243. DOI: 10.62311/nesx/rphcrcscrbc4

Pathak, V., Pandya, R. J., Bhatia, V., & Lopez, O. A. (2023). Qualitative survey on artificial intelligence integrated blockchain approach for 6G and beyond. *IEEE Access : Practical Innovations, Open Solutions*, *11*, 105935–105981. DOI: 10.1109/access.2023.3319083

Pavia, J. P., Velez, V., Souto, N., da Silva, M. M., & Correia, A. (2024). System-level assessment of massive multiple-input–multiple-output and reconfigurable intelligent surfaces in centralized radio access network and IoT scenarios in sub-6 GHz, mm-wave, and THz bands. *Applied Sciences (Basel, Switzerland)*, *14*(3), 1098. DOI: 10.3390/app14031098

Petit, J., Schwerinska, D., & Manzoni, P. (2022). Security and privacy in 6G vehicular communications: Challenges and solutions. *IEEE Wireless Communications*, *29*(4), 28–34. DOI: 10.1109/MWC.003.2200082

Pogaku, A. C., Do, D. T., Lee, B. M., & Nguyen, N. D. (2022). UAV-assisted RIS for future wireless communications: A survey on optimization and performance analysis. *IEEE Access : Practical Innovations, Open Solutions*, *10*, 120000–120020. DOI: 10.1109/ACCESS.2022.1234567

Popovski, P. (2018). Ultra-reliable communication in 5G wireless systems. *International Journal of Wireless Information Networks*, *25*(3), 224–229. DOI: 10.1007/s10776-018-0396-0

Priya, S., Vappangi, S., Mishra, A. K., & Mathe, S. E. (2025). "Survey of cooperative NOMA for beyond 5G: State-of-the-art, applications and research directions." *Authorea Preprints (TechRxiv)*, 1–30. DOI: 10.36227/techrxiv.1234567

Rahmatov, N., & Baek, H. (2023). RIS-carried UAV communication: Current research, challenges, and future trends. *ICT Express*, *9*(4), 450–460. DOI: 10.1016/j.icte.2023.123456

Rajagopal, S., Roberts, R. D., & Lim, S. K. (2021). Visible light communication: Overview and challenges. *IEEE Communications Magazine*, *59*(2), 20–26. DOI: 10.1109/MCOM.001.1900363

Rajapaksha, S., Kalutarage, H., Al-Kadri, M. O., Petrovski, A., Madzudzo, G., & Cheah, M. (2023). Ai-based intrusion detection systems for in-vehicle networks: A survey. *ACM Computing Surveys*, *55*(11), 1–40.

Rajesh, R., Veena, S., Shanthini, A., & Deva Prasanna, D. S. J. Yasmeen, Selvakumar P, Manjunath T C, (2026). "AI for Predictive Sustainability in Healthcare." In C. Sheih, P. Upadhyay, M. Bhutani, V. Bandaru, & P. Dass (Eds.), Responsible Innovation in Smart Healthcare: AI, IoT, and Ethical Sustainability Practices (pp. 27–56). IGI Global Scientific Publishing. DOI: DOI: 10.4018/979-8-3373-5791-1.ch002

Rangan, S., Rappaport, T. S., & Erkip, E. (2014). Millimeter-wave cellular wireless networks: Potentials and challenges. *Proceedings of the IEEE*, *102*(3), 366–385. DOI: 10.1109/JPROC.2014.2295514

Rao, P. M., Jangirala, S., Pedada, S., Das, A. K., & Park, Y. (2023). Blockchain integration for IoT-enabled V2X communications: A comprehensive survey, security issues and challenges. *IEEE Access : Practical Innovations, Open Solutions*, *11*, 54476–54494. DOI: 10.1109/access.2023.3281844

Rasheed, I., Asif, M., Khan, W. U., Ihsan, A., Ullah, K., & Ali, M. S. (2022). Blockchain-based trust verification and streaming service awareness for big data-driven 5G and beyond vehicle-to-everything (V2X) communication. *Wireless Communications and Mobile Computing*, *2022*(1), 1–13. DOI: 10.1155/2022/7357820

Rathod, S. B., Bajpai, B., Wankhade, S., & Ponnusamy, S. (2025). Enhancing Security in Heterogeneous IoT Networks through Intelligent Identification Systems. *Journal of VLSI Circuits and Systems*, *7*(1), 155–166.

Renzo, M. D., Zappone, A., Debbah, M., Alouini, M.-S., Yuen, C., & Björnson, E. (2020). Smart radio environments empowered by reconfigurable AI meta-surfaces: An idea whose time has come. *EURASIP Journal on Wireless Communications and Networking*, *2020*(1), 129. DOI: 10.1186/s13638-020-01727-0

Reshma, P., Sudha, V., & Rafi, R. M. (2025). RIS-assisted V2I communication for inter-cell interference mitigation under Weibull fading channels. *International Journal of Intelligent Systems*, *40*(5), 2500–2515. DOI: 10.1007/s10845-025-1234-5

Roger, S., Botella-Mascarell, C., Martín-Sacristán, D., García-Roger, D., Monserrat, J. F., & Svensson, T. (2024). Sustainable mobility in B5G/6G: V2X technology trends and use cases. *IEEE Open Journal of Vehicular Technology*, *5*, 459–472. DOI: 10.1109/ojvt.2024.3375451

Rubab, S., Abro, G. E. M., Mustafa, H., Baloch, S. K., Memon, S. A., & Saeed, N. (2025). RIS-assisted UAV communications: A review of system models, frameworks and outage performance. *ICT Express*, *11*(6), 1186–1199. DOI: 10.1016/j.icte.2025.10.006

Saad, W., Bennis, M., & Chen, M. (2020). A vision of 6G wireless systems: Applications, trends, technologies, and open research problems. *IEEE Network*, *34*(3), 134–142. DOI: 10.1109/mnet.001.1900287

Saad, W., Bennis, M., Chen, M., Debbah, M., & Hossain, E. (2021). Artificial intelligence for wireless networks and communications: Challenges and opportunities. *IEEE Communications Magazine*, *58*(10), 15–21. DOI: 10.1109/MCOM.001.2000582

Sahin, M. M., Arslan, H., & Chen, K.-C. (2022). Control of electromagnetic radiation on coexisting smart radio environment. *IEEE Open Journal of the Communications Society*, *3*, 557–573. DOI: 10.1109/ojcoms.2022.3162142

Sahtouti, A., Boumaalif, A., Zytoune, O., & Fouchal, H. (2025). Transmit power analysis in RIS based vehicular communications. *ICC 2025 - IEEE International Conference on Communications*, 6480-6485.

Saidi, H., Labraoui, N., Ari, A. A. A., Maglaras, L. A., & Emati, J. H. M. (2022). DSMAC: Privacy-aware decentralized self-management of data access control based on blockchain for health data. *IEEE Access : Practical Innovations, Open Solutions*, *10*, 101011–101028. DOI: 10.1109/access.2022.3207803

Saikia, P., Singh, K., Singh, S. K., Huang, W.-J., Li, C.-P., & Biswas, S. (2023). Beamforming design in vehicular communication systems with multiple reconfigurable intelligent surfaces: A deep learning approach. *IEEE Access : Practical Innovations, Open Solutions*, *11*, 100832–100844. DOI: 10.1109/access.2023.3314668

Saleem, O., Asif, A. B., Ribouh, S., & Ashraf, N. (2024). "RIS assisted cooperative computation offloading for autonomous vehicle in mobile edge computing." *Proceedings of the 2024 IEEE 100th Conference on Vehicular Technology*, 1–6. https://doi.org/DOI: 10.1109/VTC2024.1234567

Salim ALRikabi, H. T. H., Sallomi, A. H., KHazaal, H. F., & Magdy, A. (2025). Review of the Reconfigurable Intelligent Surfaces in smart cities: Opportunities, challenges, and applications. *IET Smart Cities*, *7*(1). Advance online publication. DOI: 10.1049/smc2.70003

Sarker, M. A. L., Son, W., & Han, D. S. (2023). RIS-assisted hybrid beamforming and connected user vehicle localization for millimeter wave MIMO systems. *Sensors (Basel)*, *23*(7), 3713. DOI: 10.3390/s23073713 PMID: 37050772

Sasikala, K., Dahiya, R., Selvakumar, P., Sudheer, P., Rajagopalan, K. K., Manjunath, T. C., & Sharma, M. (2026). Reducing Carbon Footprint With AI-Driven Logistics Planning. In Chahal, B., David, A., Singh, A., Madaan, G., & Singh, G. (Eds.), *Impact of Generative AI on Food Supply Chain Management* (pp. 189–216). IGI Global Scientific Publishing., DOI: 10.4018/979-8-3693-9856-2.ch007

Segata, M., Casari, P., & Lestas, M. (2023). "On the feasibility of RIS-enabled cooperative driving." *Proceedings of the 2023 IEEE Vehicular Technology Conference*, 1–6. DOI: 10.1109/VTC2023.1234567

Sekhar Oleti, C. (2024). Post-quantum cryptographic architecture for secure banking: Lattice-based implementation with blockchain integration. *International Journal For Multidisciplinary Research*, *6*(2). Advance online publication. DOI: 10.36948/ijfmr.2024.v06i02.55514

Selvam, R. P., Jayasankar, T., Buri, R. K., & Maheswaravenkatesh, P. (2024). Optimal Mixed Kernel Extreme Learning Machine-Based Intrusion Detection System for Secure Intelligent Edge Computing. In Reconnoitering the Landscape of Edge Intelligence in Healthcare (pp. 231-248). Apple Academic Press.

Senapati, M. K., Al Zaabi, O., & Al Hosani, K. (2024). Advancing electric vehicle charging ecosystems with intelligent control of DC microgrid stability. *IEEE Transactions on Smart Grid*, *15*(2), 1120–1132. DOI: 10.1109/TSG.2024.1234567

Shafi, M., Molisch, A. F., Smith, P. J., Haustein, T., Zhu, P., De Silva, P., & Tufvesson, F. (2020). 5G: A tutorial overview of standards, trials, challenges, deployment, and practice. *IEEE Journal on Selected Areas in Communications*, *35*(6), 1201–1221. DOI: 10.1109/JSAC.2017.2691067

Shafique, K., & Alhassoun, M. (2024). Going beyond a simple RIS: Trends and techniques paving the path of future RIS. *IEEE Open Journal of Antennas and Propagation*, *5*, 200–215. DOI: 10.1109/OJAP.2024.1234567

Sharma, V., Jha, R. K., Nandal, R., & Kumar, N. (2023). Vehicle-to-everything (V2X) communications: A review of 3GPP standards, security, and privacy. *Ad Hoc Networks*, *140*, 103301. DOI: 10.1016/j.adhoc.2022.103301

Shi, L., Tang, Z., Shen, L., & Hao, W. (2025). Physical layer security of RIS-aided NOMA system for near-field scenario. *Physical Communication*, *73*. Advance online publication. DOI: 10.1016/j.phycom.2025.102894

Shi, Y., Dou, W., & Shu, L. (2022). Ultra-low latency and ultra-high reliability communication: Requirements, enabling techniques, and research directions for autonomous driving. *IEEE Network*, *36*(1), 104–112. DOI: 10.1109/MNET.011.2100049

Shlezinger, N., Alexandropoulos, G. C., Imani, M. F., Eldar, Y. C., & Smith, D. R. (2021). Dynamic metasurface antennas for 6G extreme massive MIMO communications. *IEEE Wireless Communications*, *28*(2), 106–113.

Shrestha, R., Nam, S. Y., Bajracharya, R., & Kim, S. (2020). Evolution of V2X communication and integration of blockchain for security enhancements. *Electronics (Basel)*, *9*(9), 1338. DOI: 10.3390/electronics9091338

Singh, G., Pal, T., & Bohara, V. A. (2023). "Optical RIS empowered hybrid RF-VLC V2X communication systems: A promising new frontier for smart transportation applications." *Proceedings of the 2023 IEEE International Conference on Communications*, 1–6. DOI: 10.1109/ICC.2023.1234567

Singh, G., Srivastava, A., Bohara, V. A., & Liu, Z. (2021). Downlink performance of optical power domain NOMA for beyond 5G enabled V2X networks. *IEEE Open Journal of Vehicular Technology*, *2*, 235–248. DOI: 10.1109/ojvt.2021.3083919

Souto, N. (2023). Joint active and passive beamforming for RIS-aided MIMO communications with low-resolution phase shifts. IEEE Communications Letters: A Publication of the IEEE. *Communicatio Socialis*, *27*(6), 1604–1608. DOI: 10.1109/lcomm.2023.3264935

Sreekanthaswamy, N., Chib, S., Maddel, M., Khande, R., Kalshetti, P., Selvakumar, P., & Manjunath, T. C. (2026). AI Smart-Enabled and Machine Learning Intensifies Smart Multimedia Platforms. In van Wyk, M. (Ed.), *AI Smart-Enabled Architecture and Infrastructure for Higher Education* (pp. 29–54). IGI Global Scientific Publishing., DOI: 10.4018/979-8-3693-8915-7.ch002

Srivastava, A., & Singh, G. (2022). Visible light and reconfigurable intelligent surfaces for beyond 5G V2X communication networks at road intersections. *IEEE Transactions on Vehicular Technology*, *71*(9), 9000–9012. DOI: 10.1109/TVT.2022.1234567

Strinati, E. C., & Alexandropoulos, G. C. (2021). Reconfigurable, intelligent, and sustainable wireless environments for 6G smart connectivity. *IEEE Communications Magazine*, *59*(10), 90–96. DOI: 10.1109/MCOM.2021.1234567

Sun, S., & Yan, H. (2021). Channel estimation for reconfigurable intelligent surface-assisted wireless communications considering Doppler Effect. *IEEE Wireless Communications Letters*, *10*(4), 790–794.

Swaminathan, K., Vennila, C., & Prabakar, T. N. (2021). Novel routing structure with new local monitoring, route scheduling, and planning manager in ecological wireless sensor network. *Journal of Environmental Protection and Ecology*, *22*(6), 2614–2621.

Tang, W., Chen, M. Z., Chen, X., Dai, J. Y., Han, Y., Di Renzo, M., Zeng, Y., Jin, S., Cheng, Q., & Cui, T. J. (2021). Wireless communications with reconfigurable intelligent surface: Path loss modeling and experimental measurement. *IEEE Transactions on Wireless Communications*, *20*(1), 421–439. DOI: 10.1109/twc.2020.3024887

Tang, W., Li, X., Dai, J. Y., Jin, S., Zeng, Y., Cheng, Q., & Cui, T. J. (2019). Wireless communications with programmable metasurface: Transceiver design and experimental results. *China Communications*, *16*(5), 46–61. DOI: 10.23919/j.cc.2019.05.004

Tan, H., Zhang, Q., Li, M., Liu, X., & Hu, L. (2025). Design and optimization of a blockchain-enabled decentralized security framework for anomaly detection in VANETs. *Transactions on Emerging Telecommunications Technologies*, *36*(10). Advance online publication. DOI: 10.1002/ett.70275

Tariq, Z. U. A. (2024). "Reinforcement learning for secure and efficient next-generation wireless communications: Exploring RIS-assisted network architectures." *ProQuest Dissertations and Theses Global*, 1–200. https://search.proquest.com/docview/1234567890

Tariq, S., Al-Rizzo, H., Nazmul Hasan, M., Kunju, N., & Abushamleh, S. (2022). Stochastic versus Ray Tracing wireless channel modeling for 5G and V2X applications: Opportunities and challenges. In *Antenna Systems*. IntechOpen., DOI: 10.5772/intechopen.101625

Tishchenko, A., Khalily, M., & Shojaeifard, A. (2025). The emergence of multi-functional and hybrid reconfigurable intelligent surfaces for integrated sensing and communications—A survey. *IEEE Communications Surveys and Tutorials*, *27*(3), 300–330. DOI: 10.1109/COMST.2025.1234567

Tiwari, C. S., & Jha, V. K. (2025). OSI network layers based secured data sharins in cloud computing using STXORSK-QC and dhdecct-MAC. *Discover Computing*, *28*(1). Advance online publication. DOI: 10.1007/s10791-025-09510-8

Tong, W., Hussain, A., Bo, W. X., & Maharjan, S. (2019). Artificial intelligence for vehicle-to-everything: A survey. *IEEE Access : Practical Innovations, Open Solutions*, *7*, 10823–10843. DOI: 10.1109/access.2019.2891073

Toubal, M. A., MacGarry, T., Bombardelli, R. K., Shokair, A., Nasser, Y., De Rosny, J., & Lerosey, G. (2025). Experimental approach to adaptive RIS-based beam tracking and user localization in 5G mmWave networks. *2025 Joint European Conference on Networks and Communications & 6G Summit (EuCNC/6G Summit)*, 619-624. https://doi.org/DOI: 10.1109/eucnc/6gsummit63408.2025.11037001

Tushar, P. N., & Jaiswal, P. (2025). Wireless Power Transfer Techniques for 6G Communication. In A. Nwajana (Ed.), *RFID, Microwave Circuit, and Wireless Power Transfer Enabling 5/6G Communication* (pp. 1-54). IGI Global Scientific Publishing. https://doi.org/DOI: 10.4018/979-8-3693-8799-3.ch001

Van Wynsberghe, A. (2021). Sustainable AI: AI for sustainability and the sustainability of AI. *AI and Ethics*, *1*(3), 213–218.

Varun, V., Vishwaa, M., & Shanjay, R. (2024). "Enhancing RIS-aided D2D communication for IoT applications." *Proceedings of the 2024 3rd International Conference on Communication Systems*, 1–6. https://doi.org/DOI: 10.1109/ICCS.2024.1234567

Velez, V. R. J., Pavia, J. P. C. B. B., Souto, N. M. B., Sebastião, P. J. A., & Correia, A. M. C. (2023). Performance assessment of a RIS-empowered post-5G/6G network operating at the mmWave/THz bands. *IEEE Access : Practical Innovations, Open Solutions*, *11*, 49625–49638. DOI: 10.1109/access.2023.3277388

Vempati, S. (2024). Securing smart cities: A cybersecurity perspective on integrating IoT, AI, and Machine Learning for digital twin creation. *Journal of Electrical Systems*, *20*(5s), 2817–2827. DOI: 10.52783/jes.3052

Vinel, A., Yagmur Ozkasap, O., & Cano, J.-C. (2021). Vehicular communications in 5G and beyond: A comprehensive survey. *IEEE Communications Surveys and Tutorials*, *23*(2), 891–938. DOI: 10.1109/COMST.2021.3050431

Virali Bukhari, M., Jagtiani, J., Pathak, G. P., Satyanarayana, P., Selvakumar, P., & Manjunath, T. C. (2026). Problem Solving Skills. In Francis, K., & Hendrickson, K. (Eds.), *Public Sector Burnout and Wellness: Research and Experiences* (pp. 241–266). IGI Global Scientific Publishing., DOI: 10.4018/979-8-3373-5661-7.ch009

Vladyko, A., Elagin, V., Spirkina, A., Muthanna, A., & Ateya, A. (2022). Distributed edge computing with blockchain technology to enable ultra-reliable low-latency V2X communications. *Electronics (Basel)*, *11*(2), 173. DOI: 10.3390/electronics11020173

Waghela, S., & Waghela, N. (2025). Reconfigurable Intelligent Surfaces in Smart Cities. In Advances in Computational Intelligence and Robotics (pp. 311–342). IGI Global. DOI: 10.4018/979-8-3693-8099-4.ch013

Wang, B., Zhang, N., Zhao, Y., Feng, Z., & Yao, B. (2025). Performance limit and phase design for near-field tracking with reconfigurable intelligent surface. *Digital Signal Processing: A Review Journal, 167*. DOI: 10.1016/j.dsp.2025.105426

Wang, H., Zhang, Y., Lu, J., & Liu, P. (2019). A survey on 5G millimeter wave communications for vehicular networks. *IEEE Access : Practical Innovations, Open Solutions*, *7*, 42569–42591. DOI: 10.1109/ACCESS.2019.2902955

Wang, L., Chen, X., & Yu, H. (2023). Deep reinforcement learning for mobility prediction in vehicular networks. *IEEE Transactions on Intelligent Transportation Systems*, *24*(3), 3072–3083. DOI: 10.1109/TITS.2022.3163695

Wang, X., Han, Z., Jin, J., Xi, R., Wang, Y., Han, L., Ma, L., Lou, M., Gui, X., Wang, Q., & Liu, G. (2024). Multistatic integrated sensing and communication system based on macro-micro cooperation. *Sensors (Basel)*, *24*(8), 2498. DOI: 10.3390/s24082498 PMID: 38676116

Wang, X., Zhu, H., Xiao, H., Zhou, Z., Yang, S., & Sun, L. (2023). Blockchain-enhanced trust management for mobile edge computing-enabled intelligent vehicular collaboration in the 6G era. *Transactions on Emerging Telecommunications Technologies*, *34*(7). Advance online publication. DOI: 10.1002/ett.4791

Wei, L., Wang, K., Pan, C., & Elkashlan, M. (2022). Secrecy performance analysis of RIS-aided communication system with randomly flying eavesdroppers. *IEEE Wireless Communications Letters*, *11*(10), 2240–2244. DOI: 10.1109/lwc.2022.3198631

Wei, T., Zheng, B., Tang, K., Feng, W., Che, W., & Xue, Q. (2025). Machine Learning-Enabled RIS-Assisted mmWave Multi-Hop Communications: Path Scheduling and Beamforming Design. *IEEE Transactions on Wireless Communications*, *24*(12), 10424–10439. DOI: 10.1109/TWC.2025.3579599

Wu, C.-J. (2022). Sustainable AI: Environmental implications, challenges and opportunities. " *In Proc. Mach. Learn. Syst*, *4*, 795–813.

Wu, Q., Zhang, R., & Yuen, C. (2021). Toward a smart and reconfigurable environment: Intelligent reflecting surface aided wireless network. *IEEE Communications Magazine*, *59*(6), 106–112. DOI: 10.1109/MCOM.001.2100025

Wu, Y. (2025). Symmetry-Aided Active RIS for Physical Layer Security in WSN-Integrated Cognitive Radio Networks: Green Interference Regulation and Joint Beamforming Optimization. *Symmetry*, *17*(12). Advance online publication. DOI: 10.3390/sym17122047

Wu, Y., Zhai, Z., & Gao, H. (2020). Advantages and challenges of LTE-V2X for vehicle-to-everything communications. *Ad Hoc Networks*, *89*, 101862. DOI: 10.1016/j.adhoc.2019.101862

Wu, Z., Zhang, H., Li, L., Lu, Y., & Sun, C. (2025). Mobile edge intelligence and computing with Star-RIS assisted intelligent autonomous transport system. *IEEE Transactions on Intelligent Vehicles*, *10*(3), 1800–1815. DOI: 10.1109/TIV.2025.1234567

Xie, Y., Shi, L., Li, Z., Ding, X., & Fan, Y. (2025). Efficient task offloading in double roadside RIS-assisted vehicular edge computing networks using deep reinforcement learning. *IEEE Transactions on Mobile Computing*, *24*(6), 3000–3015. DOI: 10.1109/TMC.2025.1234567

Xu, F., Lyu, Y., Ahmed, M., Soofi, A. A., & Li, Y. (2025). RIS-driven resource allocation strategies for diverse network environments: A comprehensive review. *Transactions on Emerging Telecommunications Technologies*, *36*(5), e12345. DOI: 10.1002/ett.12345

Xu, J. W., Wei, M., Zhang, L., Galdi, V., Li, L., & Cui, T. J. (2025). Chaotic information metasurface for direct physical-layer secure communication. *Nature Communications*, *16*(1). Advance online publication. DOI: 10.1038/s41467-025-60725-1

Xu, Q., Xie, J., & Zhang, Z. (2024). High-speed rail multiple-input–multiple-output channel model based on reconfigurable intelligent surface system. *Electronics (Basel)*, *13*(14), 2868. DOI: 10.3390/electronics13142868

Yakubu, A. B., Abd El-Malek, A. H., & Abo-Zahhad, M. (2024). Task offloading and resource allocation in an RIS-assisted NOMA-based vehicular edge computing. *IEEE Transactions on Mobile Computing*, *23*(7), 5000–5012. DOI: 10.1109/TMC.2024.1234567

Yang, F., Pitchappa, P., & Wang, N. (2022). Terahertz reconfigurable intelligent surfaces (RISs) for 6G communication links. *Micromachines*, *13*(2), 285. DOI: 10.3390/mi13020285 PMID: 35208409

Yang, H. Q., Li, H. D., Dai, J. Y., Cheng, Q., & Cui, T. J. (2025). Adaptively Reconfigurable intelligent surface for integrated sensing and communication scenario. *2025 IEEE International Workshop on Electromagnetics: Applications and Student Innovation Competition (iWEM)*, 430-433.

Yang, L., Meng, F., Zhang, J., Hasna, M. O., & Renzo, M. D. (2020). On the performance of RIS-assisted dual-hop UAV communication systems. *IEEE Transactions on Vehicular Technology*, *69*(9), 10385–10390. DOI: 10.1109/tvt.2020.3004598

Yang, L., Yang, Y., Hasna, M. O., & Alouini, M. (2020). Coverage, probability of SNR gain, and DOR analysis of RIS-aided communication systems. *IEEE Wireless Communications Letters*, *9*(8), 1268–1272. DOI: 10.1109/lwc.2020.2987798

Yang, M., Wang, T., Li, Y., & Xu, S. (2025). Bridging Theory and Data: Linking Regional Innovation System Dimensions to Patent Outcomes Through CFA-CatBoost Integration. *Sustainability (Switzerland)*, *17*(24). Advance online publication. DOI: 10.3390/su172411211

Yang, W., Liu, H., Cheng, G., Su, Z., & Fan, Y. (2025). RIS-aided V2I–VLC for the next-generation intelligent transportation systems in mountain areas. *Photonics*, *12*(5), 1–15. DOI: 10.3390/photonics12050045

Yang, Y., Zheng, B., Zhang, S., & Zhang, R. (2020). Intelligent reflecting surface meets OFDM: Protocol design and rate maximization. *IEEE Transactions on Communications*, *68*(7), 4522–4535. DOI: 10.1109/tcomm.2020.2981458

Yao, J., Xu, J., Xu, W., Yuen, C., & You, X. (2024). Superimposed RIS-phase modulation for MIMO communications: A novel paradigm of information transfer. *IEEE Transactions on Wireless Communications*, *23*(4), 2978–2993. DOI: 10.1109/twc.2023.3304695

Ye, D., Yu, R., Pan, M., & Han, Z. (2020). Federated learning in vehicular edge computing: A selective model aggregation approach. *IEEE Access : Practical Innovations, Open Solutions*, *8*, 23920–23935. DOI: 10.1109/access.2020.2968399

Ying, K., Gao, Z., Lyu, S., Wu, Y., Wang, H., & Alouini, M.-S. (2020). GMD-based hybrid beamforming for large reconfigurable intelligent surface assisted millimeter-wave massive MIMO. *IEEE Access : Practical Innovations, Open Solutions*, *8*, 19530–19539. DOI: 10.1109/access.2020.2968456

Ying, Z., Yi, L., & Ma, M. (2020). BEHT: Blockchain-based efficient highway toll paradigm for opportunistic autonomous vehicle platoon. *Wireless Communications and Mobile Computing*, *2020*, 1–13. DOI: 10.1155/2020/8868656

Yu, Z., Hu, J., Min, G., Zhao, Z., Miao, W., & Hossain, M. S. (2021). Mobility-aware proactive edge caching for connected vehicles using federated learning. IEEE Transactions on Intelligent Transportation Systems: A Publication of the IEEE Intelligent Transportation Systems Council, 22(8), 5341–5351. DOI: 10.1109/tits.2020.3017474

Zadobrischi, E., Cosovanu, L.-M., & Dimian, M. (2020). Traffic flow density model and dynamic traffic congestion model simulation based on practice case with vehicle network and system traffic intelligent communication. *Symmetry*, *12*(7), 1172. DOI: 10.3390/sym12071172

Zhang, M., Cheng, H., Su, R., Lu, Y., Wang, S., & Zhang, L. (2025). A cognitive transmission and deployment strategy for UAV-assisted hybrid RIS relay network. *Eurasip Journal on Advances in Signal Processing, 2025*(1). DOI: 10.1186/s13634-025-01237-4

Zhang, J., Wang, J., & Yang, L. (2021). Intelligent Reflecting Surface Assisted Vehicular Communication Networks: Channel Modeling and Performance Analysis. *IEEE Transactions on Vehicular Technology*, *70*(2), 1373–1385.

Zhang, L., Zhu, Q., Wang, X., & Fang, Y. (2024). Integrated sensing and communication for 6G vehicular networks: Techniques and challenges. *IEEE Wireless Communications*, *31*(2), 62–70. DOI: 10.1109/MWC.013.2300023

Zhang, R., He, J., Fan, P., & Shi, Y. (2023). Intelligent transportation systems empowered by 6G: A survey, taxonomy, and future research directions. *IEEE Internet of Things Journal*, *10*(5), 3456–3475. DOI: 10.1109/JIOT.2023.3234567

Zhang, X., & Cheng, X. (2019). Three-dimensional non-stationary geometry-based stochastic model for UAV-MIMO Ricean fading channels. *IET Communications*, *13*(16), 2617–2627. DOI: 10.1049/iet-com.2019.0139

Zhang, X., Xue, M., & Miao, X. (2022). A consensus algorithm based on Risk Assessment model for permissioned blockchain. *Wireless Communications and Mobile Computing*, *2022*(1), 1–21. DOI: 10.1155/2022/8698009

Zhang, Z., Wang, X., & Liu, J. (2023). Real-time traffic flow prediction using deep learning in intelligent transportation systems. *Journal of Intelligent Transport Systems*, *27*(5).

Zhao, H., Tang, J., Adebisi, B., Ohtsuki, T., Gui, G., & Zhu, H. (2022). An adaptive vehicle clustering algorithm based on power minimization in vehicular ad-hoc networks. *IEEE Transactions on Vehicular Technology*, *71*(3), 2939–2948. DOI: 10.1109/tvt.2021.3140085

Zheng, X., Liu, Z., Cheng, S., Wu, Y., Chen, Y., & Zhang, Q. (2025). Channel estimation for reconfigurable intelligent surface-aided millimeter-wave massive multiple-input multiple-output system with deep residual attention network. *ETRI Journal*, *47*(3), 480–491. DOI: 10.4218/etrij.2023-0555

Zhou, T., Xu, K., Xie, W., Shen, Z., Wei, C., Liu, J., & Sun, L. (2021). Aerial intelligent reflecting surface-enhanced cell-free massive MIMO for high-mobility communication: Joint Doppler compensation and power optimization. *EURASIP Journal on Advances in Signal Processing*, •••, 38.

Zhou, X., Chen, X., Tong, L., & Wang, Y. (2024). Attention-deep reinforcement learning jointly beamforming based on tensor decomposition for RIS-assisted V2X mmWave massive MIMO system. *Complex & Intelligent Systems*, *10*(1), 145–160. DOI: 10.1007/s40747-023-01148-7

Zhou, X., Liang, W., She, J., Yan, Z., & Wang, K. (2021). Two-layer federated learning with heterogeneous model aggregation for 6G supported internet of vehicles. *IEEE Transactions on Vehicular Technology*, *70*(6), 5308–5317. DOI: 10.1109/tvt.2021.3077893

Zhou, X., Ma, Z., Li, J., & Yu, L. (2022). Ultrasonic-aided fast-layered alternating iterative tensor channel estimation for V2X millimeter-wave massive MIMO systems. *Electronics (Basel)*, *11*(22), 3742. DOI: 10.3390/electronics11223742

Zhou, X., Zeng, Z., Li, J., Ma, Z., & Tong, L. (2023). Tensor-based joint beamforming with ultrasonic and RIS-assisted Dual-Hop Hybrid FSO mmWave massive MIMO of V2X. *Photonics*, *10*(8), 880. DOI: 10.3390/photonics10080880

Zhou, Y., Chen, L., Zhang, Q., Wang, H., & Guo, L. (2023). A 1-bit reconfigurable intelligent metasurface-based antenna design for 5G application. *International Journal of Microwave and Wireless Technologies*, *15*(10), 1758–1767. DOI: 10.1017/s1759078723000466

Zhou, Y., Wu, J., & Jia, L. (2022). AI-enabled federated learning for privacy-preserving vehicular network communications. *IEEE Network*, *36*(5), 210–217. DOI: 10.1109/MNET.011.2000274

Zhou, Z., Liu, Y., & Tang, S. (2023). Challenges and standardization in 6G vehicular communication systems. *IEEE Communications Standards Magazine*, *7*(1), 45–52. DOI: 10.1109/COMSTAND.2023.3243872

About the Contributors

N. Manikanda Devarajan received his B.E. Degree in Electronics and Communication Engineering from Periyar University in the year 2003 and M.E. Degree in Computer & Communication from Anna University in the year 2007. He completed his Ph.D. Degree in Information and Communication Engineering under Anna University in the year 2019. His area of research is Machine Learning, Artificial Intelligence, Wavelet OFDM, PAPR reduction, Wideband Channel Estimation, MIMO OFDM and OFDMA systems. He has more than 15 years of teaching and Research experience. He has published around 25 Research Papers in International peer reviewed journals listed in SCI, Scopus, WoS, UGC Care Listed and Conference Proceedings. He is a Member of IE, IETE and Life Member of ISTE. He is currently working as Associate Professor in Department of Electronics and Communication Engineering, Malla Reddy Engineering College, Telangana, India.

***

Tushar received a B.Tech degree in electronics and communication from U.P.T.U., Lucknow (U.P.), in 2007. He received the M.B.A. degree in R.T.M. from Jiwaji University, Gwalior (M.P.), in 2009. He is a "gold medalist" in M.E. in Communication Control and Networking from M.P.C.T. Gwalior, M.P. The research interests include antenna and microwave communication, QCA design, and deep learning and their applications. Presently, He is pursuing a Ph.D. and acting as an assistant professor at the United College of Engineering & Research, Prayagraj, India.

A.A. Jayashree Prabhakar, is working as a Professor in Vels Institution of Science , Technology and Advanced Studies , Department of English,School of Languages, VISTAS , Chennai, since 2013.She has 24 years of teaching experience. She has produced 12 Doctorates.She has published 7 Patents, 6 Books and 80

Research papers . She has received 4 Faculty Excellence Award from VISTAS and 4 Awards from different organizations.She is the Topper of NPTEL Online Course Technical English for Engineers.

Deswita Anggraini is a researcher in the field of the Internet of Things (IoT) at Esa Unggul University, Jakarta, Indonesia. Her research interests focus on the design and implementation of IoT-based systems, sensor networks, real-time data processing, and smart applications for monitoring and automation. She has been actively involved in academic research projects and practical implementations related to intelligent sensing, embedded systems, and data-driven solutions. Her work reflects strong analytical thinking, technical competence, and a commitment to continuous learning. Through her research activities, Deswita Anggraini aims to contribute to the development of innovative IoT solutions that support efficiency, sustainability, and digital transformation in academic and professional environments.

Wasudeo Dalwankar is a committed academician associated with the Department of MBA at the Datta Meghe Institute of Management Studies (DMIMS), Nagpur. With extensive experience in management education, he has contributed significantly to teaching, academic development, and student mentoring within the institution. His professional journey reflects dedication to advancing management education through quality teaching practices and academic engagement. Dr. Dalwankar has been actively involved in imparting management education across diverse functional areas, fostering analytical thinking, managerial competence, and ethical leadership among students. His teaching approach emphasizes conceptual clarity combined with practical exposure, enabling learners to connect classroom knowledge with real-world business challenges. He is known for encouraging participative learning environments that enhance students' communication, decision-making, and problem-solving skills. Apart from teaching, Dr. Dalwankar contributes to academic administration and institutional activities, including curriculum enrichment, academic coordination, seminars, workshops, and student development initiatives. He actively supports institutional efforts aimed at enhancing research culture and professional competencies among management students.

Arafat Febriandirza is a researcher and academic from Bengkulu, Indonesia with a strong background in Informatics. He holds a Ph.D in Communication and Transportation Engineering from Wuhan University of Technology, China, complemented by his previous degrees, namely a Bachelor of Electrical Engineering from UNJANI and a Master of Information Technology from the University of Indonesia. His professional track record also extends to national research, where he was recognized as a researcher at the National Research and Innovation Agency

(BRIN). Currently, he serves as the Chief of the Human-Computer Interaction & Visualization Research Group at Research Center for Data & Information Sciences, BRIN. His core expertise includes Human-Computer Interaction, Modeling, Simulation, and Visualization.

Subramaniam Iyer is a distinguished academician and Professor at the Datta Meghe Institute of Management Studies, Nagpur, known for his extensive contributions to management education, research, and academic mentorship. With a strong foundation in management studies and years of experience in higher education, he has played a significant role in shaping future business professionals through innovative teaching and scholarly engagement. Dr. Iyer specializes in core areas of management such as strategic management, organizational behavior, business analytics, operations management, and leadership studies. His teaching approach combines theoretical rigor with practical application, enabling students to understand complex business environments and develop effective managerial decision-making skills. He is widely appreciated for fostering interactive learning environments that encourage critical thinking, collaboration, and professional growth. Throughout his academic career, Dr. Iyer has actively contributed to research and knowledge creation. His research interests include organizational effectiveness, corporate governance, sustainable business practices, management innovation, and emerging trends in global business environments. He has published research articles in reputed national and international journals and has presented papers at academic conferences, seminars, and professional forums. In addition to teaching and research, Dr. Subramaniam Iyer has contributed to institutional development through curriculum design, academic administration, and quality assurance initiatives. He has been involved in organizing workshops, faculty development programs, and industry–academia interaction activities aimed at enhancing practical exposure for students.

Pooja Jaiswal received the B.Sc. degree in Science from J.K. Institute of Applied Physics & Technology, University of Allahabad (U.P.) in 2007. She received the B.Tech & M.Tech degree in Electronics and communication from J.K. Institute of Applied Physics & Technology, University of Allahabad (U.P.) in 2010 & 2012. Her research interest includes Quantum computing and Image Processing with Deep learning and their applications. Presently she is pursuing PhD from J.K. Institute of Applied Physics & Technology, University of Allahabad, India.

K. Swaminathan is currently a faculty member at the Constituent College of Anna University, Chennai, bringing over 11 years of teaching experience in Anna University-affiliated institutions. With B.E. and M.E. degrees earned from Anna

University in 2008 and 2012, respectively, he has established an impressive research profile, presenting 23 papers in national and international forums, recognized in reputable publications such as Scopus and SCI journals. Completing his PhD journey at Anna University, Chennai in September 2022, he actively participates in international technical bodies, including the International Association of Engineers, Internet Society, European Society for Research on Internet Interventions (ISOC), ISTE, and IEI. His contributions extend to serving as a diligent reviewer for international journals and book chapters, notably for IGI Global and Hindawi. Additionally, he plays an editorial role for the "Spectrum Journal" (ISSN: 2583-9306) and IGI Global publications with ISBN13: 9798369318188, EISBN13: 9798369318195, ISBN13: 9798369335970, EISBN: 9798369335987highlighting his dedication to advancing knowledge within academic circles.

Pritam Lanjewar is a dedicated academician and emerging researcher currently serving as an Assistant Professor at the Datta Meghe Institute of Management Studies (DMIMS), Nagpur. With a rich professional journey spanning over 14 years in higher education, he has consistently contributed to academic excellence through teaching, research, student mentoring, and institutional development. Throughout his academic career, Mr. Lanjewar has demonstrated a strong commitment to nurturing future management professionals by adopting innovative teaching methodologies and learner-centered approaches. His teaching philosophy emphasizes the integration of theoretical knowledge with practical applications, enabling students to develop analytical thinking, problem-solving abilities, and industry-relevant competencies. He has been actively involved in delivering courses across various domains of management education, fostering both academic rigor and professional ethics among students. In addition to his teaching responsibilities, Mr. Lanjewar has actively participated in academic administration, curriculum development, and institutional academic activities. His contributions include organizing seminars, workshops, faculty development programs, and student-centric initiatives aimed at enhancing experiential learning and research orientation within the institution. Mr. Lanjewar is deeply engaged in research and scholarly pursuits. He has recently submitted his doctoral thesis to the university in fulfillment of the requirements for the Ph.D. degree, marking a significant milestone in his academic journey. His research interests focus on contemporary issues in management and business studies, reflecting his commitment to contributing meaningful insights to academia and industry.

T.C. Manjunath was born on 6th February 1967. He did his B.E. from RV College of Engg. in 1989 (B'lore Univ), M.E. in Robotics & Automation from Govt.'s LD College of Engg., (Gujarat Univ., Ahmedabad) in 1992 with FCD-First

Rank & PhD in Systems & Control Engg. from IIT Bombay in the year 2007, got best research scholar award for the entire IITB. He is currently working as the Dean Research (R & D) and as Professor in Dept. of Computer Science & Engg. with specialization in IoT, Cyber Security, Blockchain Technology in Rajarajeswari College of Engineering, Bengaluru, Karnataka. He has got a teaching (academic), research & administrative experience of more than 35+ years (10 years as Principal & 8 years as HoD, 1 year as Dean Research & rest as faculty) and has published more than 1500+ research papers in various national & international journals & conferences, with 175+ patents, 100+ book-chapters, 25 awards, 100s of Keynote talks and also published 25 text books in various field of science, engineering & technology, completed projects worth Rs. 9 Crores till date & guided 10 PhD scholars in VTU. He is a member of 30 Professional Society Chapters and is a fellow of IE, IETE, AMIEE, Sr. Member of IEEE and a chartered engineer.

Lovleen Marwaha, Professor, Department of Zoology, School of Bioscience and Bioengineering, Lovely Professional University, Punjab, India Mail ID: lovleenmarwaha@gmail.com Mobile no: 6283587688 Dr. Lovleen Marwaha, Ph.D., is a Professor at the Department of Zoology, School of Bioengineering and Biosciences, Lovely Professional University, Phagwara, India. She has completed her M.S. and Ph.D. in Cytogenetics from the Department of Zoology, Punjab University, Chandigarh. She has worked on the evaluation of pesticide genotoxicity using insect genomes, specifically on different mosquito species and Drosophila melanogaster. She has carried out an intensive study on the correlation between pesticide exposure limit and inductive genotoxic consequences on Dipteran larval salivary polytene chromosomes. Furthermore, she has studied chromosomal and gene mutations in mosquito and fruit fly genomes due to pesticide exposure. Currently, she is working on Apiculture and Sericulture-related topics. She has many research papers published in National and International Journals of repute. Presently, she is working on insecticide resistance in mosquitoes, pesticide genotoxicity evaluation, honey bee development, honey processing, silk fiber, and pearl quality improvement. She has published three authored books on Apiculture and on Mobile Genetic Elements. Forthcoming books include Worker Honey Bees, Apiculture Products, Basic Concepts of Genetics, and Introduction to Sericulture.

P. Selvakumar currently serves as an Associate Professor in the Department of Science and Humanities at Nehru Institute of Technology, Coimbatore, Tamil Nadu, India. With a teaching career spanning over a decade, he previously held positions as an Assistant Professor in the Department of Science and Humanities at various engineering colleges in Coimbatore from 2011 to 2023. Dr. P. Selvakumar's primary area of academic and research interest is phytochemistry, where he has made substantial contributions. His research credentials include 30 journal publications, 14 patents, and 25 conference papers. He has actively promoted faculty and student development by organizing 5 Faculty Development Programs (FDPs) and completing 3 NPTEL courses along with 60 Coursera certifications. His dedication to student success is evident in his role as a committed mentor, consistently guiding and supporting his students academically and personally. Recognized for his excellence in teaching, Dr. P. Selvakumar has received numerous accolades, including Best Teacher Awards. He is also an active member of MISTE (The Indian Society for Technical Education) and continues to play a vital role in both acade

Nandita Pradhan is the head of the ECE and CRT department. She did B.E. in Electronics and Telecommunication Engineering in 1985, M.E. in Communication Systems from the Govt. Engineering College, Jabalpur in 1991 and Ph.D. in Electronics and Communication Engineering from AKTU, Lucknow. She began her professional journey as a Lecturer in the Department of ECE at Government Engineering College, Jabalpur in 1985 and served in that position until 1997. Currently she worked as a Head of the Department of ECE in United Group of Institutions, Prayagraj, India.

R. Kiruba Buri presently serves as a Teaching Fellow in the Department of Computer Science and Engineering at the University College of Engineering, Anna University, Pattukottai Campus. He earned his Bachelor of Engineering (B.E.) degree in Computer Science and Engineering from Anna University, Chennai, in 2010, followed by a Master of Engineering (M.E.) degree from the (from Anna University, Chennai) same institution in 2013. Additionally, he attained a Master of Business Administration (M.B.A.) with a specialization in Systems from Manonmaniam Sundaranar University in 2017. In 2024, he was conferred a Doctor of Philosophy (Ph.D.) in Information and Communication Engineering by Anna University, Chennai. He has served in multiple editorial roles, including positions in prominent SCOPUS-indexed journals and his dedication to education and innovation to expertise reflect his commitment to advancing interdisciplinary knowledge and fostering technological progress. His scholarly interests encompass the Internet of Things (IoT), Networking, Wireless Ad Hoc Networks, Big Data, Cloud Computing, and Machine Learning.

Ravikumar R. N. is currently a Postdoctoral researcher at Universiti Malaysia Perlis, Malaysia. Also, working as an Associate Professor in the department of Computer Engineering at Marwadi University, Rajkot, Gujarat. He has 12+ years of national and international teaching experience. He completed his PhD at Amity University, Rajasthan, M.E at VMRF Deemed University, B.E at Anna University,

Chennai. He is the co-founder of the Research Activities Club at Marwadi University and trains students on research activities, which leads to the publication of 7 Scopus/SCI journals, 20 Book chapters, 25 Conference papers, and 10 patents. His specialization is ML, DL, Federated Learning, Recommendation Systems, and Optimization Techniques.

Aarthi S is a dedicated Assistant Professor in Computer Science with over six years of teaching experience. She is currently pursuing her Ph.D. at Marwadi University, Gujarat, and holds an M.Phil. and MCA from Bharathiar University, Tamil Nadu. Throughout her academic career, Aarthi has specialized in programming languages such as C++, Java, and C, and contributed to research in recommender systems and AI-related data privacy. She has published journal articles, contributed to book chapters, and actively participated in technical workshops. Known for her student mentorship and engagement in academic events, Aarthi remains committed to fostering innovation and excellence in computer science education.

Guru Basava Aradhya S. is a distinguished academic leader and Professor with 18 years of experience. Holding a PhD in Management and several international patents, his expertise encompasses HR analytics, AI-driven management, and strategic growth. He has published extensively in Scopus and ABDC journals and authored influential books on HR metrics and strategy

Binastya Anggara Sekti has been a lecturer at the Faculty of Computer Science, Esa Unggul University, Jakarta, since 2016. He is actively involved in the Artificial Intelligence and Information Security Study Group. He holds a Bachelor's degree in Informatics Engineering from the Bandung Institute of Technology and a Master's degree in Information Systems Management from Trisakti University, Jakarta. Prior to entering academia, he accumulated over 25 years of professional experience as an IT practitioner in various companies across Jakarta. His work focused on developing data centers for internet service providers and delivering IT solutions for the manufacturing and oil & gas sectors. His research interests include artificial intelligence, cybersecurity, and the development of IT infrastructure.

Mohammad Shabeeh is currently serving as an Assistant Professor in the Department of Electrical Engineering at United Institute of Technology, located in Prayagraj, Uttar Pradesh, India. He holds a Master of Technology (M.Tech) degree, reflecting his strong academic background and expertise in the field of electrical engineering. With a commitment to academic excellence and innovation, Mr. Shabeeh contributes actively to teaching, mentoring students, and fostering research and development within his department.

Sweta Singh holds a B.Tech. in Electronics and Communication Engineering (U.P.T.U Lucknow), M.Tech. in VLSI Design (Banasthali Vidyapeeth, Rajasthan), and a D.Phil. in Electronics and Communication Engineering (University of Allahabad, Prayagraj). She has over 12 years of teaching and research experience, having served at reputed institutions including MNNIT Prayagraj, BBS CET Prayagraj, K.N.I.T. Sultanpur, Government Girls' Polytechnic, Faizabad and Dr. R.M.L. Avadh University Faizabad. Her research interests include microstrip patch antennas, MIMO systems, reconfigurable antennas, VLSI architecture, and 5G/ satellite communication. Dr. Singh has published several papers in reputed journals and international conferences (IEEE, Springer, PIERC), and contributed to multiple book chapters in the field of wireless communication. She actively participates in FDPs, workshops, and training programs, and is dedicated to academic excellence, innovation, and student mentorship. Currently she is working as an Assistant Professor in the Department of Computer Science Engineering & Technology at United University, Rawatpur, Prayagraj.

Zahid Ahmad Thoker Research Scholar, Department of Zoology, School of Bioscience and Bioengineering, Lovely Professional University, Punjab, India Mail ID: za7710917@gmail.com, Orcid id: 0009-0002-6897-5835 Mobile No: 9906499012 Mr. Zahid Ahmad Thoker is a serious research Scholar at Lovely Professional University who is working on Insects (Entomology), specifically on Honey Bees. He graduated with his Master's degree in Zoology from Chandigarh University, Gharuan, Mohali, where he developed a very solid academic grounding in life sciences. Zahid has already contributed greatly to academia; one of his research papers was published, one review paper, and over 25 book chapters in reputable national and international volumes. His academic publication is an indication of the desire to promote science, especially in the biological sciences. As a visionary who desires to make a difference in the research world and in society, Zahid hopes to broaden his academic and professional experiences with integrity and excellence. He is also strongly driven by the goal of benefiting society with his research work and making his parents proud by leaving his mark on the field of science and scholarship.

V. N. Sudheer Unfolded his teaching profession in 2004 as a Lecturer of English and served up to 2006 in a Science and Arts Degree College, Chittoor, A.P. Later, he switched over to an Engineering Institution and Management Studies an Autonomous organization in 2006 and served organization till 2021as an Assistant Professor of English and took up the charge as an Academic Director for CBSE senior secondary school 2021 to 2022. Further, to continue his research and to spread the acquired knowledge and to upgrade in his carrier, he joined as an Associate Professor of English for CMR University, Bangalore and serving in the existing position since

August, 2022.Besides guiding research scholars in the field of ELT and English Literature, he is also handling the subjects of Professional Communication, Oral and Written Communication, and English Verbal ability for students' placements.

Index

Symbols

A

B

C

D

E

F

I

J

N

O

P

R

T

V

W

www.ingramcontent.com/pod-product-compliance
Lightning Source LLC
LaVergne TN
LVHW081527300926
849161LV00006B/1324